Silicon Integrated Circuits

Part C

Applied Solid State Science

ADVANCES IN

MATERIALS AND DEVICE RESEARCH

Editor: Raymond Wolfe

       AT&T BELL LABORATORIES
       MURRAY HILL, NEW JERSEY

Supplement 1
Magnetic Domain Walls in Bubble Materials
A. P. Malozemoff and J. C. Slonczewski

Supplement 2 (in three parts)
Silicon Integrated Circuits
Edited by *Dawon Kahng*

Silicon Integrated Circuits
Part C

Edited by Dawon Kahng

AT&T BELL LABORATORIES
MURRAY HILL, NEW JERSEY

1985

ACADEMIC PRESS, INC.
(*Harcourt Brace Jovanovich, Publishers*)
Orlando   San Diego   New York   London
Toronto   Montreal   Sydney   Tokyo

COPYRIGHT © 1985, BY ACADEMIC PRESS, INC.
ALL RIGHTS RESERVED.
NO PART OF THIS PUBLICATION MAY BE REPRODUCED OR
TRANSMITTED IN ANY FORM OR BY ANY MEANS, ELECTRONIC
OR MECHANICAL, INCLUDING PHOTOCOPY, RECORDING, OR
ANY INFORMATION STORAGE AND RETRIEVAL SYSTEM, WITHOUT
PERMISSION IN WRITING FROM THE PUBLISHER.

ACADEMIC PRESS, INC.
Orlando, Florida 32887

United Kingdom Edition published by
ACADEMIC PRESS INC. (LONDON) LTD.
24–28 Oval Road, London NW1 7DX

ISSN 0194–2891

ISBN 0–12–002960–X

PRINTED IN THE UNITED STATES OF AMERICA

85 86 87 88   9 8 7 6 5 4 3 2 1

Contents

LIST OF CONTRIBUTORS ... vii

PREFACE ... ix

Transient Thermal Processing of Silicon
G. K. Celler and T. E. Seidel

I. Introduction ... 2
II. Adiabatic Annealing ... 4
III. Thermal Flux Annealing ... 23
IV. Isothermal Rapid Annealing ... 32
V. Related Rapid Thermal Processes ... 55
VI. Summary ... 66
References ... 67

Reactive Ion-Beam Etching and Plasma Deposition Techniques Using Electron Cyclotron Resonance Plasmas
Seitaro Matsuo

I. Introduction ... 75
II. Reactive Ion-Beam Etching ... 76
III. Plasma Deposition ... 95
References ... 116

Physics of VLSI Processing and Process Simulation
W. Fichtner

I. Introduction ... 119
II. Physics of Processing and Process Simulation ... 122
III. Conclusions and the Future ... 324
References ... 325

AUTHOR INDEX ... 337

SUBJECT INDEX ... 349

List of Contributors

Numbers in parentheses indicate the pages on which the authors' contributions begin.

G. K. CELLER, AT&T Bell Laboratories, Murray Hill, New Jersey 07974 (1)

W. FICHTNER, AT&T Bell Laboratories, Murray Hill, New Jersey 07974 (119)

SEITARO MATSUO, Atsugi Electrical Communication Laboratory, Nippon Telegraph and Telephone Public Corporation, Atsugi-shi, Kanagawa 243-01, Japan (75)

T. E. SEIDEL, AT&T Bell Laboratories, Murray Hill, New Jersey 07974 (1)

Preface

It has been twenty years since the first MOS transistors were demonstrated using the SiO_2–Si system. This system is unique in that thermal SiO_2 films possess high dielectric strength and its interfaces contain manageable amounts of interfacial states, both conditions being essential to a successful MOS transistor. In conjunction with Si planar technology, the MOS-based integrated circuits are impacting our daily lives on a scale not encountered since the Industrial Revolution. The MOS circuit performance has steadily improved with the advent of fine-line lithography and is expected to surpass that of bipolar transistor circuits. The *Applied Solid State Science* serial publication has followed these exciting developments through judicious selection of review articles, although they have been somewhat disjoint. The time is now ripe for presenting a package of reviews, in the form of supplementary volumes to the publication, on the current status of MOS device physics, which has shown remarkable maturity during the past five years, and of device processing technology, which is still undergoing almost daily improvement.

The first supplementary volume begins with a chapter by John R. Brews. This chapter develops the most complete theory to date of long-channel MOS transistors on good physical foundations. Important device parameters are derived in closed form, mostly compact enough to aid circuit simulations, based on sound approximations with clearly defined validity. The chapter closes with an examination of short channel effects that indicate the future direction in research. The chapter has been written in a tutorial spirit and should prove an excellent text for students in undergraduate and graduate school, as well as a guide to practicing scientists and engineers. The first volume also contains two more chapters designed to introduce readers to emerging, next-generation integrated circuits. One is a review article by Yoshio Nishi and Hisakazu Iizuka covering the recent efforts to develop nonvolatile semiconductor memories. An ideal memory stores data permanently, yet permits fast access using a minimum of energy, and is physically compact. It appears that silicon technology is evolving to finally create such an ideal memory. The readers should find this chapter both illuminating and exciting. The final chapter of the first volume, by Alfred C. Ipri, reviews the current status of silicon-on-sapphire (SOS) technology. This article assesses the future of SOS technology, which is presently at a crossroad. Long-held promises of higher circuit performance are being challenged by the evolving VLSI and non-SOS circuits on the one hand and lingering materials problems associated with silicon–sapphire interfaces on the other. Hopefully, this chapter prepares those who wish to work toward resolving the difficulties and attaining the promised land in the near future.

The main applications of MOS integrated circuits have been in low-power circuitry (i.e., memories and logic circuits). Recent movements toward high power integrated circuits promise to carve out another major domain. The second volume, therefore, deals with the special considerations needed to achieve high-power Si-integrated circuits. The first chapter of this volume, by Richard B. Fair, lays foundation for the most important operations needed for the high-power circuitry, namely, impurity diffusion and oxidation. This chapter treats these related phenomena in light of the most recent understanding of crystal defects under thermal equilibrium in silicon. The second chapter, by B. Jayant Baliga, systematically develops essential high-power device physics and associated technology. This chapter should serve the needs of practicing scientists and engineers for immediate applications. Again, it is written in a tutorial tone and should be appropriate as a text.

The third volume contains topics on ever-evolving processing technology. Since Si-integrated circuits are matured commercial entities, new technological innovations rather than new physics tend to play a major role. It is felt appropriate, therefore, to review in this volume some of the most promising new approaches along with the new understanding of processing-related areas of physics and chemistry.

The first chapter, by G. K. Celler and T. E. Seidel, is on the transient thermal processing of silicon. The second, by Seitaro Matsuo, is concerned with the use of electron cyclotron resonance plasmas in two important materials processing techniques: reactive ion-beam etching and plasma deposition. The third, by W. Fichtner, deals with the exploding area of VLSI processing and process simulation.

The Editor wishes to thank the contributing authors for their arduous efforts and personal sacrifices that made the publishing of this volume possible. Finally, the Editor acknowledges AT & T Bell Laboratories, some facilities of which were used in editing these volumes, and especially the editorial skill rendered by Ms. Denise McGrew.

Dawon Kahng

APPLIED SOLID STATE SCIENCE, SUPPLEMENT 2C

Transient Thermal Processing of Silicon

G. K. CELLER AND T. E. SEIDEL

AT&T BELL LABORATORIES
MURRAY HILL, NEW JERSEY

I.	Introduction	2
II.	Adiabatic Annealing	4
	1. Introduction	4
	2. Absorption of Photons and Electrons	4
	3. Equipment	10
	4. Microstructure and Dopant Incorporation	11
	5. Amorphization of Si and Other Rapid Regrowth Phenomena	18
	6. Summary	22
III.	Thermal Flux Annealing	23
	7. Equipment	23
	8. Diffusion Profiles	25
	9. Defects in Beam-Annealed Si	26
	10. SPE Rate Measurements	28
IV.	Isothermal Rapid Annealing	32
	11. Equipment and General Uses	32
	12. Temperature Determination and Stress Effects	35
	13. Dopant Activation	40
	14. Dopant Diffusion (Boron and Arsenic)	44
	15. Defect Removal	53
V.	Related Rapid Thermal Processes	55
	16. Fast Evaluation of Implantation	55
	17. Shallow Junctions for ICs	55
	18. Grain Boundary Diffusion	56
	19. P-Glass Flow	58
	20. Silicide Formation	58
	21. Aluminum Contact Sintering	59
	22. Recrystallization of Si on Insulator	60
	23. Laser Gettering of Impurities	64
VI.	Summary	66
	References	67

Copyright © 1985 by Academic Press, Inc.
All rights of reproduction in any form reserved.
ISBN 0-12-002960-X

I. Introduction

Very large scale integration (VLSI) of semiconductor devices has become a reality. Its needs are the driving force for developing new processing technologies. Typical processes for fabrication of semiconductor devices can be grouped into three categories: (1) pattern definition by lithography, (2) introduction of electrically active impurities (doping) and deposition of conducting or insulating films, and (3) heat treatments for oxidations, diffusions, sintering, reflowing, silicide formation, and for annealing of defects introduced by any of the preceding processes.

Although great advances have been achieved over the years in the fields of lithography, doping, and film deposition, the delivery of heat has not changed significantly since the early days of the semiconductor industry. It usually involves insertion of a stack of silicon wafers held in a quartz boat into a resistively heated quartz tube. The heating rates are a function of the insertion speed and the number of wafers in the boat. The wafers start heating from the edges and it takes several minutes to reach the final temperature and even heat distribution. If wafers were inserted rapidly, the temperature gradients would be sufficient to cause wafer slip and bow, and boat rollers could scatter particles onto silicon surfaces. To avoid these problems wafers are heated and cooled slowly over several minutes and absorb much more thermal energy than is necessary for annealing. An additional disadvantage of conventional heat treatments is the lack of any provision for localized heating of selected areas or layers. For example, since most semiconductor devices are formed in the top $1-3$-μm surface layer of 500-μm-thick silicon wafers, it would often suffice to heat this top layer only, something that cannot be done in a tube furnace.

As devices become smaller, it becomes more important to control precisely the spatial extent of the electrically active layers. By selecting a suitable acceleration energy, the stopping range of implanted ions can be controlled quite accurately. This advantage of depth control is largely compromised by thermal diffusion of impurities during furnace annealing and during other high-temperature processing steps. For that reason VLSI devices necessitate development of "low-temperature" processing. Rapid annealing encompasses several methods of reducing the heating cycles. They range from laser annealing with nanosecond pulses of light to rapid isothermal heating over several seconds.

The rapid annealing processes can be divided into three groups: (1) adiabatic, (2) thermal flux, and (3) rapid isothermal annealing. This classification, first proposed by Hill,[1] is illustrated in Fig. 1. In adiabatic annealing, the energy is deposited right at the surface, within the top 1-μm layer, in a time too short to allow any appreciable heat loss by diffusion into the material. Consequently, the near-surface layer is melted while the rest of the sample remains at room temperature. All irradiations with energy beam pulses shorter than 10^{-7} sec fall into this category. In thermal flux annealing, the thermal diffusion length is comparable to the wafer thickness. Heat is diffused into the bulk at a substantial rate and temperature gradients span the wafer thickness. Heating with scanned electron and laser beams falls into this category and so does flashlamp irradiation. During rapid isothermal annealing the entire wafer reaches a uniform temperature. This requires a heating time of at least 1 sec. Conventional furnace heating is of course isothermal as well. The difference lies in the fact that the entire chamber is isothermal in a standard furnace, whereas in rapid isothermal heating wafers absorb most of the radiation and the walls of the enclosure are usually at a lower temperature.

In this chapter we review annealing with directed-energy beams and rapid isothermal annealing. Section II is devoted to adiabatic annealing with laser and electron beams. Pulsed melting provides a unique tool for the study of very rapid solidification phenomena, some of which are discussed. Thermal flux annealing is reviewed in Section III. It allows diffusionless

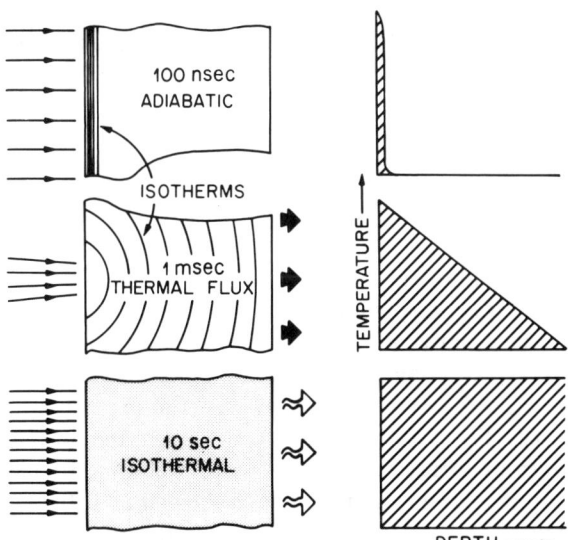

Fig. 1. Definition of three heating regimes—adiabatic, thermal flux, and rapid isothermal. (From Hill.[1] Copyright North-Holland Physics Publishing, Amsterdam, 1981.)

solid-phase regrowth of ion-implanted layers. In this section we also present some novel measurements of solid-phase regrowth rate. Section IV describes rapid isothermal annealing, from equipment and temperature measurement considerations to diffusion profiles and defect removal. Section V includes several applications stemming from rapid annealing and semiconductor processing with directed-energy beams.

II. Adiabatic Annealing

1. INTRODUCTION

In the mid-1970s Russian scientists were first to report that silicon amorphized by ion implantation recovered its crystallinity when irradiated with a short, high-intensity pulse from a ruby or Nd:glass laser.[2-4] In the process the implanted impurities were incorporated into silicon lattice and became electrically active. Since the effect of irradiation was similar to that obtained in conventional furnace annealing of ion implants, the term *laser annealing* was coined. Over the next several years laser annealing has been studied in great detail and the concept of laser annealing has been broadened to encompass almost any thermal processing of semiconductors with lasers.[5] In this section we primarily consider removing implantation damage and activating impurities using short laser pulses; we also discuss the limiting case of an extremely fast solidification. Crystalline recovery with longer, millisecond irradiation is discussed in Section III.

2. ABSORPTION OF PHOTONS AND ELECTRONS

In adiabatic annealing, a pulse from a Q-switched laser is focused on an implanted Si target to supply an energy density in excess of 1 J/cm² to the surface layer. Since pulses are 10^{-7}–10^{-9} sec long, the power density has to be in the 10^8-W/cm² range. Lasers with sufficient energy for short-pulse annealing include ruby, Nd:glass and Nd:YAG, excimer lasers, CO_2 and alexandrite lasers. For better absorption in silicon, the output of Nd lasers is often shifted from infrared wavelengths into visible by frequency doubling. Some representative lasers and their uses are listed in Table I.

The thickness of a heated layer is a function of the optical absorption depth α^{-1} and of thermal diffusivity $D = \kappa/C_V\rho$. The latter determines a characteristic thermal diffusion length $d = (2Dt_p)^{1/2}$, where κ is the thermal conductivity, C_V is the specific heat per unit mass (joules per gram per degree

TABLE I
Output Parameters of Commercial High-Power Lasers

Type	Wavelength (μm)	Operating mode	Pulse duration	Maximum power or pulse energy
Solid state lasers				
Ruby	0.694	Pulsed	0.1–6 msec	100 J
		Q^a	10–60 nsec	10 J
Nd:glass	1.060	Q	10–60 nsec	30 J
	0.530	Q	10–50 nsec	5 J
Nd:YAG	1.064	cw	—	100 W
	1.064	Q	50–200 nsec	0.1–5 J
	0.532	Q	30–100 nsec	0.1 J
		Mode locked	10–30 psec	—
Alexandrite	0.7–0.815	Q	100 nsec	1 J
Gas lasers				
Argon	0.35–0.52	cw		20 W
Krypton	0.33–0.80	cw		15 W
Excimer	0.2–0.4	Pulsed	10 nsec	1 J
CO_2	10.6	cw		1000 W
	10.6	Q	100 nsec	10 J
CO	5–6	cw		20 W

[a] Q stands for "Q-switched."

TABLE II
Physical Properties of Silicon[a]

Density (solid)	2.33 g/cm^3 (at 300°K)
Density (liquid)	2.533 g/cm^3 (at 1685°K)
Melting point	1685°K
Boiling point	3540°K
Heat of fusion	1810 J/g
Heat of vaporization	15.9 × 10^4 J/g
Specific heat (solid)	0.715 J/g·°K (at 300°K)
Specific heat (liquid)	1.0 J/g·°K (at 1685°K)
Thermal conductivity (solid)	1.47 W/cm·°K (at 300°K)
Thermal conductivity (solid)	0.25 W/cm·°K (at 1685°K)
Thermal conductivity (liquid)	0.6 W/cm·°K (at 1685°K)
Surface tension (liquid)	736 dynes/cm (at 1685°K)
Linear thermal expansion coefficient	2.6 × 10^{-6}/°K (at 300°K)
Shear modulus	7.55 × 10^{11} dynes/cm^2

[a] After Jackson and Witt.[7]

Celsius), ρ is the mass density, and t_p is the pulse duration. For short-wavelength irradiation usually $d < \alpha^{-1}$ and the laser pulse heats a layer of thickness d. An average adiabatic temperature rise is[6]

$$\Delta T = \frac{(1-R)It_p}{C_V\rho(2Dt_p)^{1/2}}, \tag{2.1}$$

where I is the light intensity (in watts per square centimeter) and R the reflectivity. By substituting appropriate values into this equation from Table II,[7] we find that ~ 1 J/cm^2 is necessary to reach the melting temperature T_M of Si with a single 50-nsec pulse. If the optical absorption depth is large compared to the thermal diffusion length, a layer of thickness α^{-1} is heated and the energy required to reach T_M is greater by the radio of α^{-1}/d.

The preceding considerations assume that all the parameters are independent of temperature. In fact, the thermal conductivity κ drops by a factor of 6 between 300 and 1683°K,[8] thus reducing heat dissipation into the substrate. The absorption coefficient of light is also sensitive to temperature.

Studies of absorption versus temperature have elucidated somewhat the issue of high-temperature absorption. Ellipsometric measurements of Si at elevated temperatures were performed by Jellison and Modine,[9] yielding band-edge absorption curves from 1.6 to 4.7 eV at several temperatures between 10 and 972°K, as shown in Fig. 2. The data indicate that for photon wavelengths between 410 and 750 nm and for substrates at 300–1000°K the absorption coefficient α depends exponentially on temperature T:

$$\alpha = \alpha_0 \exp(T/T_0). \tag{2.2}$$

The values of α_0 and T_0 were determined for a number of commonly used laser wavelengths. Jellison and Lowndes[10] also measured the optical absorption coefficient of Si versus temperature at the HeNe near-infrared line ($\lambda = 1.152$ μm), and with the help of other published data[11,12] extrapolated the high-temperature absorption values to 1.064-μm wavelength of the Nd:YAG laser, as shown in Fig. 3. These data may be useful in computing the rate of energy absorption during a high-intensity laser pulse.

Optical properties of Si are also a function of its microstructure and doping. In particular, layers amorphized by implantation reflect more light in the visible part of the spectrum and are more absorbing[13]; moreover, their absorption coefficient is less dependent on temperature.[14]

Over the years, many thermal models were developed to describe short-pulse annealing.[15–20] In one such calculation, Baeri et al.[15] obtained the threshold energy for melting Si as a function of the absorption coefficient α. Their result, based on temperature-independent values of absorption, thermal conductivity (0.28 W/cm·°K), and thermal diffusivity (0.17 cm^2/sec), is shown in Fig. 4. Under these assumptions, when $\alpha^{-1} \gg d$ the threshold energy is independent of the pulse duration. In reality, since α and κ are functions of temperature, the pulse shape determines the heating rate and the total energy needed for melting in the case of low absorption.

When light penetration depth is very small compared to the thermal diffusion length (large α), the threshold energy for melting E^{th} is proportional to $\sqrt{t_p}$, since heat flow is then determined by the pulse length.

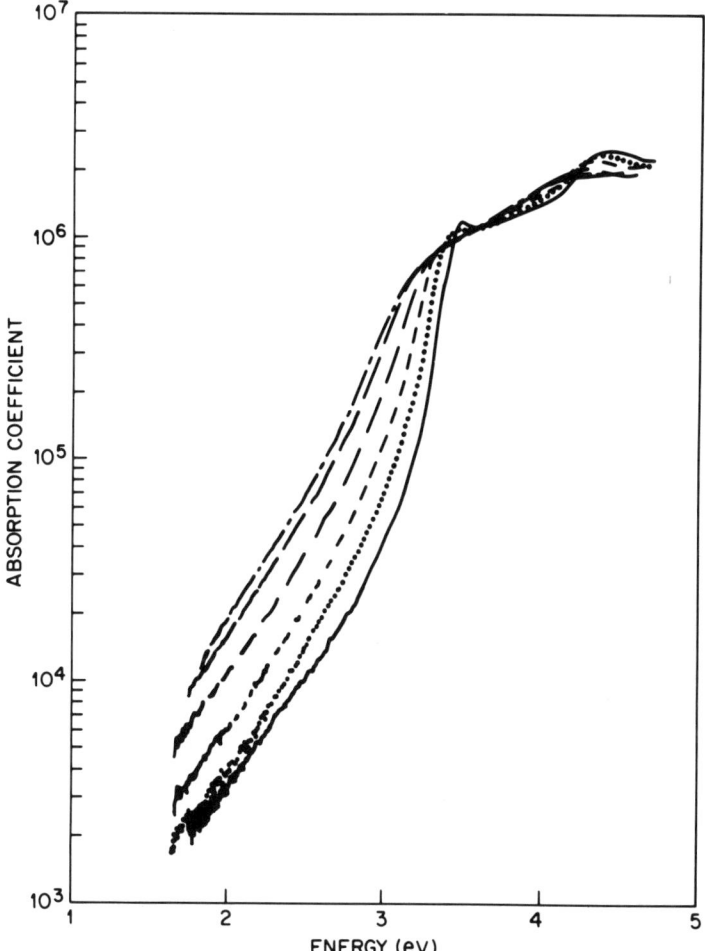

Fig. 2. Optical absorption coefficient of Si versus photon energy at several temperatures. —, 10°K; ···, 297°K; ---, 465°K, --, 676°K; ——, 874°K; —·—, 972°K. (From Jellison and Modine.[9])

For short-wavelength radiation and pulses shorter than a few nanoseconds, heating is limited to a layer < 1000 Å. Thermal gradients exceed 10^8 °C/cm and conductive cooling and solidification are so rapid that crystalline structure has no time to form. Instead, amorphous Si is formed, as discussed later.

The depth profile of energy deposited in Si by a beam of electrons is quite different from that generated by photons. The rate of energy deposition from the electron beam depends on the material density but not its crystallinity or band structure, and the reflectivity is only a function of the atomic number.

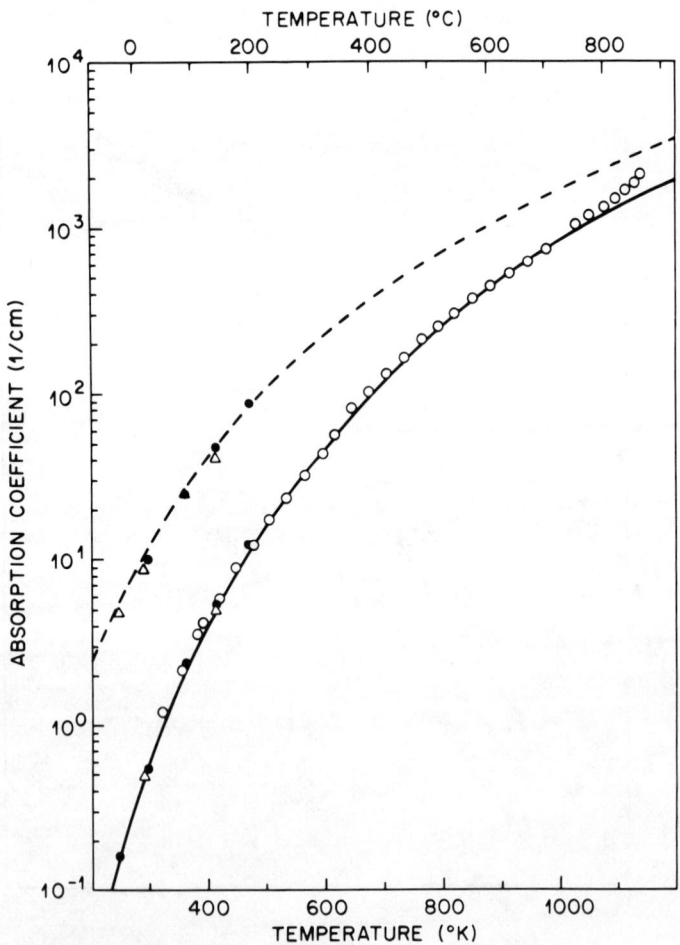

Fig. 3. Optical absorption coefficient of silicon at two wavelengths as a function of temperature. The lines represent calculated values. Solid curve, $\lambda = 1152$ nm; dashed curve, $\lambda = 1064$ nm; ○, Ref. 10; ●, Ref. 12; △, Ref. 11. (From Jellison and Lowndes.[10])

Electrons lose energy by inelastic small-angle collisions with valence and core electrons in the solid. Each high-energy incident electron leaves in its wake electron–hole pairs. The average energy expended in forming these pairs is constant for a given material and independent of the beam energy.[21] For all semiconductors, $E_{eh} \sim 3E_g$ and in particular for Si $E_{eh} = 3.64$ eV. The transfer of energy from the excited electron–hole gas to the lattice proceeds just as it does for optical excitations. Nonradiative recombination and Auger interactions reduce the number of excited carriers, while their energy is dissipated into atomic vibrations.

In Fig. 5 the number of electron–hole pairs generated in Si by an average electron is plotted for several beam energies. Clearly, for monoenergetic beams the peak of energy deposition is under the surface, at a depth that is a function of the incident beam energy. Continuously operating beams are usually at a constant acceleration voltage and the plots of Fig. 5 describe them well. On the other hand, pulsed high-fluence electron beams used for

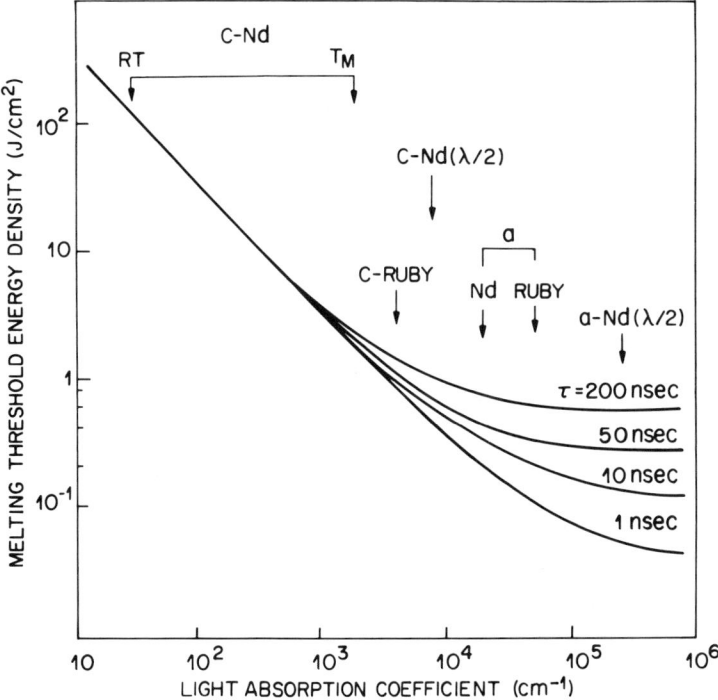

Fig. 4. Calculated energy density threshold for surface melting of silicon as a function of the absorption coefficient and pulse length. Representative absorption coefficients in crystalline and amorphous Si for a few lasers are indicated with arrows. (From Baeri et al.[15])

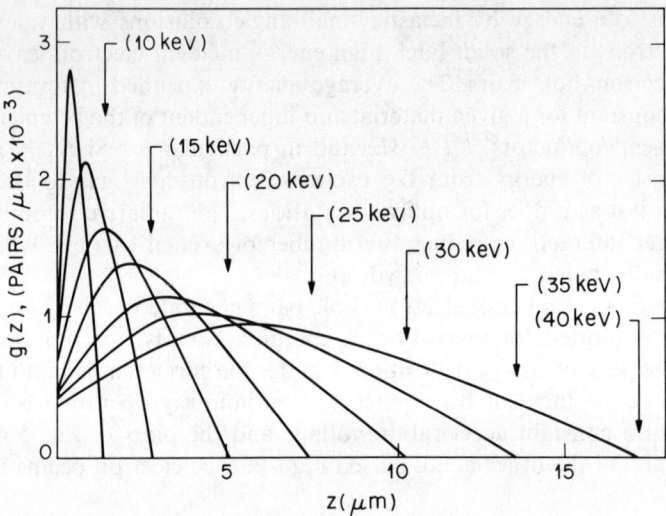

Fig. 5. Average number of electron–hole pairs generated by one electron, as a function of distance from the surface of Si, for several initial energies. (From Leamy.[21])

annealing are not monoenergetic and can provide quasi-exponential absorption spectra,[22] similar to the laser beams.

3. Equipment

Among the variety of laser types in existence only a few have a sufficient energy per pulse and a high enough duty cycle to be used for adiabatic annealing. In addition to the power consideration the wavelength has to be matched to the absorption spectrum of the target, in our case primarily that of crystalline and amorphous silicon. The early experiments were done primarily with ruby and Nd:glass lasers. They offer high-energy density in short pulses obtained by Q-switching. The absorption coefficient of silicon at the 0.69-μm wavelength of the ruby laser is $\sim 10^4$ cm^{-1} and is only weakly dependent on temperature. The main disadvantage of the ruby laser is the inherent nonuniformity of output, resulting in hot spots and damaged regions on the silicon surface.

Nd:glass systems offer higher energy efficiency but at the expense of a less suitable wavelength. Frequency doubling from 1.06 to 0.53 μm provides good absorption but lower efficiency. Glass lasers suffer also from low duty cycles necessary to cool the glass laser rod between pulses.

Excimer lasers such as XeCl (0.308 μm or 308 nm) or XeF (350 nm) appear to offer many advantages for semiconductor processing.[23,24] The laser cavity supports a very large number of longitudinal modes, reducing beam coherence and undesirable interference effects. The spatial output can be flat-topped and uniform to within a few percent, instead of gaussian as in other lasers. The pulse energy of 1–10 J per pulse is readily available and the pulse repetition rate can be high. The absorption coefficient of Si at 308 nm is $\sim 10^6$ cm^{-1}; thus, thermal gradients are determined by the pulse length. The reflectivity of Si at 308 nm wavelength of the XeCl laser is 60%, or almost double that at 532 nm, but the reflective loss of incident energy is compensated by the relatively high overall energy conversion efficiency of excimer lasers (about 3%).

Two approaches have been developed to anneal large areas of the order of a wafer diameter. The first approach has been to use high-energy beams capable of melting ~ 1 cm^2 area in a single pulse. Experiments with ruby and Nd:glass laser demonstrated feasibility but also pointed to severe non-uniformities associated with such irradiations. A significant improvement was achieved by passing the laser beam through a quartz homogenizer that allowed irradiation uniform to 5% over a 3-mm-diameter spot.[25] An alternative for large-area uniform irradiation is to use one of the excimer lasers, as they offer fairly uniform intensity profile over a rectangular output aperture.

The second approach to large-area annealing is based on a dense overlap of low-energy pulses focused into ~ 50-μm spots.[26] By translating the beam with respect to the target plane, entire wafers can be annealed with depth of melt uniform to $\sim 10\%$. If necessary, fine patterns of micrometer dimensions can be written in the amorphized surface. Nd:YAG lasers offer pulse repetition rates up to 20 kHz and a sufficient average output power to anneal 60 wafers/hr by the overlapping-spot method.

4. Microstructure and Dopant Incorporation

It has been established beyond any doubt that annealing of silicon with pulses shorter than 10^{-7} sec occurs by melting and rapid solidification of a thin surface layer. Initially, the presence of the molten zone was not obvious and some nonmelting annealing theories were proposed.[27] However, experimental evidence of melting has been accumulated that is incontrovertible. The first results that clearly indicated melting were the doping profiles after pulsed laser annealing, obtained by Celler et al.[26] These profiles, shown in Fig. 6, indicated arsenic diffusion far in excess of what could be accounted for by the solid-phase diffusivity of As in silicon at $\sim 1400°$C. The results were only consistent with the diffusivity in the molten Si, which is $\sim 10^{-4}$ cm^2/sec, about 10^7 times higher than that in the solid.

A powerful technique for detection of the molten layer, based on time-resolved surface reflectivity, was developed by Auston et al.[28] They measured the duration of the liquid phase and correlated it with the pulse energy, laser wavelength, and crystalline structure of the target. Over the years virtually all tests and experiments, described in detail in Ref. 5, proved the entirely thermal character of laser annealing and in particular showed that the energy transfer from photons through electrons into lattice vibrations took $<10^{-11}$ sec.[29-32]

Overlapped annealing allowed uniform coverage of large areas. An example of surface morphology after annealing of As-implanted Si with tightly focused pulses of 532-nm radiation from a frequency-doubled Nd:YAG laser is shown in Fig. 7.[33]

Melting of ion-implanted surface regions with pulsed-energy beams can completely remove the displacement damage caused by implantation, provided that the depth of melting exceeds that of the defective layer. When all the structural defects are obliterated by the melt, the new crystalline layer that is formed by epitaxial regrowth from the liquid phase is as good as the crystalline template on which it is formed. For that reason, the density of extended defects in pulse-recrystallized samples is often much lower than that in the equivalent samples subjected to a standard furnace anneal.[34] Defect-free material has been obtained by laser melting for a variety of

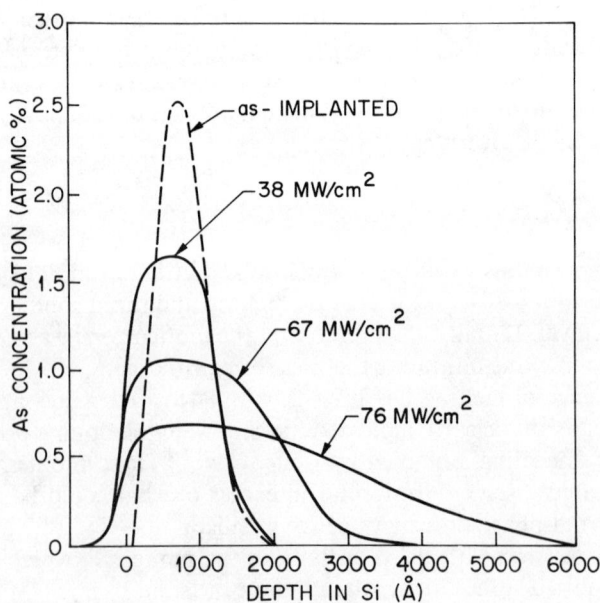

Fig. 6. Depth profiles of As implanted into (100) Si and annealed with a Q-switched Nd:YAG laser in an overlapped spot mode. (From Celler *et al.*[26])

dopants, including boron.[35] Displacement damage caused by a high-dose B implant is difficult to eliminate by solid-phase furnace annealing. In contrast, laser annealing yielded material free of dislocation loops. The ion-channeling yield that is a good measure of crystalline structure quality was also much closer to that of virgin material.

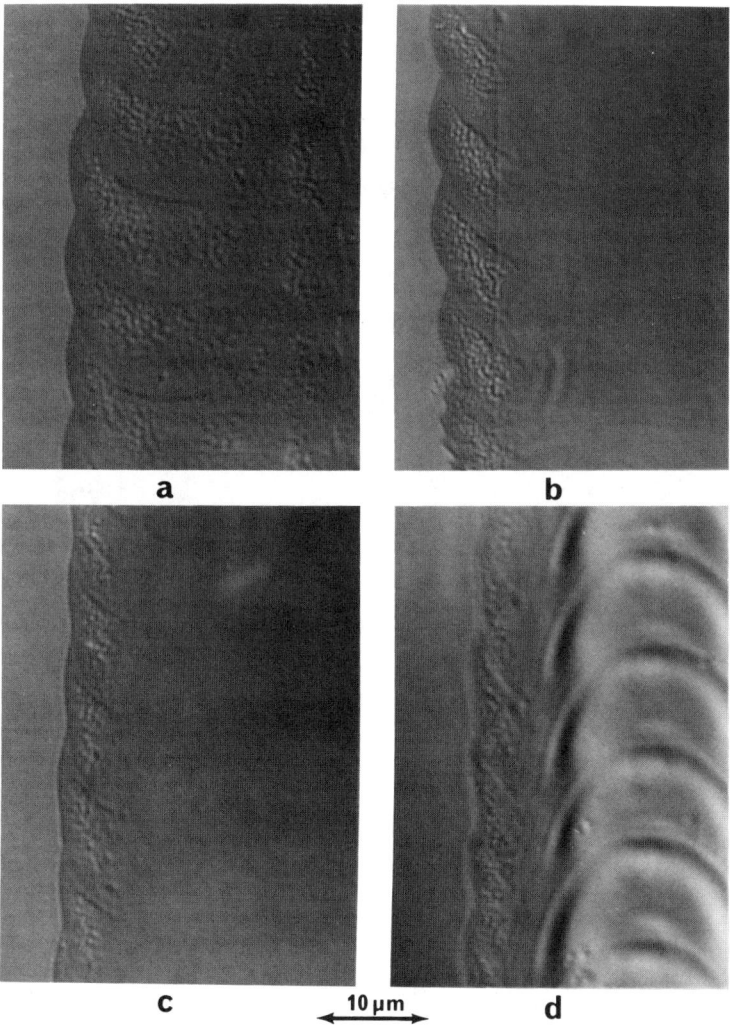

Fig. 7. Nomarski interference contrast micrographs of As-implanted samples annealed at (a) 0.6, (b) 1.0, (c) 1.5, and (d) 2.3 J/cm² with overlapping 0.532-μm laser pulses. (From Aspnes et al.[33] Reprinted with permission of the publisher, The Electrochemical Society, Inc.)

Rapid quenching associated with pulsed melting generates a high density of point defects. These were studied extensively by deep-level transient spectroscopy (DLTS) and related techniques.[36-39] Postlaser low-temperature annealing was shown[38] to reduce substantially the defect density, and passivation of defects with atomic hydrogen[40] was particularly effective, as shown in Fig. 8.

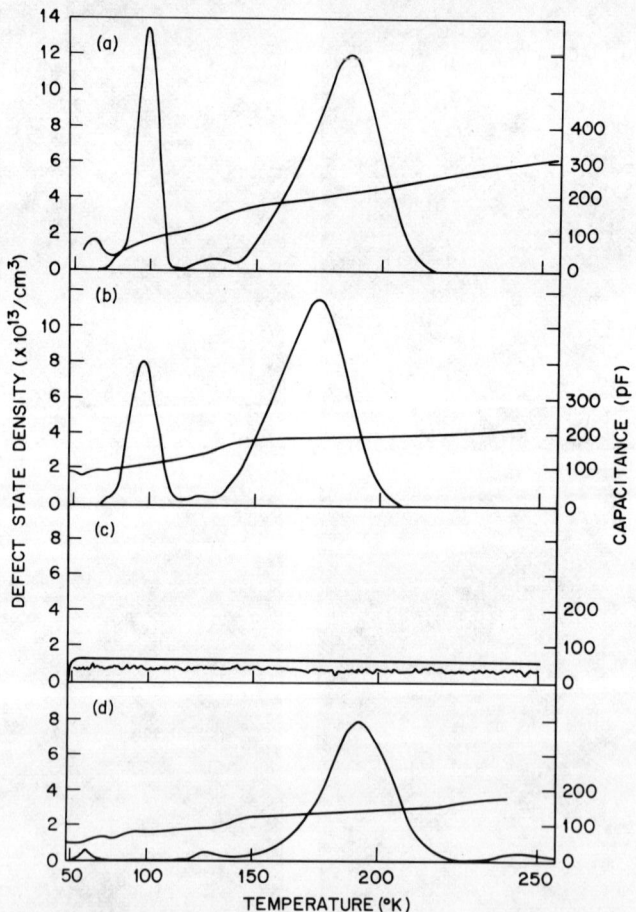

Fig. 8. (a) DLTS spectrum and capacitance versus temperature (TSCAP) scan for P-doped float-zone Si melted with a Q-switched Nd:YAG laser. (b) DLTS spectrum and TSCAP scan after 4 hr, 300°C H_2 anneal, showing no change in defect state density. (c) Same spectra showing passivation of electrically active defects after treatment in a 0.38 Torr hydrogen plasma at 200°C for 4 hr. (d) Return of defect spectra after hydrogen evolution in vacuum at 400°C for 1 hr subsequent to hydrogen plasma annealing. $n = 5 \times 10^{14}/cm^3$, $\tau_t = 9$ msec, $\lambda = 0.53$ μm. (From Benton et al.[40])

Since diffusivities of all impurities used for semiconductor device fabrication are so much higher in the molten Si than in the solid at approximately the same temperature, pulse melting can be used to obtain dopant profiles that would not be possible by conventional solid-phase diffusion. A series of approximately rectangular As profiles is shown in Fig. 9.[41] The profiles

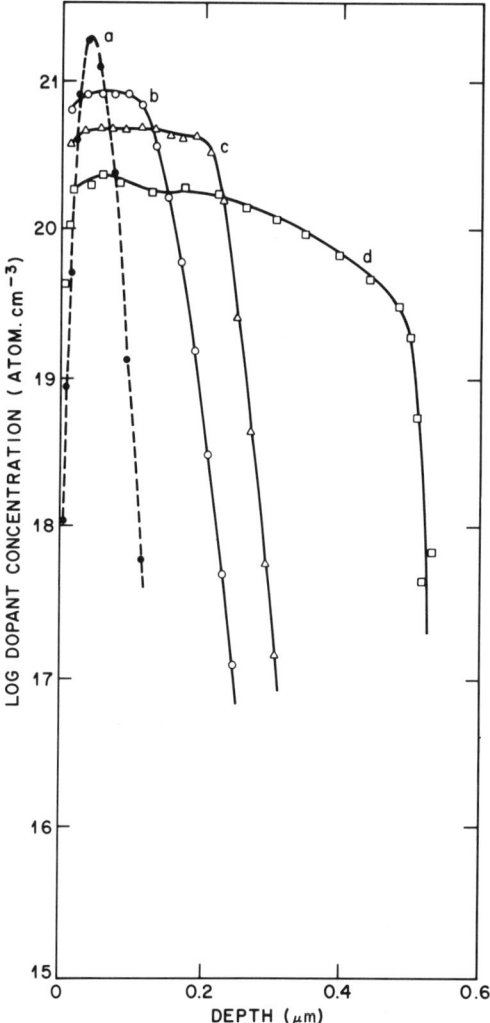

Fig. 9. Effect of laser energy density on redistribution of As implant: (a) as implanted, $10^{16}/cm^2$ at 40 keV; (b) after 1 pulse of 1 J/cm² and 9 pulses of 1.7 J/cm²; (c) after 1 pulse of 1 J/cm² and 9 pulses of 2.0 J/cm²; (d) 10 pulses of 2.5 J/cm². (After Hill et al.[41] Copyright North-Holland Physics Publishing, Amsterdam, 1982.)

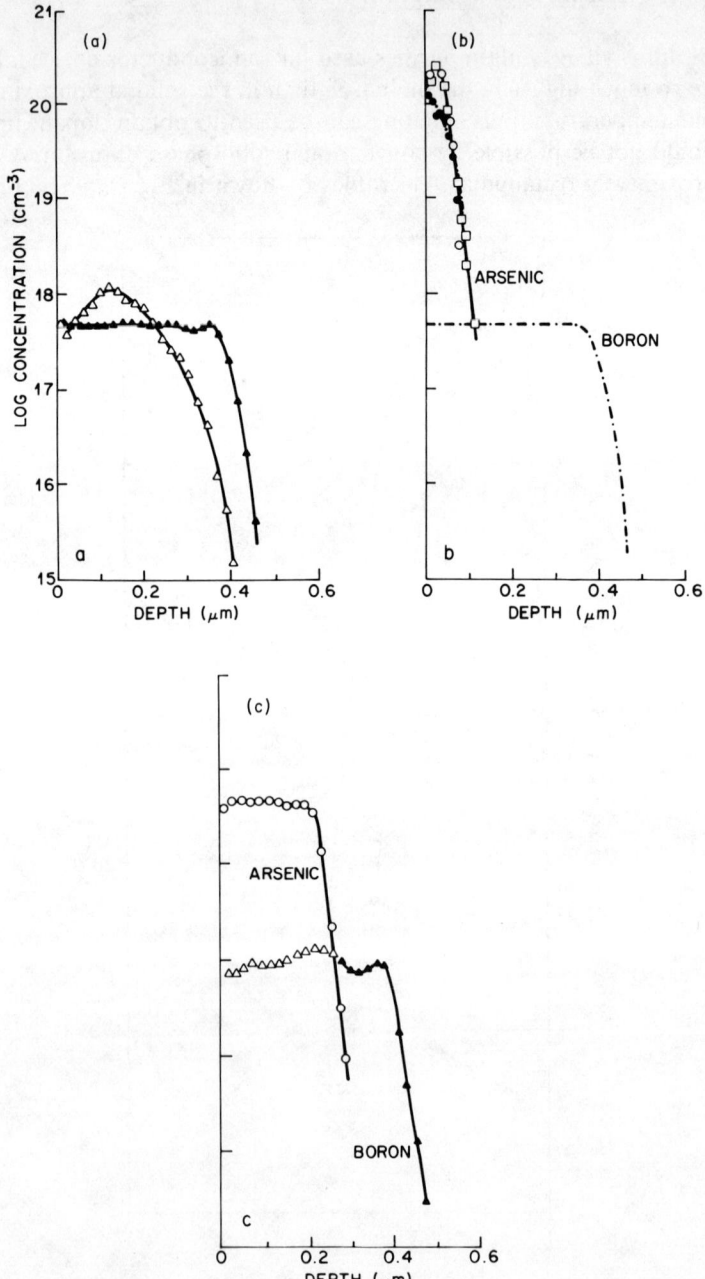

Fig. 10. Fabrication of the emitter base region of a bipolar transistor by laser processing: (a) B implant followed by 10 laser pulses (△, implanted; B, 2×10^{13}, 30 keV; ▲, redistributed, E, $1 \times 1.0, 9 \times 2.3$); (b) subsequent As implant and single-pulse annealing after arsenic implant, 1×10^{15}, 40 keV; □, arsenic; —·—·—·, boron; after laser anneal, E, 1×1.0; ○, arsenic; ●; electrons. (c) redistribution of As by nine more pulses (As, 1×10^{15}; E, 9×2.04; ○, arsenic; ▲, boron). (From Hill et al.[41] Copyright North-Holland Physics Publishing, Amsterdam, 1982.)

were obtained by repeatedly melting the same area to the same depth to allow complete diffusion of As through the liquid layer. The concentration drops sharply beyond the maximum melt depth. Bipolar transistor structures were fabricated by Hill et al.,[41] in which both boron-doped base and the As-doped emitter were formed by multiple laser pulse redistribution process, as shown in Fig. 10.

When a solidification front propagates through silicon, some impurities are rejected into the liquid while others tend to segregate into the solid. This

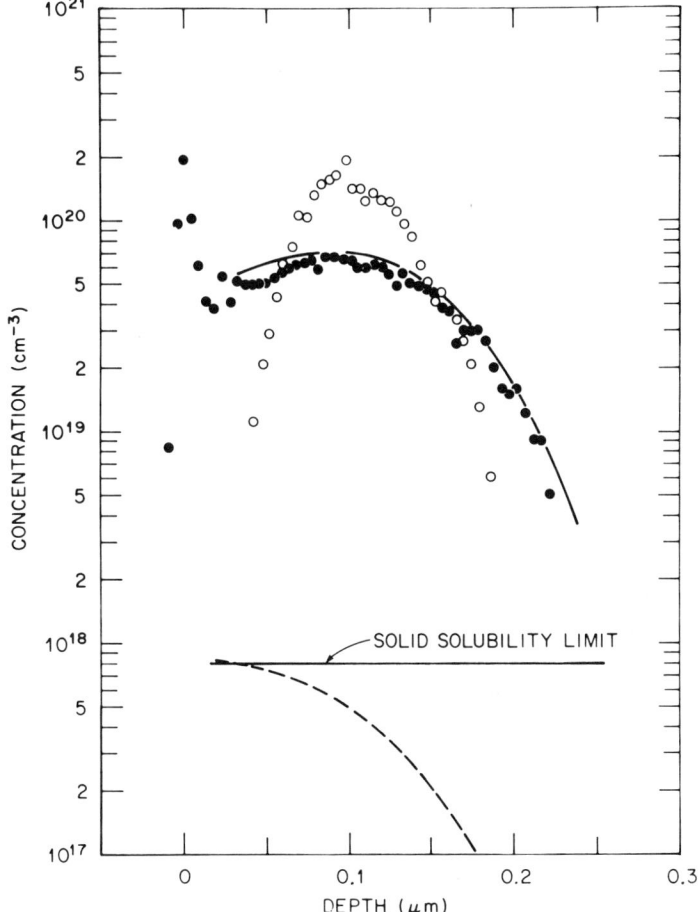

Fig. 11. Measured depth profiles of ^{209}Bi before and after laser annealing compared with profiles calculated for two values of the distribution coefficient k. ○, implanted; ●, laser annealed; —, $k = 0.4$; - - -, $k = 0.0007$, calculated; 250 keV 1.2×10^{15}/cm^2 in ⟨100⟩ Si. (From White et al.[43])

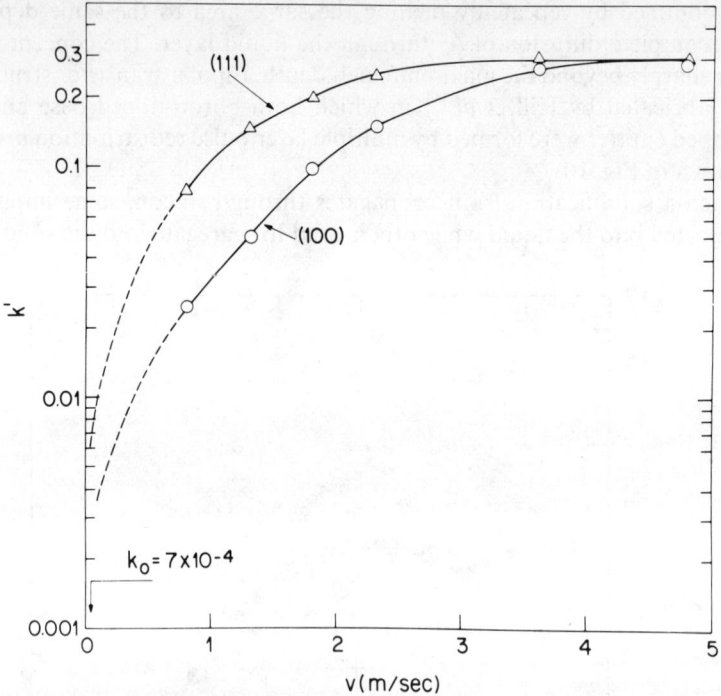

Fig. 12. Bi segregation coefficients in Si as a function of liquid–solid interface velocity for two crystal orientations. (After Baeri et al.[44])

behavior of impurities is governed by their distribution coefficients $k = C_S/C_L$, where C_S and C_L are impurity concentrations in the solid and in the liquid near the interface. Near equilibrium, values of distribution coefficients k_0 are determined from the phase diagrams, but during rapid solidification $k \neq k_0$.[42] A striking example of the distribution coefficient k exceeding its equilibrium value by more than three orders of magnitude is presented in Fig. 11, where depth profiles of Bi are shown after implantation and after pulsed laser annealing.[43] The actual dopant profile after recrystallization agrees with a calculated value of $k = 0.4$, whereas $k_0 = 0.0007$. By varying melt conditions, a range of regrowth velocities can be obtained and allows measuring k as a function of the solidification velocity, as shown in Fig. 12.[44]

5. Amorphization of Si and Other Rapid Regrowth Phenomena

When solidification velocity exceeds a threshold value of ~15 m/sec epitaxial growth can no longer occur. Instead, an amorphous Si layer is

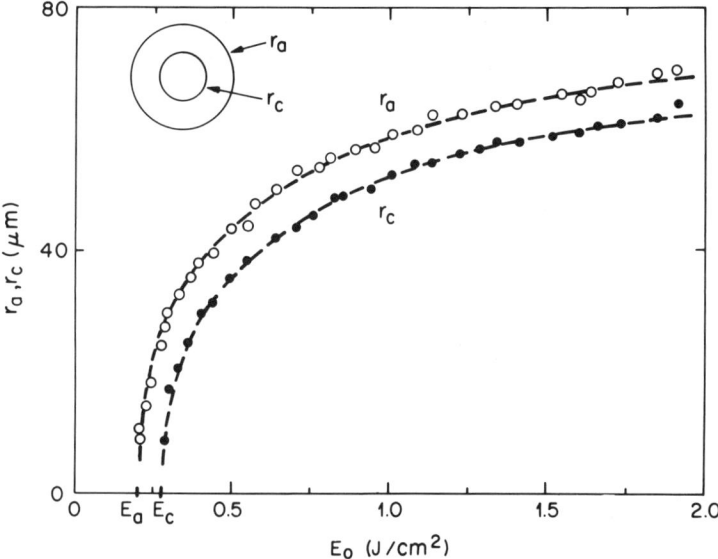

Fig. 13. Inner and outer radii of the amorphous ring following the 20-psec, 530-nm laser pulse, as a function of energy fluence at the center. (From Yen et al.[47] Copyright North-Holland Physics, Publishing, Amsterdam, 1982.)

formed. This phenomenon was first discovered by Harvard and IBM groups in 1979.[45,46] Rings of amorphous Si were observed after solidification of extremely shallow molten zones in crystalline silicon. In one case the melt was formed with 30-psec pulses at 532 nm[45]; in another, 10-nsec pulses at 266 nm gave a similar result.[46] Short pulses and large absorption coefficients are necessary for maximum temperature gradients that can drive the solidification front at velocities exceeding the epitaxial growth limit. Yen et al.[47] determined the range of energy densities required for amorphization with 20-psec pulses at 532 nm, as plotted in Fig. 13. Since too high energy density slows regrowth enough to cause crystallization, gaussian laser beams usually form an annulus of amorphous Si, since only at the beam perimeter is the melt shallow enough and consequently the solidification rapid enough to suppress crystallization. Cullis et al.[48] obtained large amorphized regions by irradiating silicon with a beam of a frequency-doubled ruby laser, shuttered to emit 2.5-nsec pulses and scrambled with a fused silica beam homogenizer to yield light intensity uniform to within a few percent across a 3-mm molten spot. TEM cross sections of amorphized Si layers are shown in Fig. 14.[49] They illustrate the difference in amorphization behavior of (100) and (111) surface regions.

Baeri et al.[50] correlated the onset of amorphization with the velocity of the solid–liquid interface. The threshold values of 18 and 15 m/sec were calculated for the $\langle 100 \rangle$ and $\langle 111 \rangle$ freezing directions, respectively. Measurements of the interface velocity by a transient conductivity technique gave a slightly lower value of 15 m/sec for the $\langle 100 \rangle$ direction.

Explosive crystallization occurs when amorphous material transforming into crystalline structure releases more latent heat than can be conducted away, thus heating the adjacent amorphous region to crystallization temperature and propagating the crystallization front some distance from the initial stimulus. Explosive crystallization phenomena have been associated most often with laser irradiation of amorphous films on thermally insulating substrates such as fused silica.[51-53] Depending on the initial substrate temperature, thickness of silicon or germanium films, thermal properties of the substrate, and energy density stored in the film, self-propagating crystallization could have a limited extent or be of a runaway nature, encompassing the whole sample.[54,55]

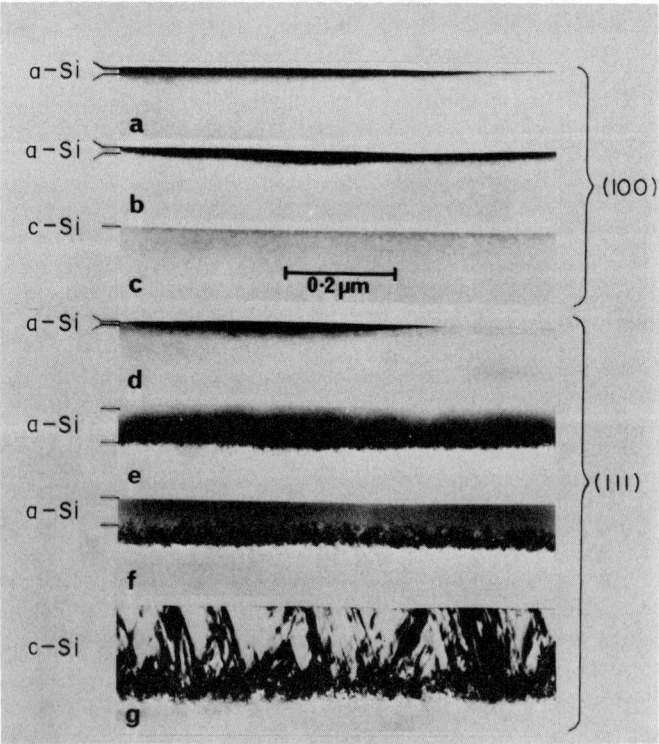

Fig. 14. TEM cross sections of Si layers irradiated with 2.5-nsec uv pulses. Si (100); (a) 0.20, (b) 0.27, and (c) 0.40 J/cm^2; Si (111); (d) 0.20, (e) 0.5, (f) 0.55, and (g) 0.9 J/cm^2. (After Cullis et al.[49])

Very shallow melting of implant-amorphized Si with short pulses, combined with transient conductivity measurements, have made it possible to observe the simultaneous existence of two melt fronts in the samples.[56] Specifically, laser energy density was sufficient to melt only about 14 nm of the surface layer of the 300-nm a-Si film. As this layer began to solidify, it released enough heat of crystallization to launch a narrow molten zone in previously unmelted material. This zone propagated through the amorphous material at 10–20 m/sec, driven by the latent heat, leaving in its wake fine-grained polycrystalline silicon. The sequence of events is illustrated in Fig. 15.[57] Incidentally, these observations of the self-propagating thin liquid

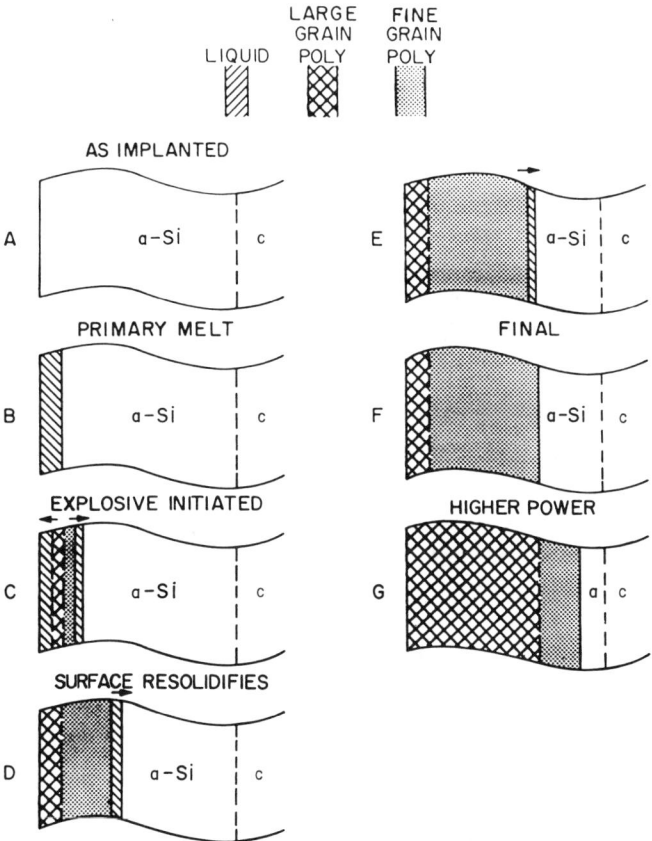

Fig. 15. A schematic sequence of events during explosive crystallization of amorphous Si under laser-melted thin layer. (From Brown.[57] Copyright North-Holland Physics Publishing, Amsterdam, 1984.)

layer confirmed earlier predictions that explosive crystallization requires the liquid phase as an intermediary.[58,59]

Further investigations of rapid regrowth phenomena are in progress in many laboratories. Laser annealing has opened for study an entire new range of crystal growth and solidification rates, and a new understanding of the rapid growth processes is emerging.

6. Summary

Adiabatic annealing has some unique advantages over other annealing methods. Whereas the near-surface layer of a typical thickness <1 μm is at or above the melting temperature, the rest of the substrate is at room temperature. In other words, thermal energy is delivered only to the implantation-damaged layer, where it is needed for annealing and none is wasted to heat the rest of the wafer. Since wafers remain at a low temperature, contamination of bulk silicon with impurities is eliminated and minority-carrier lifetimes in the material are not degraded. Since the defective layer is melted, the material loses all memory of the defects, provided that no defects extend beyond the molten region. Rapid solidification with the associated solute trapping effects permit a very high concentration of dopants to be incorporated into the lattice, although in a metastable configuration. Since diffusion rates in the liquid are 10^7 times higher than those in the solid, dopant profiles that are very different from the conventional gaussian and erfc functions can be obtained. Such profiles appear advantageous for solar cells and some bipolar devices. Selective annealing of small areas and even writing of crystalline patterns in amorphized Si are easily accomplished. Finally, adiabatic annealing can be adapted easily for cassette-to-cassette processing.

Adiabatic annealing also has some disadvantages in comparison to furnace annealing. First, it is not a batch process; the beam addresses one point at a time. In this respect, it is like ion implantation, and similar methods of moving the beam relative to the wafer can be used. Since radiation comes in pulses, additional consideration has to be given to the overlap of the irradiated spots. Second, energy density and irradiation uniformity have to be controlled carefully, a task more difficult than controlling temperature in a furnace. In particular, absorption of light depends on the optical properties of silicon and on thickness of any transparent layers on the surface. Third, rapid dopant diffusion within the molten layer cannot be avoided, limiting the utility of adiabatic annealing to devices where such redistribution can be tolerated or is desirable, as in solar cells and some bipolar structures. Finally, although extended defects are easily eliminated by melting, point defects are introduced below the molten layer by rapid quenching, requiring further low-temperature annealing or some other postmelting procedure such as hydrogen plasma annealing.

III. Thermal Flux Annealing

Silicon amorphized by ion implantation can be recrystallized by heating to temperatures well below the melting point. Regrowth then occurs by solid-phase epitaxy (SPE), a process in which atoms move into regular positions of the diamond lattice to reduce the Gibbs free energy of the system. Furnace annealing studies in the 400–600°C range established that reordering proceeds by a planar motion of the amorphous–crystalline interface. Rutherford backscattering (RBS) and ion channeling were used by Csepregi et al.[60] to determine the kinetics of SPE. The velocity of the interface v was found to depend on temperature through an Arrhenius expression

$$v = v_0 \exp[-E_a/kT],$$

with the activation energy $E_a = 2.35$ eV and the parameter $v_0 = 3.2 \times 10^6$ cm/sec for regrowth in the $\langle 100 \rangle$ direction. Csepregi et al.[60] also determined that among the high symmetry directions regrowth was fastest along the $\langle 100 \rangle$ axis and the slowest in the $\langle 111 \rangle$ direction.

7. Equipment

Equipment for thermal flux annealing includes cw (continuous wave) lasers such as argon- or krypton-ion, CO_2, and cw Nd:YAG lasers. Many scanning electron-beam systems are also suitable for this type of annealing, as are flashlamp based heaters. In all cases the effective dwell time is between 10^{-5} and 10^{-1} sec; i.e., it spans the whole time spectrum between short pulses requiring the intermediate liquid phase for annealing and isothermal heating. Lasers are scanned in a similar fashion to pulsed lasers, but for best results a uniform scan rate is necessary; therefore, wafer stages powered by dc motors are often used. The alternative is to scan the beam itself at a constant rate with galvanometer-controlled mirrors.

Argon and krypton ion lasers have been most popular in scientific investigations but CO_2 and Nd:YAG lasers [61] should be better for commercial use. Both types are more rugged than the ion lasers, are widely accepted in industrial applications, and offer high-energy conversion efficiencies. Their main disadvantage is a poor coupling of light to the surface of silicon at 300°K. Any 10.6-μm radiation of a CO_2 laser is absorbed primarily by free carriers, while 1.064-μm light from a Nd:YAG laser is in the tail of the fundamental absorption edge. Preheating of samples largely alleviates this problem, since the density of free carriers increases exponentially with temperature, while the absorption edge shifts to lower energies. As a result, the absorption at both wavelengths increases rapidly with temperature, as was pointed out in Section II. Temperatures of 200–400°C were found sufficient to obtain controllable annealing of ion-implanted surface layers with cw infrared beams.

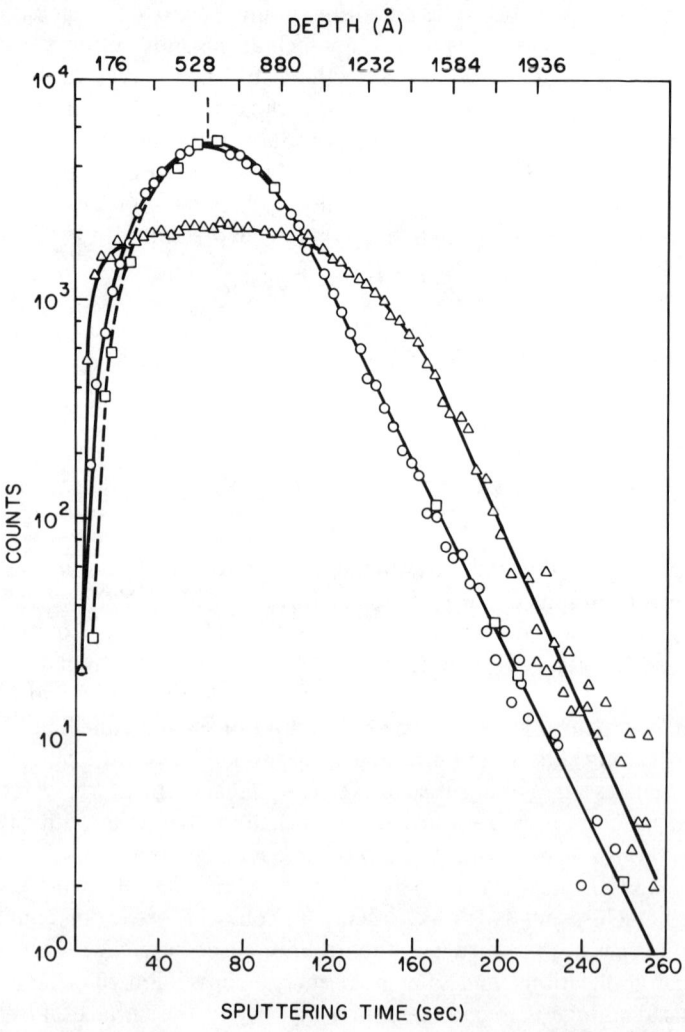

Fig. 16. Comparison of arsenic concentration profiles after solid-phase laser annealing and after furnace annealing. □ as implanted; ○, laser anneal; △, thermal anneal (1000°C, 30 min); --- PEARSON IV; distribution with LSS range statistics. (From Gat et al.[62])

8. Diffusion Profiles

The most attractive and unique feature of thermal flux annealing is suppression of dopant diffusion. Gat et al.[62] and Williams et al.[63] showed first that irradiation of ion-implanted (100) Si with a cw Ar or Kr laser caused complete solid-phase regrowth of the amorphized region and dopant activation without any redistribution.

In Figure 16, three impurity profiles obtained by SIMS are shown.[62] The as-implanted and laser-annealed profiles are essentially identical, whereas the profile after furnace annealing is considerably different. There has been no diffusion of the implanted species during the laser anneal, a consequence

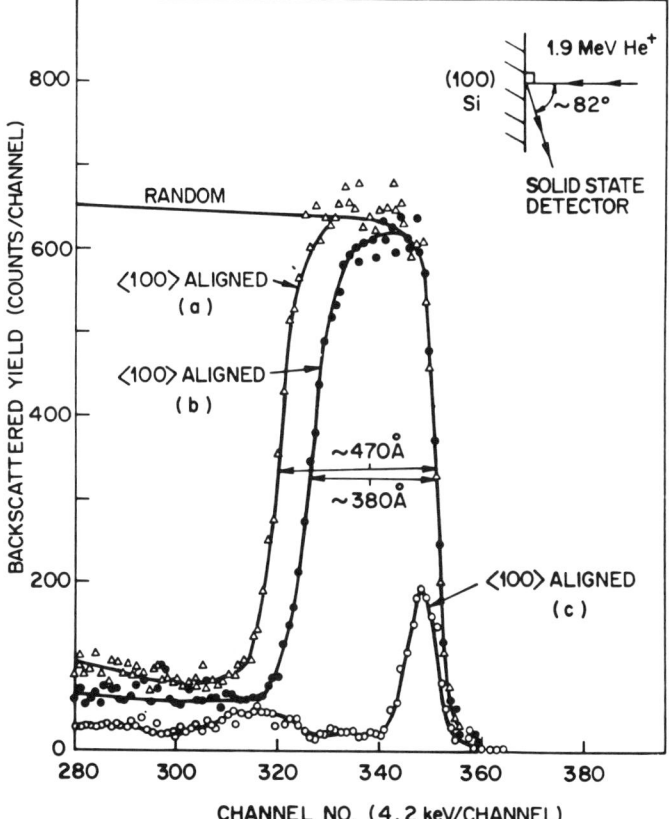

Fig. 17. Glancing-angle RBS random and $\langle 100 \rangle$ aligned spectra of As-implanted (100) Si. The aligned spectra show progressive regrowth of amorphous layer (a) before laser annealing, (b) at an early stage of regrowth, and (c) completely recrystallized. 30-keV As → Si (100). Dose: 3×10^{15} As/cm², cw argon laser annealed. (After Williams et al.[63])

of the short dwell time (1–10 msec) of the heat source at any surface location. Grazing angle He-ion channeling provided additional insight into the regrowth mechanism. Aligned spectra before annealing and after partial and complete anneal, in Fig. 17,[63] show a progression of the crystalline–amorphous interface toward the surface, consistent with solid-phase epitaxial regrowth.

Recovery of crystalline structure by solid-phase epitaxy and dopant substitutionality do not necessarily indicate complete electrical activation. This is particularly true for implants of As. It has been observed that although the chemical solubility limit is $\sim 3 \times 10^{21}/cm^3$, the maximum stable electrically active concentration is only $3 \times 10^{20}/cm^3$.[64] A higher electrical activation can be achieved by cw laser annealing, or even by furnace annealing,[65] but this concentration is metastable and relaxes to a lower value on subsequent thermal annealing.[66]

Laser annealing preserves the as-implanted depth profiles of dopants, and this may be advantageous in designing integrated circuits with submicrometer design rules. Moreover, beam writing of crystalline patterns in amorphized surface is possible, and if followed by wet chemical etching to remove the remaining amorphous region would result in nonlithographic pattern definition.

9. DEFECTS IN BEAM-ANNEALED Si

Although ion-implantation-amorphized Si layers can be recrystallized rapidly by SPE with lasers and electron beams, the complete removal of extended and point defects remains more elusive. Rapid quenching from the high temperature always leads to a high density of vacancies and interstitials, and these frequently condense into dislocation loops. Since large lateral temperature gradients exist during scanned beam annealing, the distribution of defects is inhomogeneous. A spatial map of the minority carrier recombination rate typically forms a striped pattern, with alternating defect-free and defective regions replicating the path of the annealing beam. Such a pattern, as detected with an SEM in the electron-beam-induced current (EBIC) configuration is shown in Fig. 18, for a Si sample scanned at room temperature with a cw Ar laser.[67] Similar inhomogeneity in the carrier collection efficiency was seen after CO_2 laser annealing[67] and after electron beam annealing with incomplete overlap of scan lines.[68]

Although in all these cases the annealing conditions were far from optimum, the results point to an inherent problem associated with sequential scanning. Interstitial atoms created at the high temperature agglomerate into dislocation loops on temperature quenching when the beam moves

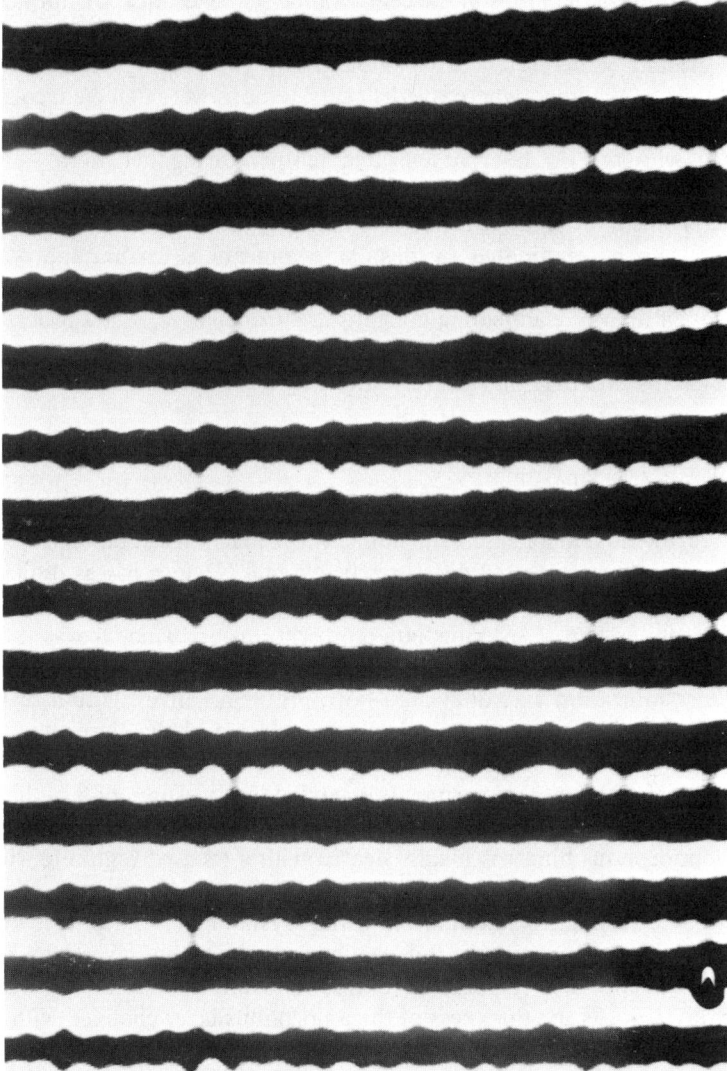

Fig. 18. EBIC-contrast SEM micrograph of Si surface annealed with a cw argon laser. The black stripes represent reduced charge collection. (After Baumgart et al.[67] Copyright North-Holland Physics Publishing, Amsterdam, 1982.)

away. The interstitial loops in the center of the scanned track are annihilated by the thermally activated climb processes, and these areas corresponding to the bright regions in the EBIC photographs provide high carrier collection. Edges of the lines, where the dislocation loops were detected, correlate with black regions in the EBIC contrast. When tracks of width >1 mm were scanned with a CO_2 laser at elevated temperatures, the lateral thermal gradients were greatly reduced and no dislocation loops were detected.[61]

The problem of inhomogeneous annealing associated with sequential scanning can be eliminated by flashlamp annealing. In this approach, a pulse of light, 50–500 μsec long, from an arc lamp floods the whole surface and heats it up to the annealing temperature. Cohen et al.[69] and others[70–72] demonstrated annealing of implanted Ge and Si by this method and flashlamp systems capable of large-area annealing have become available.[73]

10. SPE Rate Measurements

Before the advent of laser annealing, all furnace measurements of SPE kinetics were limited to temperatures below 600°C, since it was difficult to raise the temperature rapidly enough to avoid substantial crystallization of the amorphous film at intermediate temperatures and since it was difficult to monitor rapid crystallization. Laser heating and optical *in situ* monitoring of the crystallization rate alleviated both problems. Investigation of laser-induced crystallization rate by Olson et al.[74] proved that laser annealing in the solid phase occurs through the same SPE process as furnace annealing. More significantly, these experiments extended the measurements of the crystallization rate up to 1000°C and in some cases above 1300°C.

The amorphous film was locally heated with a focused high-intensity cw Ar laser beam, and time-resolved reflectivity of an auxiliary He–Ne laser beam served to monitor the motion of the crystalline interface. It is known that amorphous surface layers and even buried amorphous layers give rise to interface effects, since the refractive indices of amorphous and crystalline Si are different. As the thickness of the amorphous layer changed with time, interference between light reflected from the surface and the advancing interface caused oscillations in the net reflectivity, as shown in Fig. 19. Since each consecutive interference maximum corresponds to a change in amorphous layer thickness of $\lambda/2n$ (652 Å in Si for 6328-Å He–Ne radiation), it is possible to extract directly the rate of interface movement. The temperature during crystallization is determined from the change in reflectivity of fully recrystallized silicon on removing the intense heating beam, and the known dependence of silicon reflectivity on temperature, as shown in Fig. 19b. An example of SPE rate obtained by measuring time-resolved reflectivity during crystallization of As-implanted (100) Si is shown in Fig. 20. From such measurements Olson et al.[74] obtained more accurate values of the activation energy and of v_0.

One intriguing aspect of the SPE measurements above 1300°C is the absence of melting of the amorphous layer. Pulsed laser melting experiments indicate that amorphous silicon melts as much as 200–300°C below the melting temperature of crystalline Si. Such a depression of the melting point has been predicted on the basis of thermodynamical considerations by Bagley and Chen[75] and by others.[76] The shape of the SPE curve versus temperature also indicates that $T_{M_c} \neq T_{M_a}$, but there is no trace of melting in the data. This issue has not been resolved yet, although it has been speculated that in cw laser heating, which is slower than that induced by pulsed

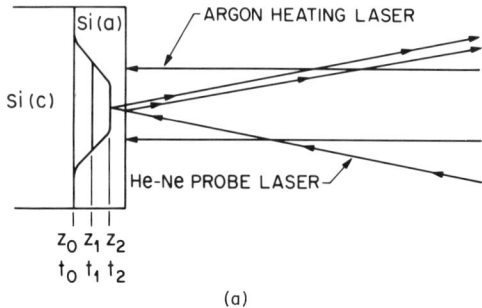

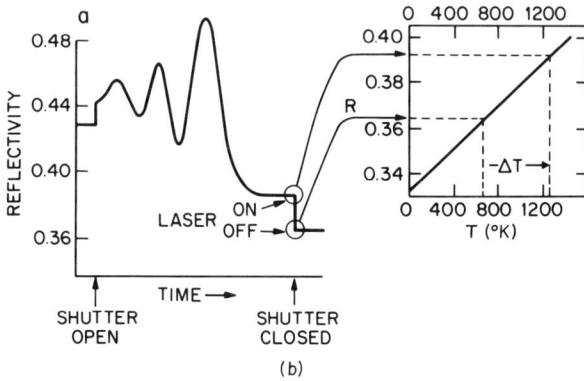

Fig. 19. (a) Schematic view of SPE growth induced with an argon laser and monitored with a He–Ne laser. (b) Time and temperature dependence of the reflectivities. Interface positions z_i are obtained from the analysis of the oscillatory reflectivity at times t_i. Temperature is determined from the change in reflectivity caused by removal of heating beam and the known dependence of optical parameters on temperature. (From Olson et al.[74] Copyright North-Holland Physics Publishing, Amsterdam, 1983.)

irradiation, amorphous Si can be superheated and transform directly in crystalline Si.

It is well known that many electrically active dopants, such as As, B, or P, enhance the SPE rate, whereas nonactive impurities, such as O, C, and noble gases, retard the rate. Technologically important BF_2 implants con-

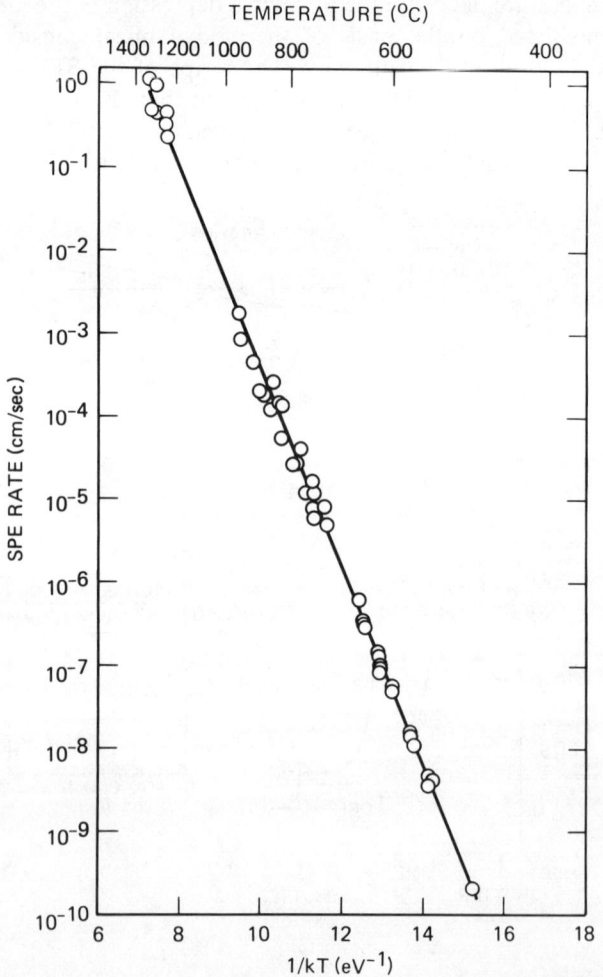

Fig. 20. Dependence of SPE rate on temperature in arsenic-implanted Si(100). All rates measured by time-resolved reflectivity. As → Si(100); $5 \times 10^{14}/cm^2$, 150 keV, $E_a = 2.76 \pm .05$ eV, $v_0 = 3.68 \times 10^8$ cm/sec. (From Olson et al.[74] Copyright North-Holland Physics Publishing, Amsterdam, 1983.)

tain both a rate-enhancing and a rate-retarding species. Olson *et al.* measured separately the regrowth kinetics of boron and fluorine implanted into preamorphized Si, as shown in Fig. 21.[77] The fluorine reduced the SPE rate over the whole range of temperatures, but in particular at low temperatures, since the activation energy of F-implanted Si was 3.06 eV, considerably

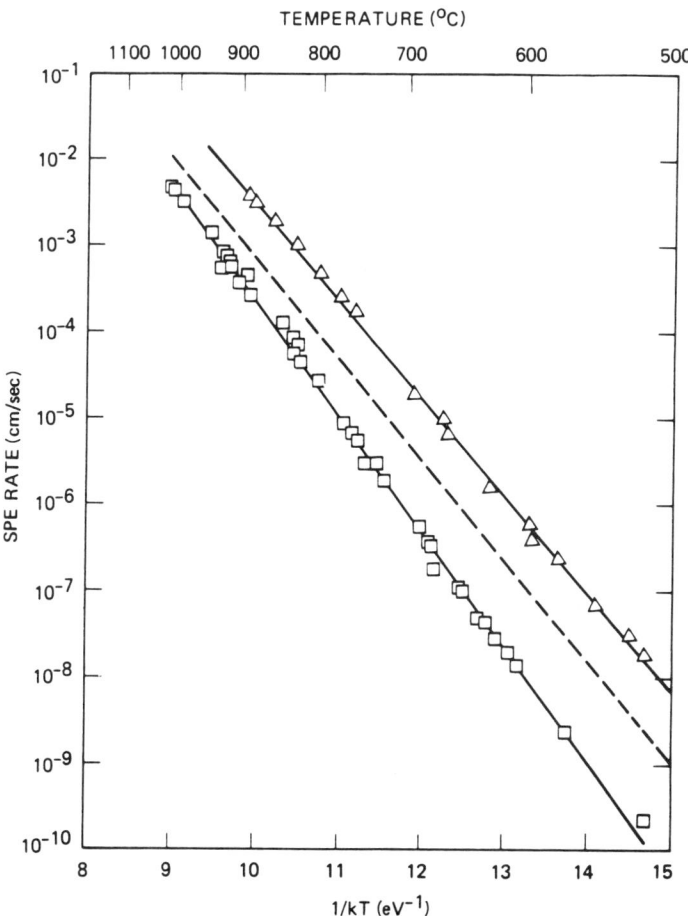

Fig. 21. Effect of temperature on SPE rate in fluorine- (18 keV, 6×10^{15} cm^{-2}) and boron-implanted (10 keV, 3×10^{15} cm^{-2}) samples. Si(100) substrates were preamorphized by Si implantation. Dashed line represents intrinsic SPE rate. △, B + Si → Si(100), $E_a = 2.59$ eV, $v_0 = 5.35 \times 10^8$ cm/sec. □, F + Si → Si(100), $E_a = 3.06 \pm .05$ eV, $v_0 = 4.35 \times 10^9$ cm/sec. ---, Si → Si(100), $E_a = 2.68 \pm .05$ eV, $v_0 = 3.07 \times 10^8$ cm/sec. (From Olson *et al.*[77] Copyright North-Holland Physics Publishing, Amsterdam, 1984.)

higher than the 2.59-eV value of the B-doped layers. Combining B and F together in the BF_2 implant caused a complex rate dependence on temperature and depth of the interface, but on average the rate was between that for B and F alone.

IV. Isothermal Rapid Annealing

Rapid thermal annealing (RTA) techniques include broad-band incoherent spectral sources with programmable anneals.[1,78] These incoherent sources allow large-area wafer processing (with larger throughput than lasers), keep the diffusion-limited aspect of rapid thermal annealing, and tend to avoid interference effects during the annealing of layered structures. The first practical applications of rapid thermal annealing are expected to use incoherent heating. Since the thermal conductivity and thermal diffusivity of silicon are relatively large, the wafer is nearly isothermal during the anneal.

11. Equipment and General Uses

Several commercial systems are now available for annealing using incoherent sources. Programmable heating durations typically range from 1 to 500 sec. High-temperature applications allow processes that are reaction-rate limited or limited by a thermal activation energy, while a relatively small and controlled thermal diffusion occurs simultaneously. For example, processes such as glass flow, attainment of impurity solubility, or defect removal after implantation require high temperatures. However, the requirement of shallow junctions for VLSI fabrication requires a limited diffusion process.

An extreme example of limited diffusion is the electrical activation of implanted boron in a preamorphized layer of silicon. Electrical activation takes place simultaneously with the solid-phase epitaxy of an amorphous layer on $\langle 100 \rangle$ oriented silicon during a ~ 10-second anneal at $\sim 800°C$. In this case the diffusion length $\sqrt{Dt}$ of boron in the solid state is only a few angstroms.[79] (This is discussed in Subsection 14.)

A variety of heating sources is available: tungsten–halogen lamps, graphite heater elements, and arc lamps. Each of these sources has a slightly different blackbody distribution; the arc lamps have relatively high intensities in the ultraviolet while the graphite sources are rich in the infrared. Power levels of the different commercial systems range from 20 to 150 kW, corresponding to a maximum silicon wafer temperature ranging from ~ 1200 to $\sim 1400°C$.

Thus, almost all temperatures used in normal silicon wafer processing are obtainable, including applications requiring melting of silicon.

The type of ambients available are determined by the design of the enclosed furnace cavity. In an open system a flush of neutral gases (e.g., N_2, Ar, O_2) is possible. In a closed system any gas, including reactive gases, may be used. A vacuum environment is also possible, to degas surfaces containing volatile components. Vacuum environments are not generally useful for silicon processing, which requires the maintenance of stabilized surfaces. One would not usually wish to lose dopant from P-glass or doped polysilicon during thermal processing.

The drive mechanism of the power source may be by open-loop (no feedback), or closed-loop methods (e.g., thermocouple feedback).[80] In the open-loop arrangement the source is driven independently of the temperature response of the wafer. In the closed loop method the temperature is controlled by feedback from a sensing thermocouple on a "control wafer." Unfortunately, the temperature of the sensing-control thermocouple may not be the same as that of the wafer to be annealed. Control by feedback from a sensing optical pyrometer appears to solve this problem.[81] Another approach is to let the lamps run continuously and expose the wafer by a mechanical shutter mechanism or by mechanical movement of the wafer.[82] The drive mechanism is critical in determining the character of short (1-sec) anneals.

Heating rates of wafers placed in a blackbody cavity depend on the optical coupling of the wafer to the source radiation. Heavily doped wafers (n^+, p^+, nn^+, or pp^+ wafers) heat up more rapidly than n^- or p^- wafers.[83]

Several examples of annealing systems are shown in Fig. 22a and 22b. In Fig. 22a the schematic shows a two-sided heating system using tungsten–halogen lamps and a water-cooled reflective enclosure.[80] A quartz isolation tube that can be sealed is placed between the lamps. The surfaces of the quartz are sandblasted to diffuse light and improve the uniformity of irradiation. The surfaces of the enclosure cavity reflect light and are water-cooled. Thus both lamps and the wafer are placed inside a reflecting blackbody cavity, leaving primarily the wafer to absorb the energy. Cooling occurs by "non-ideal" losses to the supporting quartz posts, convection, conduction, and finite absorption of the reflecting surfaces. Also shown in Fig. 22a is a thermocouple attached to a monitor wafer, for feedback to control the power to the lamps.

In Fig. 22b the schematic shows a "one-sided" lamp-illuminated system.[84] Here the lamps are cooled by flowing air, and the wafer's temperature can be measured by an optical pyrometer that views the wafer side opposite the lamps. It is common to place the polished, highly reflecting side of the wafer toward the pyrometer, if temperature calibrations were done for a

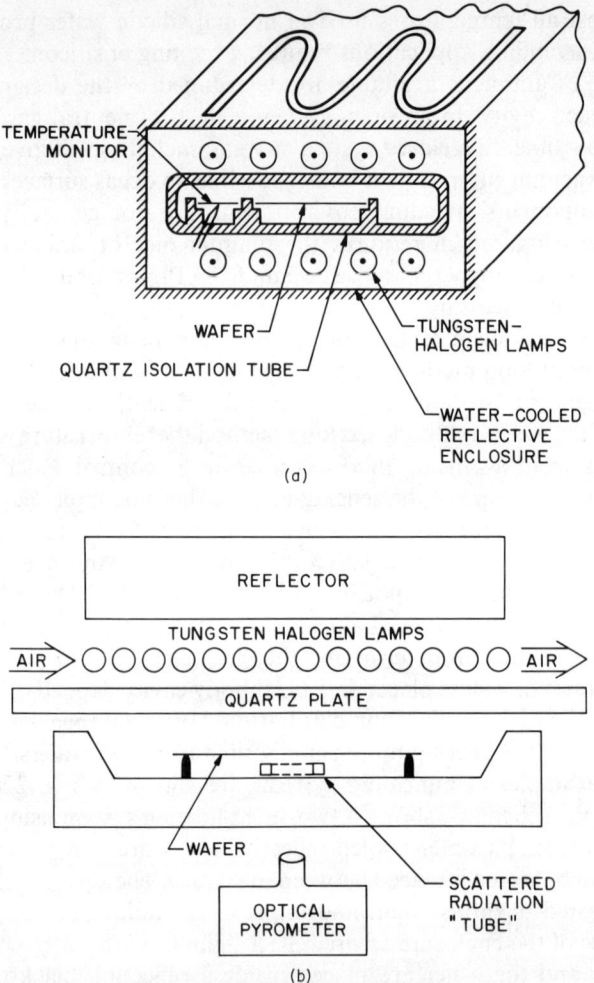

Fig. 22. (a) Two-sided heating system using tungsten-halogen lamps and a water-cooled reflective enclosure. (After AG Associates.[96]) (b) One-sided heating system, the wafer temperature is measured with an optical pyrometer. (After Lischner and Celler.[84] Copyright North-Holland Physics Publishing, Amsterdam, 1982.)

polished silicon surface. It is important to keep scattered radiation out of the pyrometer, so it is good practice to place a quartz tube between the silicon surface and the pyrometer.[85] This tube prevents scattered radiation from getting into the pyrometer. In this way, the pyrometer has "tunnel vision," to see only the wafer's emission.

Some of the commercially available incoherent rapid thermal annealing

TABLE III

RTA EQUIPMENT

Manufacturer	Source	Ambient	Max. power (kW)	1- or 2-side exposure	T_{max} (°C)	Temperature measure/control
AG2101 (AG)	W-Hal.	Closed[a]	23	2	1250	T-C/feedback[b]
IA-200 (VARIAN)	Graphite	Vac	20	1	1400	PYRO/mechanical shutter[c]
ROA-400 (EATON)	Arc lamp	Closed[d]	100	1	>1400	PYRO/feedback
180-RIP (TAMARAK)	W-Hal	Closed[e]	~20	1	1400	IR PYRO/feedback

[a] Closed quartz-walled chamber allows various gases.
[b] IR pyrometer with feedback in development.
[c] Lamps run cw, wafer is exposed by mechanical shutter.
[d] Closed metal-walled chamber.
[e] Chamber walls are gold, or quartz sputtered on gold.

systems are described in Table III (RTA equipment). The manufacturer and model numbers are listed in the first column. The reader may study the table columns to compare the detailed information. We also need to mention that the isothermal furnace equipment has cold walls, and therefore is inherently clean. The wafer temperture uniformly is limited by edge losses but with proper design can be reduced to $\pm 1°$ at 1200°C. This is an important consideration to be discussed along with its consequences in the next section.

12. TEMPERATURE DETERMINATION AND STRESS EFFECTS

Temperature measurements are by thermocouples or by pyrometers. Both infrared and visible wavelengths sensitive pyrometers are used to view the nonilluminated side of the wafer. In such an arrangement, the infrared sensitive pyrometer initially "sees" the source filament temperature until intrinsic free carriers in the silicon wafer make the wafer opaque. The infrared pyrometer gives a signal that first goes through a maximum (seeing the filament) and then gives a signal corresponding to the actual emission temperatures of the silicon. When thermocouples are used they must be small so that the temperature is determined by the wafer and not the heat capacity and properties of the thermocouple. The heating rate of the thermocouple itself should be faster than that of the wafer. Other methods are necessary to insure that one knows the temperature of the sample being heated. One method that may prove useful involves infrared pyrometers that are sensitive only for $\lambda > 4$ μm. [The detector views infrared emissions from the wafer surface, while a quartz diffuser plate (opaque to $\lambda > 4$ μm) is placed between the other wafer surface and source filaments.] Thus the detector-pyrometer will not respond to the emission of the filaments. Other methods may use annealing phenomena (a thermally activated process), oxide thickness detection,[86] lattice expansion, and so on.

Measurement of the temperature of a calibration wafer followed by a physically different wafer or even by a quite similar second wafer may lead to a temperature assignment error. This is the case if the second wafer is not identical to the calibration wafer in thickness, optical properties, and thermal losses. On the other hand, it is impractical to attach a thermocouple to each wafer that is being heated. Development continues to solve this difficult measurement problem. It is likely that long-wavelength pyrometers will provide the most reliable temperature control.

Alternatively, if one is interested in a "dedicated" furnace use, then it makes sense to use the actual annealing effect to calibrate and keep the equipment calibrated. An example may be the activation of an implanted layer such as boron, where progressively higher temperatures result in lower sheet resistance. The activation of arsenic (high dose, $>5 \times 10^{14}/cm^2$) is relatively *insensitive* to temperature changes for temperatures somewhat above that required to get solid-phase epitaxial regrowth of the implant-induced amorphous layer. After 800°C for a few seconds the resistivity of the activated arsenic layer is not much different than it is after 1050°C and 5 sec. Above 1050°C some diffusion starts to occur and the mobility is increased over the profile. Implanted boron (which does not produce an amorphous layer), on the other hand, changes its resistance during activation over a wider range of temperatures. Oxide thickness detection is useful to characterize radial thermal gradients in a RTA system.[86]

One aspect of temperature measurement is the heating rate, since the "early-time" temperatures depend on the optical absorption and reflection of the wafer being heated. When a RTA system is driven with a fixed cycle, the temperatures of n^- or p^- wafers and n^+ heavily doped wafers are different, as shown in Fig. 23. Here temperatures of ~ 900°C are reached after ~ 20 sec, but after 10 sec differences of several hundred degrees exist between lightly and heavily doped wafers.[83]

The doping effect on heating rates is due to optical absorption in the infrared for high extrinsic free carrier concentrtions. For a uniformly doped wafer the heating rate initially depends on the power absorbed. We have

$$\frac{dT}{dt} = \frac{(1-R)}{C} \int_0^\lambda d\lambda \, I_\lambda (1 - e^{-\alpha_\lambda d}), \tag{4.1}$$

where R is the surface reflectivity, which we take to be a constant; $C = C_V d$ is the specific heat per unit area (C_V is the specific heat per unit volume); λ is the wavelength; I_λ is the incident intensity at λ; α_λ is the absorption coefficient; and d is the wafer thickness, I_λ and α_λ are plotted in Fig. 24 for a source temperature of 2200°K and several doping concentrations. The free carrier coupling is significant for doping above middle $10^{18}/\text{cm}^3$ levels. It is also important for annealed (activated) high-dose implantations.[83] An adverse effect of the supporting posts is to introduce a heat sink and reduce the temperature in the wafer just around the point of contact. Better uniformity is obtained if the posts are tapered to a point.

The radial temperature gradients are a major practical problem. A wafer radiates from the perimeter region out to cold surfaces and results in gradients that in turn—if not eliminated—will introduce slip dislocations and possibly warpage in the wafers.[87,88] Slip occurs at the perimeter and propagates toward the center of a $\langle 100 \rangle$ wafer along $\langle 110 \rangle$ directions with the dislocations in $\{111\}$ planes. Initial approaches to reduce the radial gradients include surrounding the heated wafer with a donut-shaped ring of silicon, a heat source shaped as a ring, or some other extra sources near the wafer edges. Material preparation is also expected to play a role: perfection (smoothness) of wafer edges[89] and oxygen content,[90] which is related to the yield stress in the wafer, are important. The vertical gradients in temperatures are minimal ($\lesssim 1°C$) for two-sided heating for times greater than 1 sec because of the high thermal diffusivity of silicon. Vertical gradients of $\sim 100°C/$

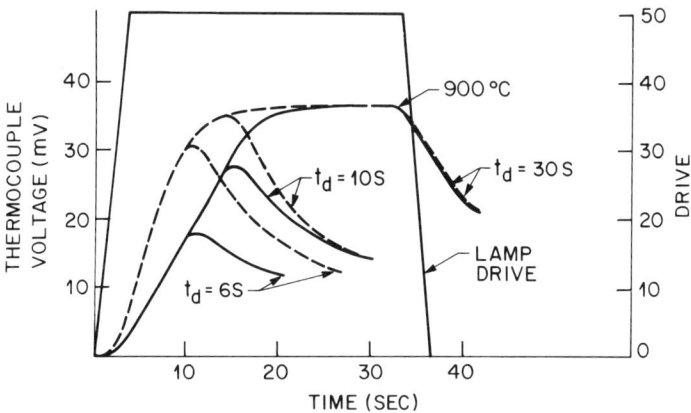

Fig. 23. Thermocouple voltage (temperature) for lightly doped n^- or p^- (—) and heavily doped n^+ (- - -) wafers as a function of time. The drive is relative power to the lamps and t_d is the duration of the constant drive. (After Seidel et al.[83] Reprinted by permission of the publisher, The Electrochemical Society, Inc.)

cm for one-sided heating are obtained that are useful for thermal gradient zone melting (TGZM) applications.[91]

Materials other than silicon couple light differently and hence have different heating rates. The absorption by metal layers is usually higher than that by silicon (even heavily doped silicon), although metals have a higher reflectivity than Si, while absorption of undoped GaAs with its larger energy gap and absorption by thin silicon-on-sapphire will be less than that of standard silicon wafers. In general, the temperature during heat-up of material placed in an adiabatically supported holder and exposed to blackbody radiation depends on the optical coupling, which is different for different absorbing media.

In many single-sided lamp systems for rapid isothermal annealing, temperature up to the melting point of silicon can be obtained. Detection of the melt at $T_M = 1685°$K can help in calibration of the temperature sensors. It should be noted that radiative melting of crystalline silicon produces rather unusual melt patterns. Although heating is uniform, any oxidized single-crystalline Si surface breaks up into a pattern of faceted molten and solid areas, with a typical side of an individual molten zone ~ 100–500 μm.[92] The shape of the molten segments is very regular and is defined by the (111)

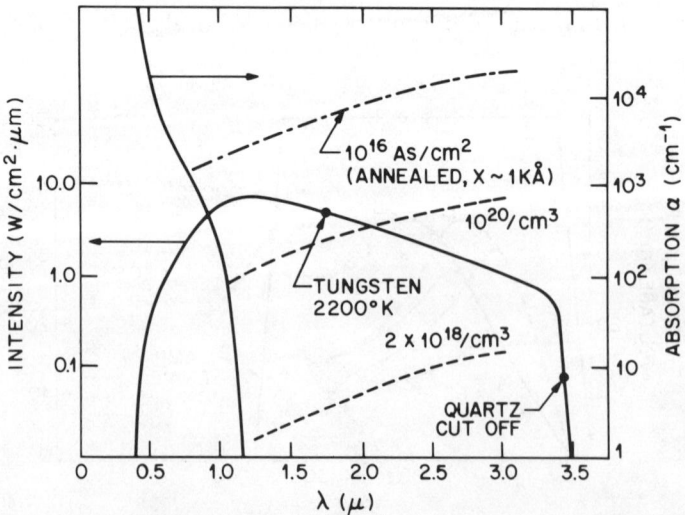

Fig. 24. Normalized blackbody-like spectrum for tungsten lamps at 2200° (solid curve and left ordinate). Absorption coefficient for silicon (right ordinate) showing fundamental edge and free carrier components (dashed curves). (After Seidel et al.[83] Reprinted by permission of the publisher, The Electrochemical Society, Inc.)

planes intersecting the surface, as the melt is bounded by these planes. For a (100) Si surface the molten areas are rectangular (often approximating squares) with edges parallel to the (110) directions, as shown in Fig. 25. Similarly, for the (111) surface plane the melt forms equilateral triangles on the surface. Within each recrystallized area many small ridges of 1–3-μm height and $\sim$50-μm spacing can be noticed. They are imprints of solid lamellae coexisting with the liquid Si.[93,94]

The coexistence of solid and liquid areas on the surface in the steady state is the consequence of a large difference in reflectivity between the two phases. Solid silicon near T_M reflects about 38% of incident light, while liquid Si, which has metallic properties, reflects as much as 72%. If melting were uniform across the surface, so much power would be lost by reflection that the melt could not be sustained. By breaking up into an array of solid and liquid zones (an "intermediate state") the surface attains an average reflectivity value that is a function of incident power and that leads to steady-state conditions.

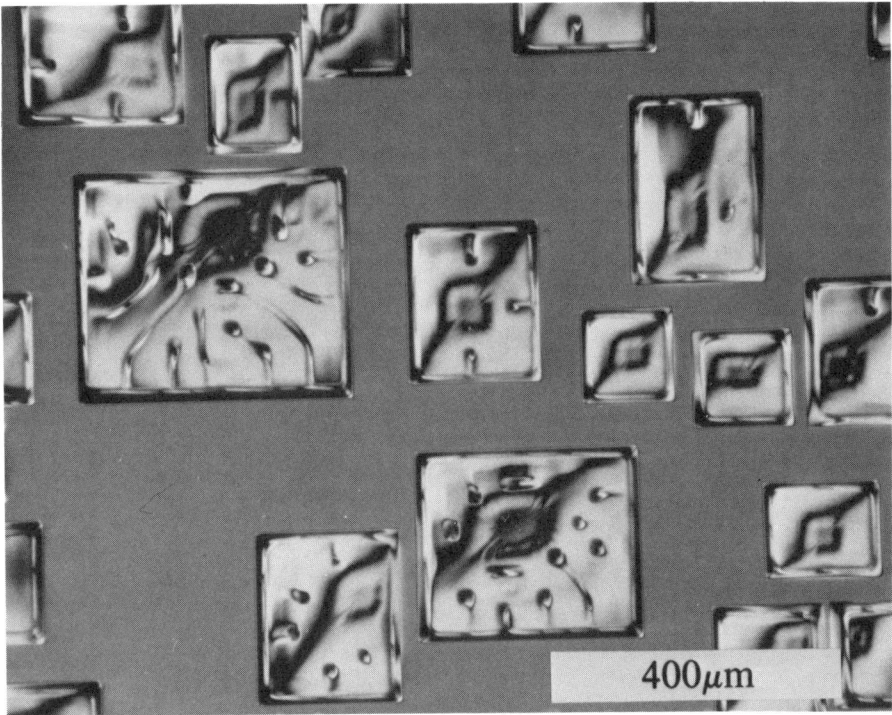

Fig. 25. Polished, single-crystalline (100) Si surface, covered with 1000 Å of oxide, after melting with a uniform radiative flux and recrystallization. (From Celler et al.[92])

Melting starts at discrete nuclei and each molten puddle expands into a faceted form because the rate of melting is a function of the crystallographic direction, with the melt moving most slowly in the [111] direction.

13. Dopant Activation

Activation of implanted dopants to form shallow junctions is potentially one of the most useful applications of rapid isothermal annealing. There is

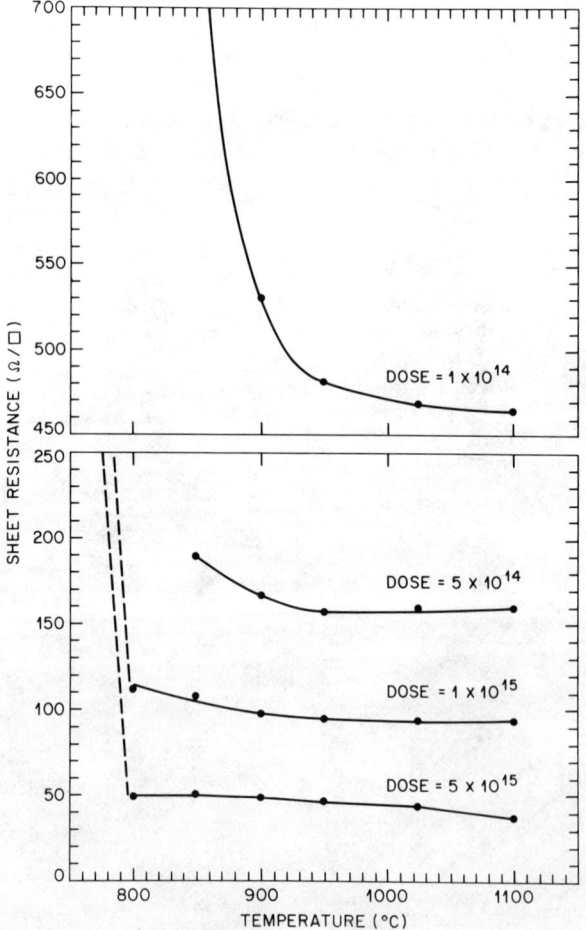

Fig. 26. Sheet resistance versus temperature for 100-keV arsenic implanted into $\langle 100 \rangle$ Si, annealed for 10 sec. The arsenic is amorphous before annealing and anneals by solid-phase epitaxy (dashed curves). (From AG Associates.[96])

no doubt that fully electrically active layers can be formed, with dopant atoms on substitutional sites, and defect trap concentrations are relatively small compared to the dopant concentrations, with very little thermal diffusion. However, residual extended defects that may cause junction leakage often remain and are the subject of continuing studies. The RTA advantage lies in the fact that activation of implanted dopant impurities and their solubility both increase with temperature.

Annealing of implanted dopants is best classified according to dopant incorporation: (1) by the solid-phase-epitaxial (SPE) process of recrystallization of amorphous silicon on a crystalline substrate or (2) by the removal

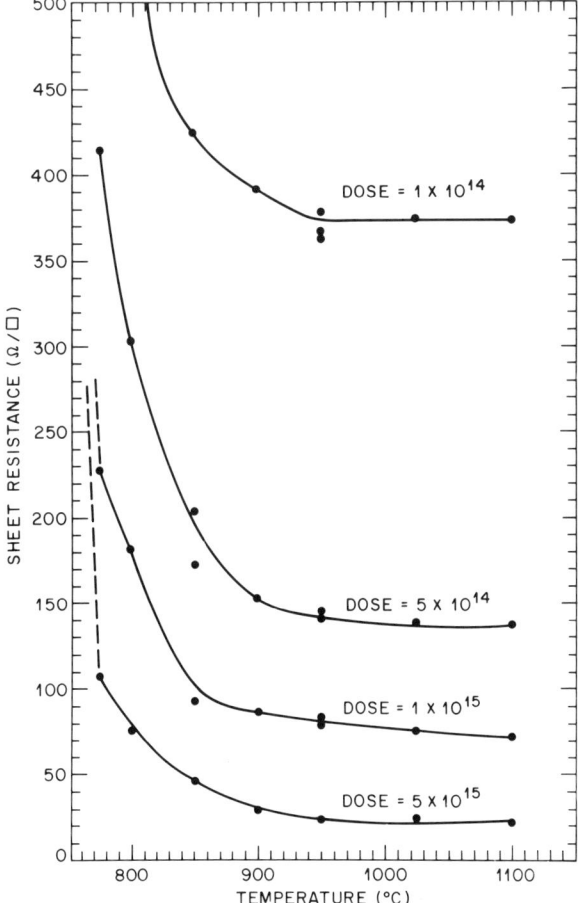

Fig. 27. Sheet resistance versus temperature for 100-keV phosphorus implanted into $\langle 100 \rangle$ Si, annealed for 10 sec. (From AG Associates.[96])

of point defects and dislocations that represent carrier traps or precipitation sites and the placement of dopant atoms on substitutional sites.[95]

When they are implanted above $\sim 10^{14}/\text{cm}^2$ into room temperature targets, P, As, or Sb produce amorphous layers over most of the dopant depth distribution. The dopant activation is then obtained mainly by the SPE recrystallization process. Activation is usually indicated by Hall or resistivity measurements that average the active concentration over the profile. Resistivity generally decreases to some limiting value. When diffusion significantly

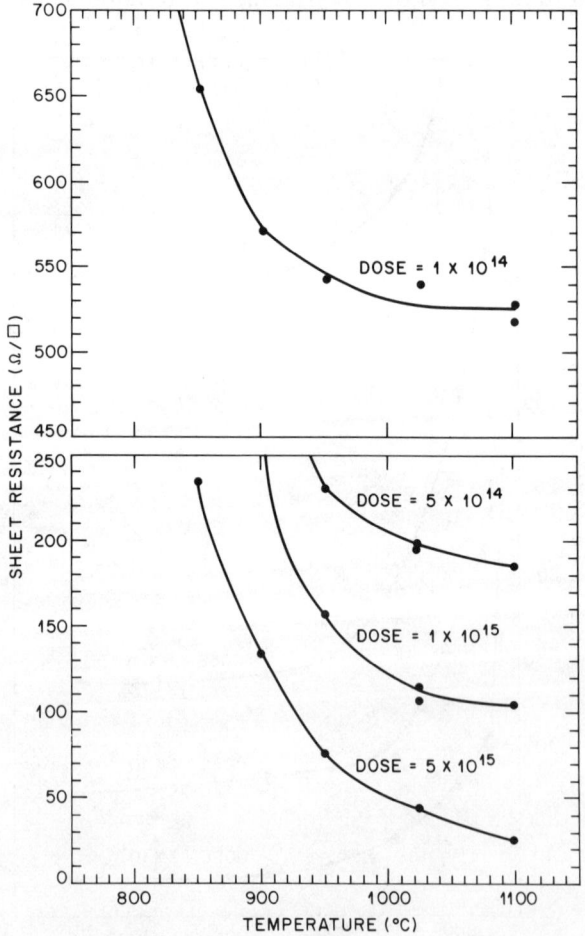

Fig. 28. Sheet resistance versus temperatures for 35-keV boron implanted into ⟨100⟩ Si, annealed for 10 sec. (From AG Associates.[96])

broadens the profile, the resisitivty decreases somewhat further since the mobility increases.

The annealing behavior is shown as sheet resistivity plotted against annealing temperature for 10-sec RTA for As, P, and B implants in Figs. 26, 27, and 28, respectively.[96] The sheet resistivities obtained at high temperatures are similar to those obtained after standard furnace anneals using longer times; thus, nearly full activation has occurred. At the lower temperatures the dashed curves indicate a solid-phase epitaxy. Systematic sheet

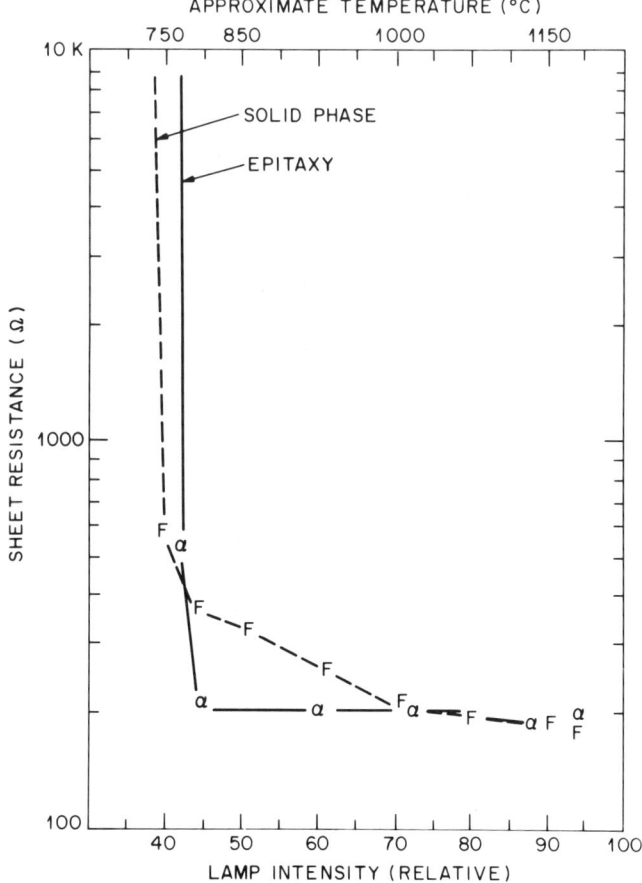

Fig. 29. Sheet resistance versus temperature for 50-keV BF_2 implanted with beam intensity $1 \times 10^{15}/cm$ into crystalline silicon (points marked F) and into preamorphized silicon (points marked α). Anneals are for 10 sec. (From T. E. Seidel.[79] © 1983 IEEE.)

resistivity data can also be found in the literature in which furnace and tungsten–halogen lamp anneals are compared for B, BF_2, and As.[97]

Hall effect measurements for As^+ and B from BF_2^+ also show full electrical activation to about $\pm 10\%$ (the Hall sheet coefficient equals the dose) after 2-sec anneals at 1100 and 950°C for $10^{15}/cm^2$ doses of arsenic and boron, respectively.[98]

Sheet resistance is shown for boron (from BF_2^+) implanted into crystalline and preamorphized silicon and annealed by solid-phase epitaxy (SPE)[79,99] using RTA in Fig. 29. When the depth of this preamorphized layer is larger than the boron distribution, complete electrical activity occurs after SPE.

14. DOPANT DIFFUSION (BORON AND ARSENIC)

Rapid thermal annealing results in profile broadening (diffusion) when temperatures are high enough or times are long enough. The question of enhanced diffusion associated with RTA of damage regions is discussed critically in this section. In order to interpret diffusion data in a quantitative way to obtain diffusion constants and activation energies, one must—as a minimum requirement—actually measure the temperatures of the *same* sample for which the diffusion analysis is carried out. Since much of the early data existing in the literature was generated without this type of temperature measurement, the early diffusion data may only be indicative of general behavior and trends.

Boron profiles (5×10^{13} B/cm^2, 50 keV) are shown in Fig. 30. Hodgson *et al.*[100] used an arc lamp for ~ 1-sec exposure and measured the temperature–time behavior on the same samples that were analyzed for diffusion. They concluded that the general diffusion behavior is consistent with normal diffusion, within experimental error. In Fig. 30 a channel tail exists in the as-implanted profile. The enhanced diffusion (ED) for the tail, according to Hofker,[101] is associated with as-implanted interstitial positions of boron and exists even for the case of standard furnace anneals (700°C, 30 min). Hodgson demonstrates the enhanced tail diffusion using RTA, as shown by the 990°C, 1-sec curve. Wilson *et al.*[99] also show the channel tail diffusing at 1150°C, 10 sec, and then a progressively increasing diffusion for longer times (see Fig. 31).

Lasky[102] has evaluated the electrically active boron profiles ($2.8 \times 10^{15}/cm^2$, 30 keV) using spreading resistance and observed a relatively large (enhanced) diffusion (Fig. 32). A graphite RTA heating source was used. A SUPREM model predicts only a 0.02-μm increase in the junction depth, whereas a 0.13-μm increase is observed after a 10-sec anneal. The SUPREM model includes the boron's concentration-dependent diffusivity but not the effects of radiation-enhanced diffusion. The spreading resistance method was checked by activating the implanted profile using a cw laser, which is known to activate the B profile electrically without thermal broadening. Lasky measured the temperature of the same sample used in the profile studies, using an optical pyrometer, and quotes a temperature accuracy of ±5°C.

Boron produced from BF_2^+ implantation may play an important role in shallow junction formation. Unlike B, the BF_2^+ implants into room temperature target silicon result in an amorphous layer over about half the depth of the entire profile.[103] During annealing solid-phase epitaxy takes place and the SPE process plays a role in the dopant activation, residual defects,

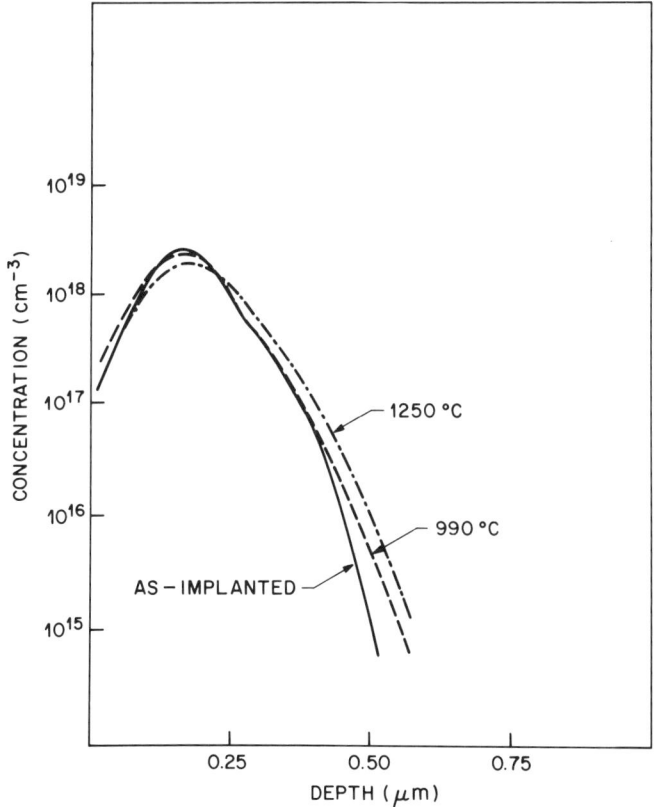

Fig. 30. Boron (SIMS) profiles for 50-keV -5×10^{13} B/cm² annealed using arc lamp pulses of $\sim$1-sec duration. (After Hodgson et al.[100] Copyright North-Holland Physics Publishing, Amsterdam, 1984.)

and subsequent diffusion behavior. The B ions stopping beyond the amorphous–crystalline (a/c) interface come to rest in a crystalline medium, and these atoms undergo channeling effects. These implants were done through a thin SiO_2 layer, which (as expected) did not eliminate the channel tail in the as-implanted profile. The SiO_2 was stripped for the SIMS analysis. Seidel et al.[103] used tungsten–halogen sources; results at nominal temperatures of 770, 925, 1150, and 1200°C are shown in Fig. 33. The 925°C curve shows an enhanced tail diffusion, as also seen in Figs. 30 and 31, and a segregation of B into a damaged region associated with the a/c interface region. (Temperatures were assigned from a calibration on different samples and are only approximate.) At higher temperatures the B breaks away from the a/c dislocation–disorder and undergoes "normal" diffusion.

Silicon can be preamorphized to greater depths than a subsequently implanted B profile. When this is done, no ion-channeling effects are possible.

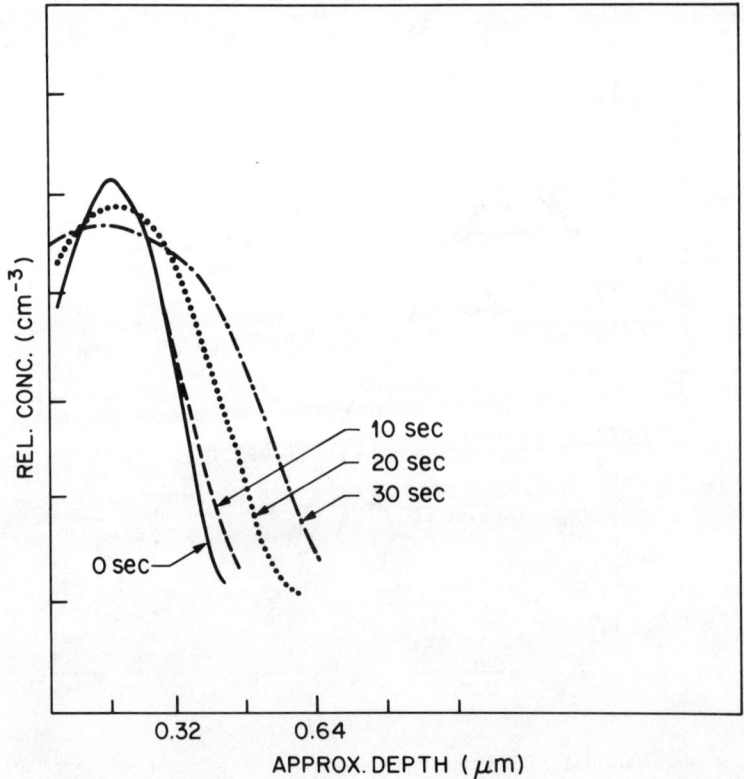

Fig. 31. Boron concentration versus depth for 50-keV -1×10^{14} B/cm^2 annealed at 1150°C using graphite-RTA. (From Wilson et al.[99] © 1983 IEEE.)

Preamorphization followed by B implants and conventional furnace anneals was studied[104] and rather shallow junctions were obtained (0.24 μm).[105] Use of BF_2 implants, followed by RTA, results in $\lesssim 0.15$-μm junction depths, as shown in Fig. 34.[103] Here profiles undergo essentially normal diffusion (without channel tail or enhanced channel tail diffusion) at 1150 and 1200°C within the accuracy of the data.

During the solid-phase regrowth of an amorphous layer, dislocations always remain localized just beyond the original a/c interface. These dislocations require rather severe thermal annealing to be removed (e.g., 1100°C, 30 sec); the detailed temperature and time depends on dose, ion species, and so on. However, the annealing of the amorphous layer also can generate other dislocations, which span between the a/c interface and the surface.[103,106] These can be avoided by using minimal damage conditions to produce the amorphous layer (e.g., by Ge^+ or Si^+ implants at lower doses into liquid nitrogen targets).[103,106-108] The dislocations that span between the original a/c interface and the surface can be removed thermally

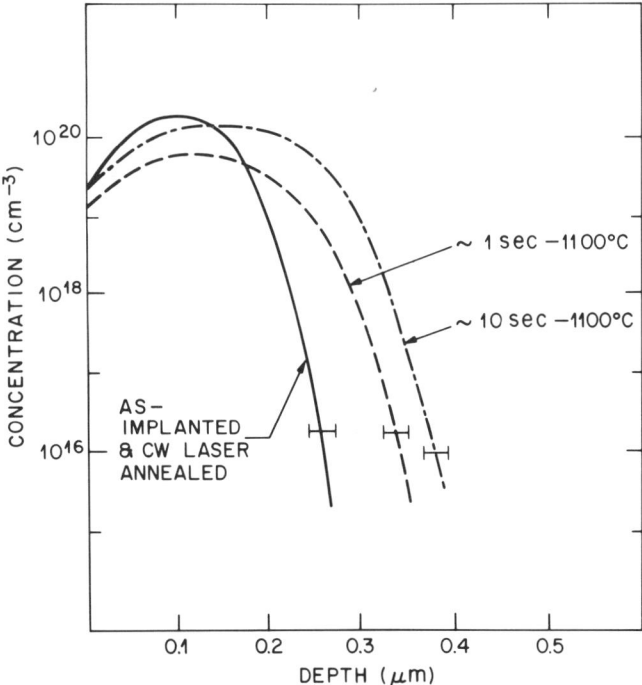

Fig. 32. Electrically active hole-concentration versus depth for implanted boron. Solid curve is fast-laser anneal that activates but does not diffuse the profile. 2.8×10^{15} B/cm², 30 keV. (From Lasky.[102])

using less severe annealing than is required for the removal of the dislocations localized beyond the a/c interface.

Arsenic RTA diffused profiles were surveyed by Benton et al.[109] using tungsten–halogen lamps with a one-sided heating arrangement. Temperatures were measured on the samples used for profile measurements using an optical pyrometer. The authors considered the diffusion to be consistent with diffusion in the solid state as opposed to very large liquid-state diffusion but did not analyze the data for possible enhanced solid-state diffusion. A

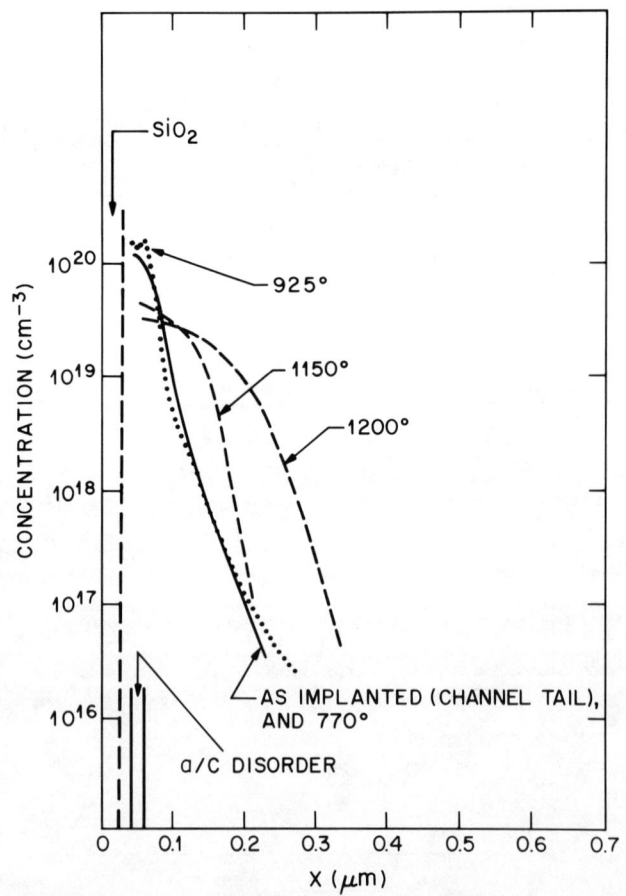

Fig. 33. Boron (SIMS) profiles for 40-keV BF_2 implanted into crystalline silicon through a thin SiO_2 layer. The channel tail shows enhanced diffusion and there is a small retrograde diffusion of B to the α/c disorder at 925°C. 1×10^{15} BF_2/cm^2, 40 keV, 10 sec RTA. (From Seidel et al.[103] Reprinted by permission of the publisher, The Electrochemical Society, Inc.)

rough analysis gives a magnitude of D [7×10^{-13} cm²/sec (at 1200°C), and 1×10^{-13} cm²/sec (at 1050°C)] consistent with concentration enhanced values of As diffusion coefficient. (See data in Fig. 35.) They also show that 1200°C for 5 sec gives a high electrical activity and solubility for 10^{16}/cm² As and B. (See Table IV.) Note the low sheet resistance values for 10^{16}/cm² doses after 1200°C, indicating greater solubility for arsenic (temperature limited) and for boron greater solubility and probably higher carrier mobility due to thermal diffusion.

The work (circa 1983–1984) of some groups stated that As diffusion coefficients under RTA are ~100 times greater than those expected. Conclusions

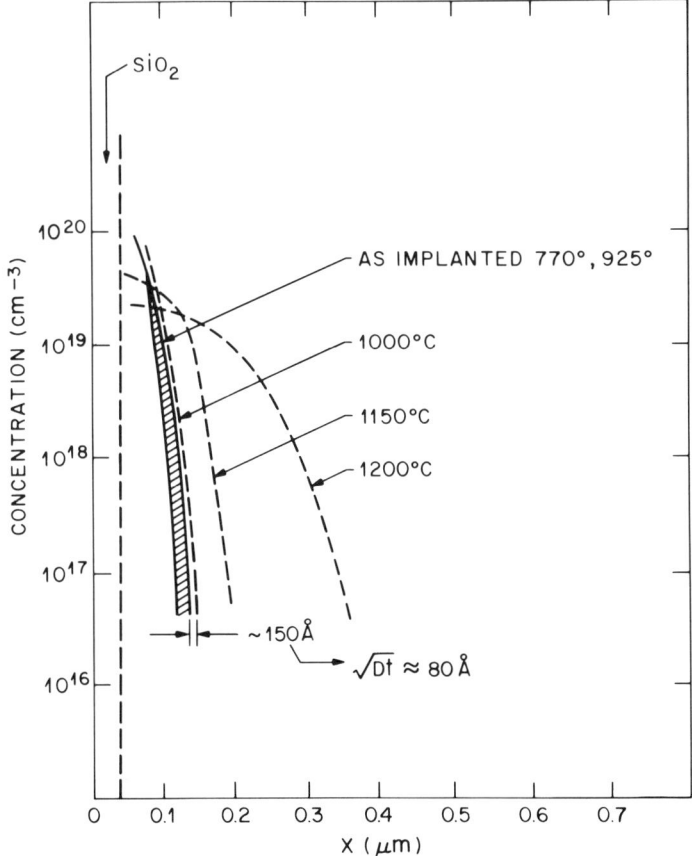

Fig. 34. Boron (SIMS) profiles for 40-keV BF$_2$ implanted into preamorphized silicon through a thin SiO$_2$ layer. Data for no anneal, 770°, and 925°C lie in the hatched band. 1×10^{15} BF$_2$/cm², 40 keV, 10 sec RTA. (After Seidel et al.[103] Reprinted by permission of the publisher, The Electrochemical Society, Inc.)

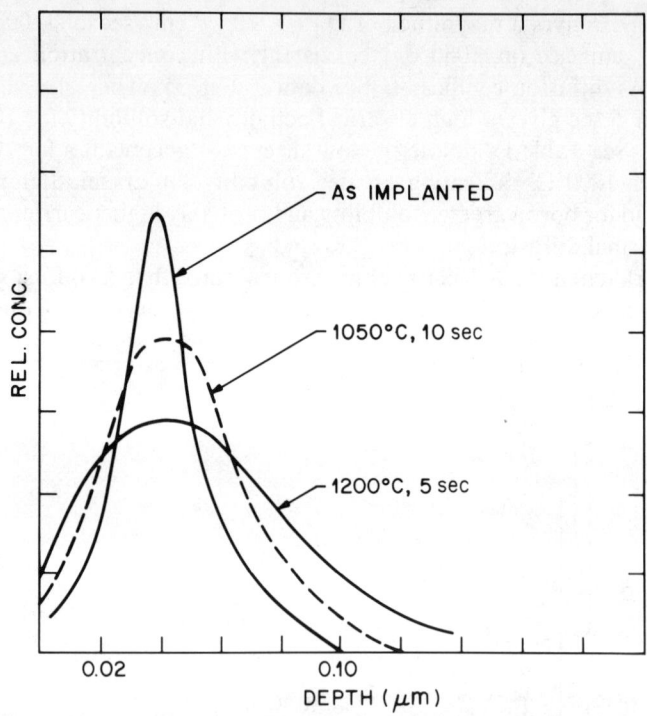

Fig. 35. Arsenic (RBS) distributions following RTA using a one-sided tungsten-halogen furnace. 10^{15} As/cm^2, 50 keV. (After Benton et al.[109] Copyright North-Holland Physics Publishing, Amsterdam, 1982.)

TABLE IV

SHEET RESISTANCE OF As- AND B- (50 keV) IMPLANTED SILICON[a]

Dose (ions/cm^2)	Sheet resistance of As ($\Omega/\square$)		Sheet resistance of B ($\Omega/\square$)	
	1050°C (10 s)	1200°C (5 s)	1050°C (10 s)	1200°C (5 s)
10^{13}	2360	2390	3166	—
10^{14}	520	555	531	511
10^{15}	113	111	101	97
10^{16}	70	24	31	14

[a] From Benton et al.[109]

were based in part on comparison with low concentration values of the As diffusion coefficient, while the actual concentrations were high ($\sim 10^{21}$/cm^3). After using a proper comparison the magnitude of the diffusion constants is only ~ 1–3 times higher than expected for comparable As concentrations. Furthermore, some groups did not measure the temperatures on the *same* samples on which RTA and diffusion analyses were carried out. Thus, temperature assignments are somewhat uncertain and no quantitative conclusions can be drawn regarding diffusion characterization.

Fair et al.[110] have associated an early time- (transient) enhanced diffusion with point defect that anneal out early. The enhanced diffusion is modeled by "overlaying a transient point defect model on the steady state, diffusion models." For boron, the channel tail atoms are viewed as being fed by point defects generated from the high concentration region of the B profile. The tail diffusion is connected to the RTA time frame of seconds and point defects generated from the high-concentration B region.[110] However, the tail diffusion occurs for standard furnace conditions (700°C, 15 min)[101] as well as RTA conditions (1000°C 10 sec). Another (alternate ad hoc) possibility is that the heavily dislocated, high-concentration region of the B is associated with *low* diffusion relative to defect-free silicon and that the tail region is relatively enhanced because of a high interstitial B concentration and/or local ion damage in the tail region as a result of the implantation.

Hodgson et al.[100] reported "normal As diffusion," within knowledge of temperature, and concentration-dependent diffusion. They have measured the temperature of the same samples on which they do diffusion analysis. Results are shown in Fig. 36, where repetitive pulses were applied to 10^{15}/cm^2 As-implanted silicon. A monotonic increase in diffusion is observed. Fair et al.[110] have reported that enhanced As diffusion (assumed to exist) is eliminated by preannealing the amorphous As-implanted (10^{16}/cm^2) layer at 550°C for 30 min. However, Kwor et al.[98] and Sedgwick et al.[111] have reported no difference in the As profile for direct RTA and preannealed As-implanted layer (5×10^{15}/cm^2). They measured the temperatures on the same samples used for diffusion analysis. Thus, the insight to the mechanism for transient-enhanced As diffusion remains obscure, even *if* RTA transient enhanced diffusion for As exists at all.

The most apparent convincing direct experimental evidence of transient-enhanced As diffusion under RTA in early reports showed the *same* diffused profile for 1 sec as for 10 sec (at ~ 1100°C). Such a result would be a definition of transient-enhanced diffusion. More recent results by Seidel and Lischner[112] were unable to show such a result at both 1095°C and 1180°C and used optical pyrometer temperature measurements on the same samples that were analyzed for diffused profiles. The results show monotonically increasing values of Dt for $\sim 1, 4,$ and 9 sec in apparent disagreement with the early experimental, greatly enhanced result. Additional recent results by Sedgwick et al.[113] also show the monotonic increase. Furthermore, a near constant D could be calculated for limited times from the data at each temperature. (Actually, D is known to be concentration dependent, and since the concentration decreases with time, D should decrease somewhat. A self-consistent analysis is needed.) With the magnitude of the As-enhanced diffusion coefficient perhaps only ~ 3 times higher than literature values for high-concentration As diffusion, the absence of an effect of the preanneal, and a value of D essentially independent of limited time at a given tempera-

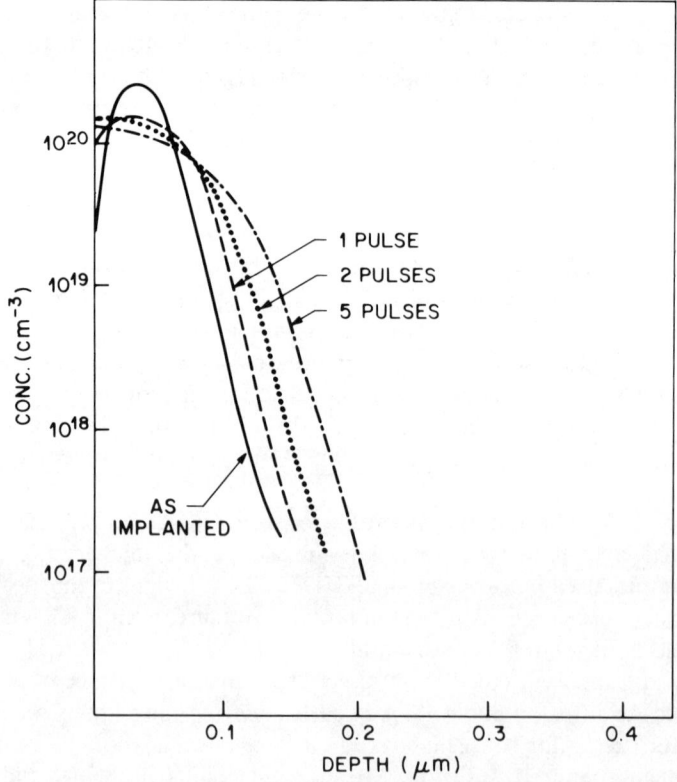

Fig. 36. Arsenic (SIMS) distribution following RTA using an arc lamp, 10^{15} As/cm^2, 50 keV, 1250°C. (From Hodgson et al.[100] Copyright North-Holland Physics Publishing, Amsterdam, 1984.)

ture, one is tempted to discard the idea of transient-enhanced diffusion of arsenic associated with RTA.

There is some experimental disagreement in the case of arsenic, but general agreement exists for a transient-tail diffusion of boron. The tail was produced by channeling during implantation. The question of enhanced early time transient diffusion may be very important in practical applications. If early time diffusions (~ 1 sec) were vastly enhanced, we would not be able to take advantage of a potentially important property of RTA, namely, the control of shallow junction depths. Both experiment and theory are likely to improve in the near future.

15. Defect Removal

Dislocation modification and removal are a key issue for RTA. Many researchers are addressing the questions, "What minimal RTA thermal cycle clears out the residual dislocation damage, and is the junction depth less (or different) for RTA under these conditions than in standard furnace annealing?" The alternate but equivalent question is whether for the same junction depth obtained by RTA and standard furnace anneals, there is a difference in the residual dislocation disorder, and if the disorder is less for RTA.

Sadana et al.[114] have surveyed the structural damage by TEM for B (at 5×10^{14} and $5 \times 10^{15}/cm^2$, 35 keV) and for As ($2 \times 10^{15}/cm^2$, 80 keV) implantations followed by RTA. For 5×10^{15} B/cm^2 they report a diffused profile with a concentration of $\sim 10^{16}/cm^3$ at 0.8 μm (the would-be "junction depth") for both a 1000°C, 30-min standard furnace anneal and a 1100°C, 30-sec RTA. Although the junction depth is identical for the two anneals, the boron profile is still peaked in the damaged region for the furnace-annealed case, while it is broadly diffused for the RTA. This may be taken as evidence for a physical difference between the residual damage for the two kinds of anneals; more boron is held in the damaged region with the furnace anneal (Fig. 37). For arsenic the dislocations are removed at 1100°C in 10 sec.

Lasky[102] has also published TEM cross sections of residual dislocation that still are not removed for 2.8×10^{15} B/cm^2 30 keV implant after 1100°C, 10-sec anneal. However, the disorder extends to only ~ 0.25 μm from the surface, which is less than the 0.40-μm junction depth. Junction leakages of ~ 2 nA/cm^2 are obtained. Lasky[102] measures the electrically active concentrations and sees no peak in the damaged region, implying that the peaked boron in the Sadana et al. paper[114] is not all electrically active. This condition is qualitatively the same for a standard furnace anneal under conditions of incomplete activation.

Hodgson et al.[100] have reported that residual defects from 5×10^{15} As/cm^2 at 50 keV are removed for temperatures above 1250°C in 1 sec. Junction depths of the order of 2000 Å are obtained. It is becoming increasingly important to establish the minimal anneal conditions for defect removal, accompanied by comparison of junction depth or SIMS profiles.

If dislocations are removed with an activation energy that is *different* from the dopant diffusion activation energy, then special annealing conditions

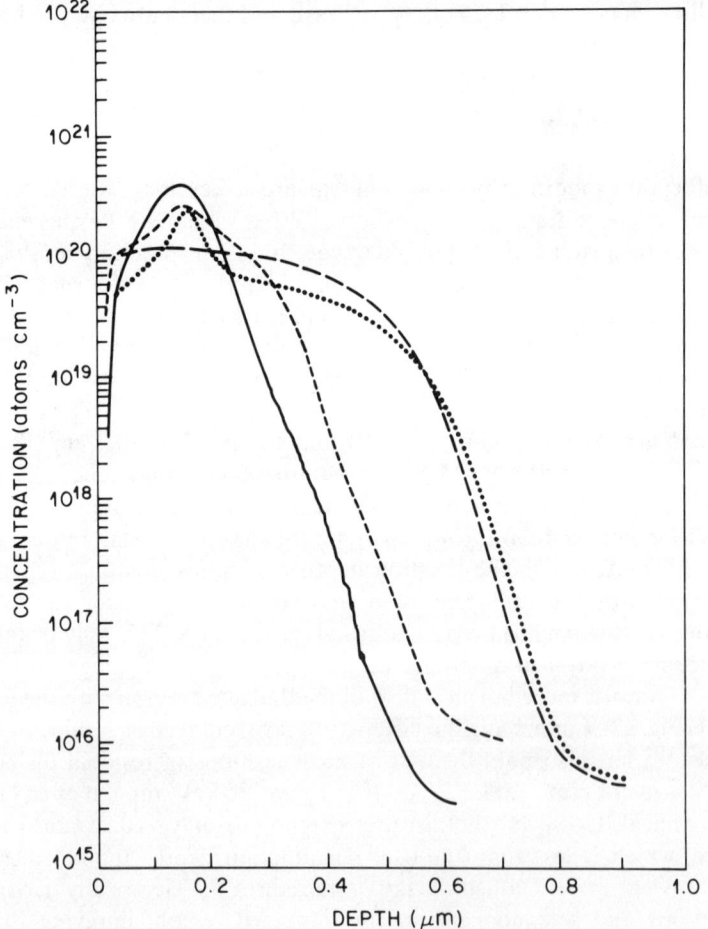

Fig. 37. Boron (SIMS) profiles for 5×10^{15}/cm^2 35 keV showing the removal of precipitation effects in the maximum damage region by use of RTA. The junction depths for 1000°C 30-min furnace and 1100°C, 30 sec RTA are the same. —, as implanted; - - -, heat pulse annealed 1100°C, 10 sec; --, heat pulse annealed 1100°C, 30 sec; ···, furnace annealed 1000°C, 30 min. (From Sadana et al.[114] Copyright © 1983 The Institute of Physics.)

may be found for which all dislocations are removed, while minimal dopant diffusion occurs. Thus, it may be advantageous to anneal at short times at very high temperatures. This should be the situation for removal of disorder that requires 5-eV activation (cooperation of Si self-diffusion), while dopant diffusion requires 3–4 eV.[112]

V. Related Rapid Thermal Processes

16. Fast Evaluation of Implantation

Presently, the most widespread commercial use of RTA is the annealing of implantations to check the dose immediately after implantation. In this manner errors in dose, and sometimes implant species, can be found rapidly and the decision to rework or junk the lot of wafers can be made without wasting valuable time on processing wafers with the incorrect implants. For best results one needs to calibrate sheet resistance against anneal time and temperature, as shown in Section 13 for a given dose and ion species. Monitor wafers or device wafers with test sites can be annealed in a relatively small one-wafer-at-a-time RTA furnace (e.g., AG-210M). Evaluations can be made by the implantation operator to confirm his own work. Suitably used, the RTA furnace may be used to evaluate implant uniformity quickly, although one must first evaluate the uniformity of the anneal process itself.

17. Shallow Junctions for ICs

Shallow junctions for ICs in the 0.2–0.4-μm range have been reported by many workers using RTA.[115–117] Use of RTA may be perceived to offer some advantage in control and defect removal, even though the same junction depths may be obtained with standard furnace anneals. For example, 30 min at 950°C gives a 0.2-μm junction for a 50-keV, 5×10^{15} arsenic dose and a 0.35-μm junction for a 10-keV, 10^{15} boron dose, which are typical MOS source–drain doses.

To obtain a very shallow junction <0.2 μm, several issues need to be addressed, such as ion-channeling tails in the as-implanted profiles, enhanced diffusion in the tail region, residual disorder that may influence junction leakage [both adversely or positively (gettering)], a suitable contact technology, and compatibility with a process that seals the devices—for example, by use of a glass barrier for sodium. These are very active areas of research and development and technology packages are considered proprietary by most laboratories at the present time. Therefore, very little is now published on specific processes that make use of RTA for very shallow junctions.

18. Grain-Boundary Diffusion

Large-grained polycrystalline silicon layers on SiO_2 can be fabricated by laser recrystallization and other melt techniques as described in the following section. MOS devices and circuits can be made in such films and they have the usual performance advantages of silicon-on-insulator (SOI) structures. Grain boundaries in the channel regions of MOS transistors affect device performance in two ways.[118] First, the electron surface mobility is reduced by the grain boundaries that intersect the current flow, and the threshold voltage is shifted by these boundaries as well. Second, dopant diffusion in the grain boundaries is enhanced by about two orders of magnitude compared to the bulk diffusion.[119,120] This means that in conventional device processing, short channel devices are likely to be shorted if there are grain boundaries that connect the source and drain regions. Indeed, experiments of Ng et al.[118] indicated that such shorting occurred for transistors with channels <2.5 μm when thermal processing subsequent to the source-and-drain As implant included a total of 90 min at 900°C. Recently, a new scheme has been tried to minimize the deleterious effects of grain boundaries on SOI transistor performance. Silicon films on SiO_2 were recrystallized with an elliptical laser beam to produce approximately parallel grain boundaries that were aligned with the current flow direction in MOSFETs.[121] This greatly reduced the influence of the grain boundries on the average electron mobility and threshold voltage. This also would maximize shorting out of transistors by enhanced diffusion if standard thermal processing were used. Instead, the processing sequence was modified to eliminate all high-temperature steps and to replace them with a single 10-sec annealing step in a lamp furnace at 900°C. This annealing activated As that was implanted into the source and drain regions and also activated As implanted into polysilicon interconnects. Electrical activation was complete, as determined from sheet resistivity measurements. However, defects were not completely removed, since control bulk transistors exhibited substantial reverse p–n junction leakage. The leakage in SOI transistors was small in spite of these residual defects, since the junction area in the SOI structures is greatly reduced.[122] Most significantly, working transistors with 1.5- and even 1-μm channels were obtained. The advantage of rapid annealing is clearly demonstrated in Fig. 38, where output characteristics of furnace and RTA-annealed devices with aligned grain boundaries are compared.

Combination of SOI structures, electron mobilities optimized by grain boundary alignment, and RTA permitted fabricating functional 19-stage ring oscillators with an effective channel length of 1.5 μm and propagation delay per stage of only 115 psec/stage.[122] This is the shortest delay achieved so far in any silicon-on-SiO_2 devices and at least twice as short as that in equivalent bulk Si ring oscillators.[123]

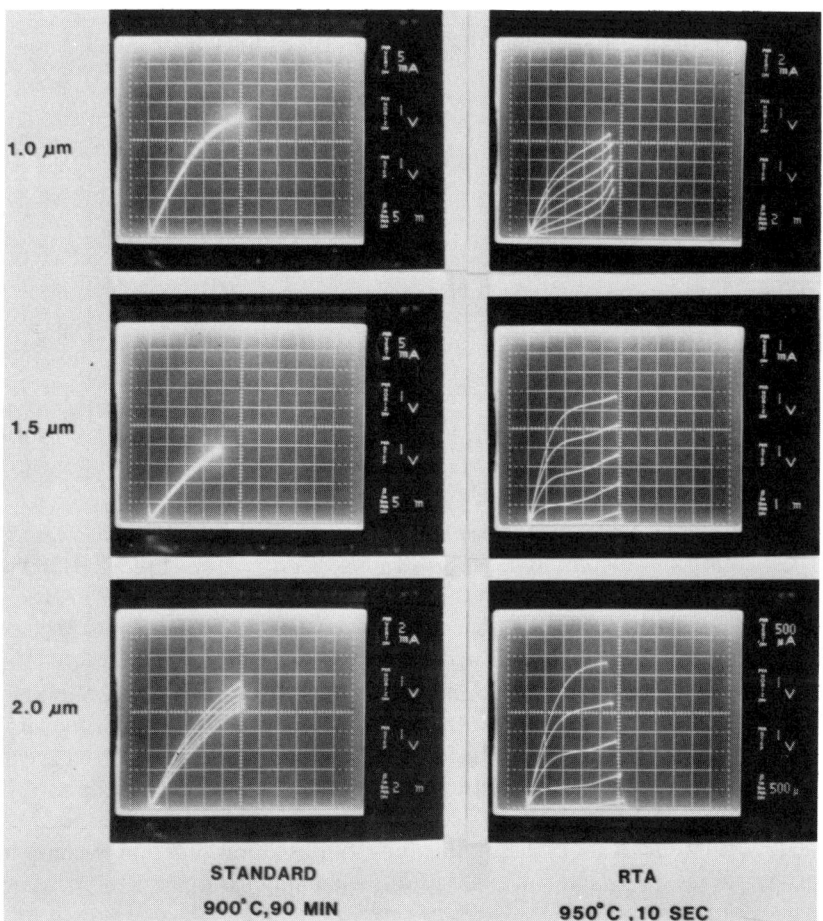

Fig. 38. Comparison of output characteristics after standard and RTA processing of MOSFETs with grain boundaries aligned parallel to the current flow. (After Celler et al.[121] Copyright North-Holland Physics Publishing, Amsterdam, 1984.)

19. P-Glass Flow

Phosphorus glass flow is used to smooth out sharp steps to the topology of integrated circuits before metallization. The use of RTA for the glass flow is an attractive idea because high temperatures $\sim 1100°C$ are needed, but a short time is sufficient to get smooth surfaces. Junction depths and all impurity redistribution would be limited. To use RTA one must obtain very good temperature uniformity and avoid degassing of the surfaces (loss of phosphorus) with vacuum environments.

The industry is developing the use of boron–phosphorus silicate glass (BPSG), which flows at lower temperatures, ~ 900–$1000°C$. This may obviate the need for using RTA to flow the glass. If BPSG is sufficient from a reliability point of view (e.g., as a sodium barrier), then there may be no need for a short high-temperature RTA.

However, one should keep in mind that the inherently clean RTA environment may prove to be attractive near the end of process in order to avoid junction degradation by contaminants.

Glass flow can be obtained by irradiation with a CO_2 laser (9.26 μm) to heat PSG selectively by exciting the stretching modes of SiO_2.[124] The glass flowed without impurity diffusion in the active devices, although junction depths were ~ 0.8 μm.

20. Silicide Formation

A number of silicides have been formed by using rapid thermal annealing. For the case of co-sputtered $MoSi_2$, Fulks et al.[125] have reported that a $1000°C/20$ sec RTA gives $5\Omega/\square$ for 2500 Å on $\langle 100 \rangle$ substrates. This sheet resistance is comparable to that which can be produced using furnace anneals requiring 15 min. Electron diffraction and TEM studies show that the grain size and structure for RTA and furnace anneals are similar, although the anneal times are very different. The implications are twofold (if the temperature measurements are correct): (1a) the RTA environment provides an enhanced reaction condition, although the specific physical mechanism is unclear, or (1b) silicide reactions occur more quickly than previously believed and (2) the limited RTA cycle should allow more shallow junction formation, corresponding to a factor of ~ 20 in diffusion time compared to standard furnace anneals.

A broader class of refractory silicides has been formed using RTA by Shatas et al.[126] Their results are shown in Table V for $TiSi_2$, $TaSi_2$, $MoSi_2$, and WSi_2. These were co-sputtered on SiO_2, except $TiSi_2$, which was metal-sputtered onto polysilicon. In all cases, $1100°C/10$ sec gave silicide resistivities

TABLE V
REFRACTORY METAL SILICIDE RESISTIVITIES[a]

Silicide	Resistivity of silicide on bulk Si	After furnace anneals[b] 900–1000°C		After RTA[c] ~1100°C, 10 sec	
		Resistivity of metal on polysilicon	Resistivity of cosputtered metal and Si	Resistivity of metal on polysilicon	Resistivity of cosputtered metal and Si
WSi_2	12	—	~70	—	48–80
$TiSi_2$	17	13–16	25	13–25	18–27
$MoSi_2$	21	—	~100	—	50–78
$TaSi_2$	46	34–45	50–55	—	46–150

[a] All resistivities in $\mu\Omega$ cm.
[b] From Murarka.[127]
[c] From Shatas et al.[126]

comparable to the furnace-annealed samples requiring 900–1000°C/30 min anneals as reported by Murarka.[127] Since diffusion of boron (3.5 eV) and arsenic (4.0 eV) are different on average by a factor of ~15 between 1000°C and 1100°C, the RTA thermal cycle would appear to give a reduced junction diffusion advantage corresponding to a factor of ~10 in diffusion time. The Shatas et al. results seem to be in qualitative agreement with the Fulks et al. experiment. Titanium silicide thin films, sputter-deposited on arsenic-ion-implanted silicon, were reacted using RTA. The silicides were uniform over the wafer and the implanted arsenic was activated during the same thermal treatment.[128]

21. ALUMINUM CONTACT SINTERING

Junction depths of integrated circuits in manufacture are presently of the order of 1 μm. Aluminum contacts are made directly to the silicon. This process is successful unless the sinter process (typically 450°C for 30 min for e-beam-evaporated Al) results in severe "spiking" effects.[129] During sintering, silicon from the contact regions diffuses through undoped aluminum connecting rails over distances of the order of tens of micrometers. The removed silicon is replaced with aluminum that forms an inverted pyramid (with $\langle 111 \rangle$ surfaces) or "spike" that may penetrate a shallow junction. Use of RTA may limit the amount of silicon removed from the contact region while still providing sintering action, to break down the native SiO_2 barriers between the Al and Si interface. Although quantitative results have not been published regarding the margin against spiking, such advantages are proposed by the commercial literature.[130]

It has been reported that large-grain Al structure can be produced by use of RTA, which may reduce electromigration effects.[131]

22. RECRYSTALLIZATION OF Si ON INSULATOR

Closely related to rapid thermal processing techniques is recrystallization of silicon films on insulating substrates such as oxidized Si or bulk fused silica. The purpose of recrystallization is to transform the deposited polycrystalline or amorphous layer into a device-quality material with few extended defects. Such material offers many advantages over bulk silicon, such as higher device switching speed and simpler device design. The latter aspect is demonstrated in Fig. 39, where CMOS inverter structure in SOI is compared with an equivalent structure in the bulk.[132]

Recrystallization can be accomplished with pulsed or cw lasers[133] and electron beams[134] and also with strip heaters scanned above the film[135] and in some cases with stationary radiative heaters.[136] To achieve a single crystalline or very-large-grained layer the entire thickness of the film has to be melted to eliminate nucleation from the existing grains; for best results large lateral gradients are necessary. These are induced by the motion of the heat source, by the structure, or by a combination of both. The pulsed-beam method allows only a moderate enlargement of the grains, since temperature gradients during solidification are primarily normal to the film thickness.

Three main configurations for crystallization of Si films on insulator are shown in Fig. 40. Continuous films on insulating amorphous substrates can be melted with any heat source mentioned previously. The recrystallized

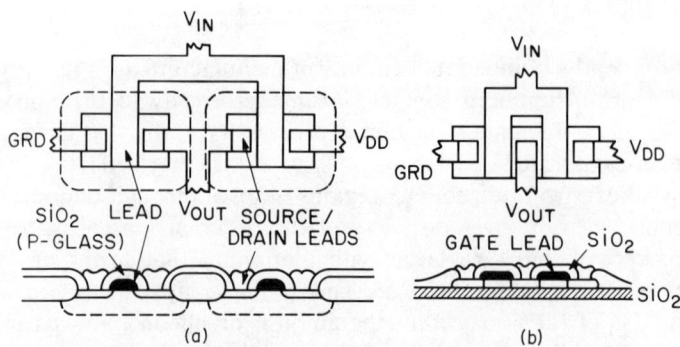

Fig. 39. CMOS inverter in (a) the bulk Si and (b) the SOI configuration. (After Leamy.[132] Copyright North-Holland Physics Publishing, Amsterdam, 1982.)

films are not entirely free of extended defects. Laser and electron-beam scanning typically forms grains that are 5–100 μm in lateral extent. Grain shape depends on the curvature of the trailing edge of the molten zone. When a line source such as a graphite strip heater is swept across the surface, interfacial energy effects often lead to highly oriented (100) Si layers with the $\langle 100 \rangle$ axis parallel to the scan direction. The defects remaining in such films are predominantly low-angle grain boundaries aligned in a characteristic wishbone pattern, as shown in Fig. 41.[137]

Patterning of Si before irradiation allows a better control of temperture gradients and nucleation conditions, often resulting in single crystalline islands. In one variant of the patterning scheme, Bridgman growth conditions are achieved on a microscale. Several silicon islands 10–100 μm in diameter and connected by narrow bridges are formed by standard lighographic techniques. As the first island in a row is recrystallized several grains may form there, but only one orientation is carried through the narrow constriction connecting this island with the subsequent ones.[138] This approach is particularly useful for display applications where an active matrix of thin-film transistors needs to be placed on a transparent substrate.

Seeded crystallization is the most effective method for eliminating extended defects when polycrystalline films are in the proximity of a single crystalline template, as in the case of films deposited on oxidized Si wafers. For best result the substrate on which a thin film is to be deposited should be planar, i.e., via openings in the oxide should contain single crystalline Si. This is usually accomplished by using local oxidation to form the vias,[139] although

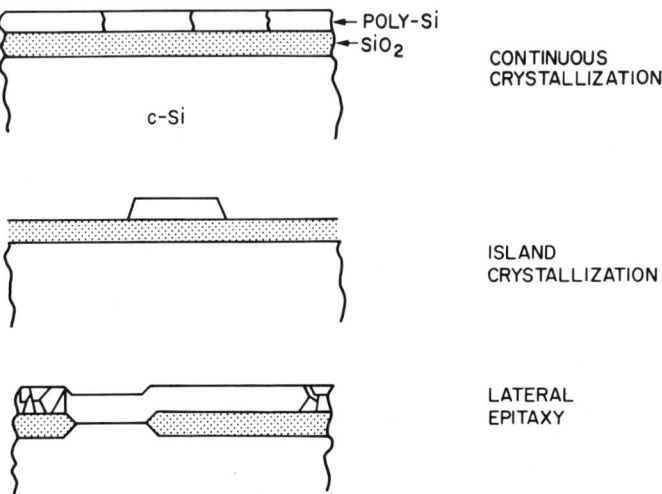

Fig. 40. Three basic approaches to crystallization of Si films on insulators.

an alternative approach of etching windows in the oxide and filling them with crystalline Si by selective epitaxy has also been demonstrated.[140] Scanned heat sources allow epitaxial crystallization to proceed for 30–100 μm from each seed window, depending on the substrate temperature and scanning conditions, after which distance the low-angle grain boundaries and even random grain boundaries reappear.[139] Fortunately, an oscillatory meltback technique allowed extending epitaxial growth from the seeds to several hundreds of micrometers.[141]

High-voltage devices require dielectric isolation of 10–100-μm-thick Si layers. Such dielectrically isolated (DI) structures have been manufactured for many years by a very expensive process that forms crystalline silicon on a polysilicon wafer, with the isolating oxide in between.[142] A new method of forming dielectrically isolated thick films has been developed by Celler et al.[143,144] that utilizes controlled melting of silicon in a lamp furnace just like the one shown in Fig. 22b. Lateral epitaxial regrowth proceeds from seeding windows that are spaced about 1 mm apart. In this case the heat source is stationary but the windows themselves act as heat sinks and introduce lateral temperature gradients necessary to carry the solidification front over the SiO_2. The recrystallized material is free of grain boundaries or subboundaries and contains few dislocations. The lack of subboundaries is associated with the film thickness and not with the method of melting.[145] The top surface of Si film recrystallized over a square SiO_2 island, 1.6 mm on each side, is shown in Fig. 42. Since lateral solidification occured from all

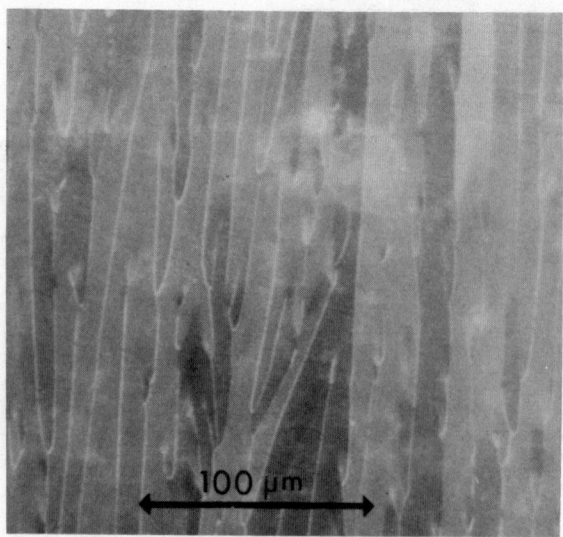

Fig. 41. Scanning electron micrograph of Si film recrystallized with a scanned strip heater. (After Leamy et al.[137])

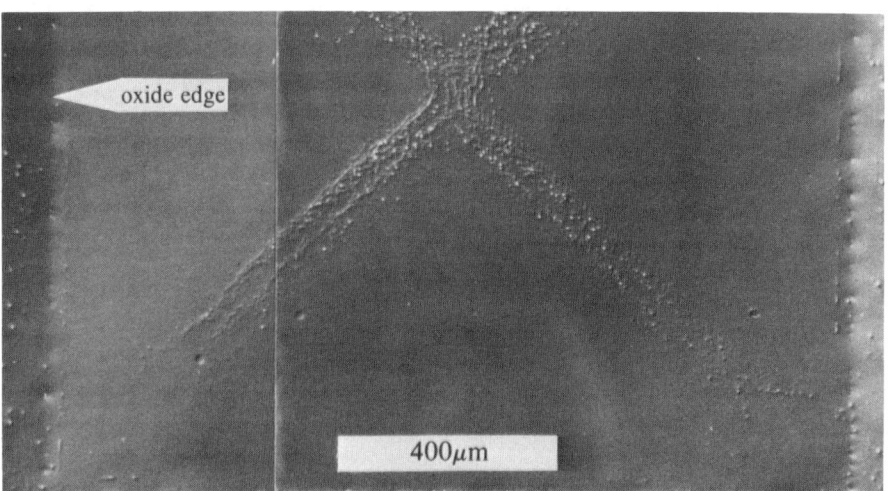

Fig. 42. Top view of lamp-recrystallized Si over a buried square island of SiO_2. The surface was etched to delineate defects. (After Celler et al.[145] Reprinted by permission of the publisher. The Electrochemical Society, Inc.)

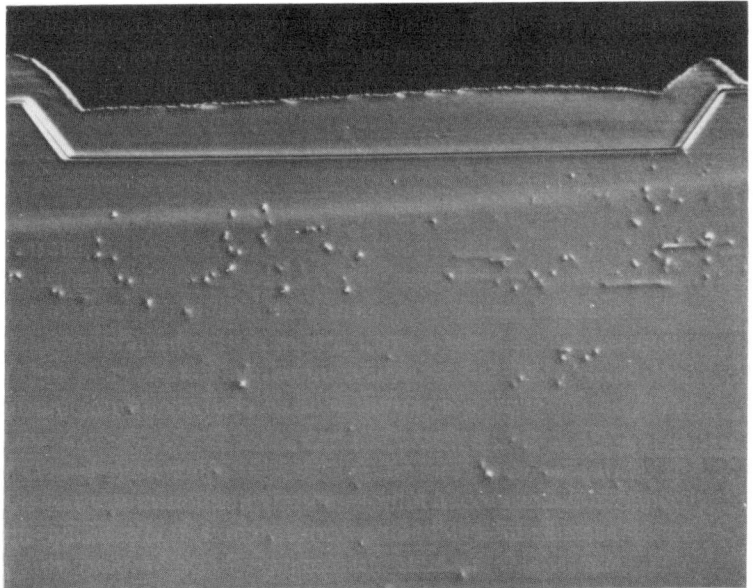

Fig. 43. Cross section of a thick Si layer crystallized over a nonplanar oxide. Seeding windows for epitaxial growth are outside the view. The surface was etched to delineate defects.

four sides, the regions along diagonals where the freezing fronts collided are more defective, with a high density of dislocations. The rest of the area is essentially defect free. A cross section of a recrystallized film over a nonplanar oxidized substrate is shown in Fig. 43. In this case seeding proceeded from narrow, slitlike openings in the SiO_2, just outside the field of view. Structures of this type are of interest for dielectrically isolated devices.

23. Laser Gettering of Impurities

In the process of fabrication of semiconductor devices it is very difficult to avoid some contamination of the silicon wafers. In particular, fast-diffusing metallic impurities are usually found in device chips, and their presence in the active regions of the devices can be very harmful. Methods of gettering these impurities away from the active parts of chips have been

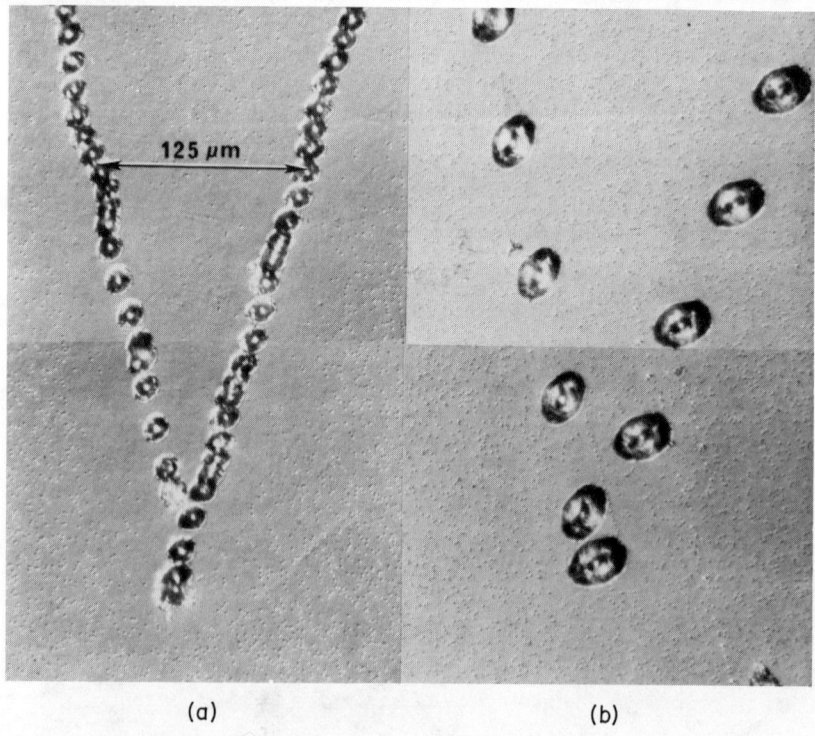

Fig. 44. Denuded zone, free of saucer pits, obtained by laser damage of Si surface. Closely spaced small craters in (a) are more effective in removing saucer pits than a few laser formed spots in (b).

used for many years. The most common approach is to diffuse a high concentration of phosphorus into the back side of device wafers, in order to enhance the solid solubility of heavy metal impurities by ion pairing. Fast-diffusing metallic impurities become bound to the ionized phosphorus atoms.[146] At higher phosphorus concentration, misfit dislocations are

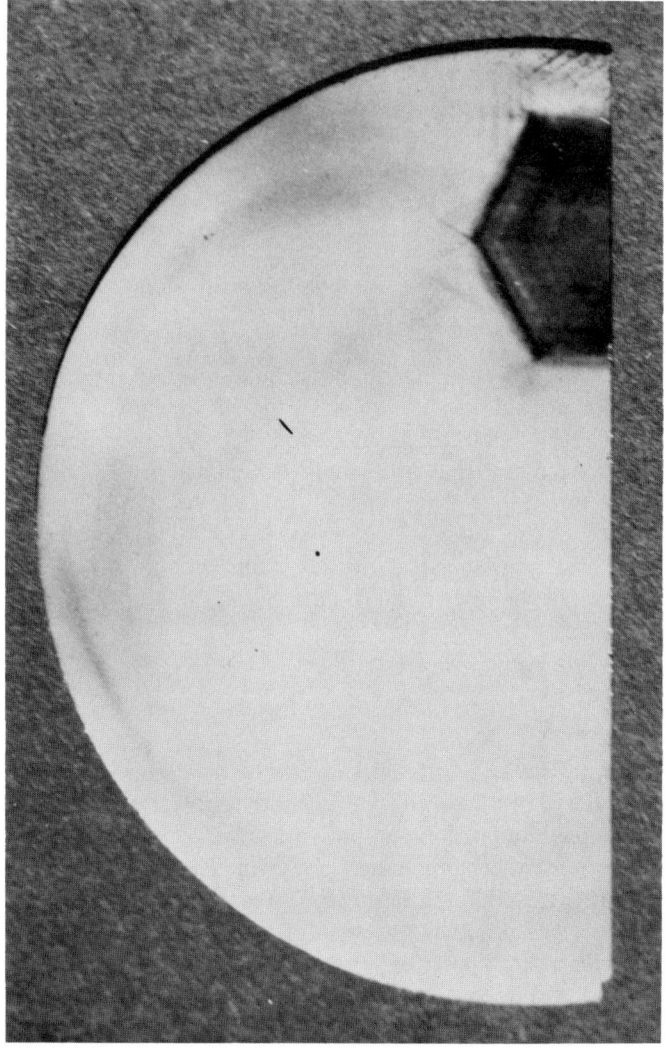

Fig. 45. A dark-field view of a front side of a 3-in. wafer with one laser-damaged segment on the back side.

formed that provide additional segregation sites for unwanted impurities. Ion implantation damage was also shown effective in impurity gettering.[147] The amount of residual damage that getters unwanted impurities depends in this case on the implanted species.

Laser-induced damage for gettering purpose was first demonstrated by Pearce and Zaleckas.[148] They showed that controlled mechanical damage could be induced on the back side of silicon wafers by irradiating that surface with tightly focused pulses of a Nd:YAG laser. Each pulse vaporized some silicon and produced a microscopic crater in silicon. The degree of damage was controlled by varying the depth of these craters and the overlap of successive pulses. The stresses induced by localized vaporization are relieved by formation of a dislocation network during heat treatment in the furnace. The damaged layer traps impurities the same way as the surface damaged by abrasion would. The advantage of laser-induced damage over any mechanical tool is that the risk of damaging the device side of the wafer is eliminated. In practical terms this means that the damage can be produced by laser at any stage of device fabrication and the refresh capability is also available. It is also possible to "write" the damage pattern on either side of the wafer. For example, damage can be easily introduced into streets between chips on the device side of the wafers. In Fig. 44 we show a typical effect of individual laser-induced craters in silicon. The region around each pit is denuded of saucer pits that are usually associated with contaminants. In Fig. 45 a half of a 3-in. wafer is viewed from the front side in the dark-field contrast. Most of the wafer surface is covered with saucer pits that scatter light into the camera. A small dark area is mirror smooth as a result of the gettering action of the laser-induced damage on the *back side* of this wafer.

VI. Summary

In this chapter we have attempted to review the fundamentals of transient thermal processing of silicon and to present the current status of applications. Rapid annealing came into being only a few years ago in the form of laser annealing. It quickly became a very powerful scientific tool for the studies of crystal growth kinetics and impurity segregation. In parallel to the efforts in basic science, the practical potential of transient annealing for electronic device fabrication has been explored. Many potential applications were identified, and a few that appeared most promising have been pursued vigorously by research and development teams. They include silicon-on-insulator, adiabatic annealing for solar cell formation, silicide formation, contact sintering, gettering, and others.

The one approach that was accepted almost immediately for device processing was rapid isothermal annealing. The main reason is the compatibility of RTA with conventional device processing practices. The equipment is scalable to produce volumes, can be automated, and is suitable for casette-to-casette wafer handling. Even more important, it fills an immediate need of the semiconductor industry for formation of shallow junctions.

The next few years should witness an implementation of rapid isothermal annealing of ion-implanted layers. RTA may also be used for reflow of P-glass, silicide formation, and contact sintering. It is already in widespread use for testing implant dose and species. At the same time the less conventional applications involving localized melting will be transferred from research stage into development. The research activities on structure modification with very short pulses will continue.

ACKNOWLEDGMENTS

We acknowledge the discussions with and the expertise of D. J. Lischner, J. M. Poate, H. J. Leamy, W. L. Brown, and L. E. Trimble. We also thank our many colleagues for data, information, and use of figures.

REFERENCES

1. C. Hill, in "Laser and Electron-Beam Solid Interaction and Material Processing" (J. F. Gibbons, L. D. Hess, and T. W. Sigmon, eds.), pp. 361–374. North-Holland Publ., Amsterdam, 1981.
2. E. I. Shtyrkov, I. B. Khaibullin, M. M. Zaripov, M. F. Galyatudinov, and R. M. Bayazitov, Sov. Phys.—Semicond. (Engl. Transl.) **9**, 1309 (1976).
3. A. G. Klimenko, E. A. Klimenko, and V. I. Donin, Sov. J. Quantum Electron. (Engl. Transl.) **5**, 1289 (1976).
4. G. A. Kachurin, N. B. Pridachin, and L. S. Smirnov, Sov. Phys.—Semicond. (Engl. Transl.) **9**, 946 (1976).
5. Proceedings of annual symposia of Materials Research Society: "Laser–Solid Interactions and Laser Processing—1978" (S. D. Ferris, H. J. Leamy, and J. M. Poate, eds.), AIP Conference Proceedings, No. 50. Am. Inst. Phys., New York, 1979; "Laser and Electron Beam Processing of Materials" (C. W. White and P. S. Peercy, eds.). Academic Press, New York, 1980; "Laser and Electron-Beam Solid Interaction and Material Processing" (J. F. Gibbons, L. D. Hess, and T. W. Sigmon, eds.). North-Holland Publ., Amsterdam, 1981; "Laser and Electron-Beam Interactions with Solids" (B. R. Appleton and G. K. Celler, eds.). North-Holland Publ., Amsterdam, 1982; "Laser–Solid Interactions and Transient Thermal Processing of Materials" (J. Narayan, W. L. Brown, and R. A. Lemons, eds.). North-Holland Publ., Amsterdam, 1983; "Energy Beam–Solid Interactions and Transient Thermal Processing" (J. C. C. Fan and N. M. Johnson, eds.). North-Holland Publ., Amsterdam, 1984.
6. N. Bloembergen, in "Laser–Solid Interactions and Laser Processing—1978" (S. D. Ferris, H. J. Leamy, and J. M. Poate, eds.), AIP Conference Proceedings, No. 50, pp. 1–10. Am. Inst. Phys., New York, 1979.

7. K. A. Jackson and A. F. Witt, Silicon Preparation—Properties—Devices. "Encyclopedia on Electronic Materials" Pergamon Press, Oxford, 1985. M. E. Glicksman and P. Voorhess, *J. Electron. Mater.* **12**, 161 (1983).
8. A. Goldsmith, "Handbook of Thermophysical Properties of Solid Materials." Macmillan, New York, 1961.
9. G. E. Jellison and F. A. Modine, *Appl. Phys. Lett.* **41**, 180 (1982).
10. G. E. Jellison and D. H. Lowndes, *Appl. Phys. Lett.* **41**, 594 (1982).
11. G. G. Macfarlane, T. P. McLean, J. E. Quarrington, and V. Roberts, *Phys. Rev.* **111**, 1245 (1958).
12. H. A. Weakliem and D. Redfield, *J. Appl. Phys.* **50**, 1491 (1979).
13. J. C. Bean, H. J. Leamy, J. M. Poate, G. A. Rozgonyi, J. P. van der Ziel, J. S. Williams, and G. K. Celler, *J. Appl. Phys.* **50**, 881 (1979).
14. M. R. T. Siregar, M. von Allmen, and W. Luthy, *Helv. Phys. Acta* **52**, 45 (1979).
15. P. Baeri, S. U. Campisano, G. Foti, and E. Rimini, *J. Appl. Phys.* **50**, 788 (1979).
16. J. C. Schultz and R. J. Collins, *Appl. Phys. Lett,* **34**, 84 (1979).
17. A. Lietoila and J. F. Gibbons, *J. Appl. Phys.* **53**, 3207 (1982).
18. C. M. Surko, A. L. Simons, D. H. Auston, J. A. Golovchenko, R. E. Slusher, and T. N. C. Venkatesan, *Appl. Phys. Lett.* **34**, 635 (1979).
19. A. Bhattacharyya, B. G. Streetman, and K. Hess, *J. Appl. Phys.* **52**, 3611 (1981).
20. R. F. Wood and G. E. Giles, *Phys. Rev. B* **23**, 2923 (1981).
21. H. J. Leamy, *J. Appl. Phys.* **53**, R51 (1982), and references therein.
22. A. R. Kirkpatrick, J. A. Minnucci, and A. C. Greenwald, *IEEE Trans. Electron Devices* **ED-24**, 429 (1977).
23. R. T. Young, G. A. van der Leeden, J. Narayan, W. H. Christie, R. F. Wood, D. E. Rothe, and J. I. Levatter, *IEEE Electron Device Lett.* **EDL-3**, 280 (1982).
24. D. H. Lowndes, J. W. Cleland, W. H. Christie, R. E. Eby, G. E. Jellison, J. Narayan, R. D. Westbrook, R. F. Wood, J. A. Nilson, and S. C. Das, *Appl. Phys. Lett.* **41**, 938 (1982).
25. A. G. Cullis, H. C. Webber, and P. Bailey, *J. Phys. E* **12**, 668 (1979).
26. G. K. Celler, J. M. Poate, and L. C. Kimerling, *Appl. Phys. Lett.* **32**, 464 (1978).
27. J. A. Van Vechten, in "Laser and Electron-Beam Interactions with Solids" (B. R. Appleton and G. K. Celler, eds.), pp. 49–60. North-Holland Publ., Amsterdam, 1982, and references therein.
28. D. H. Auston, J. A. Golovchenko, A. L. Simons, R. E. Slusher, P. R. Smith, C. M. Surko, and T. N. C. Venkatesan, in "Laser–Solid Interactions and Laser Processing—1978" (S. D. Ferris, H. J. Leamy, and J. M. Poate, eds.), AIP Conference Proceedings, No. 50, pp. 11–26. Am. Inst. Phys., New York, 1979; D. H. Auston, J. A. Golovchenko, A. L. Simons, C. M. Surko, and T. N. C. Venkatesan, *Appl. Phys. Lett.* **34**, 777 (1979).
29. N. Bloembergen, H. Kurz, J. M. Liu, and R. Yen, in "Laser and Electron-Beam Interactions with Solids" (B. R. Appleton and G. K. Celler, eds.), pp. 3–12. North-Holland Publ., Amsterdam, 1982.
30. J. M. Liu, H. Kurz, and N. Bloembergen, in "Laser–Solid Interactions and Transient Thermal Processing of Materials" (J. Narayan, W. L. Brown, and R. A. Lemons, eds.), pp. 3–12. North-Holland Publ., Amsterdam, 1983.
31. P. H. Bucksbaum and J. Bokor, in "Laser–Solid Interactions and Transient Thermal Processing of Materials" (J. Narayan, W. L. Brown, and R. A. Lemons, eds.), pp. 51–56. North-Holland Publ., Amsterdam, 1983.
32. C. V. Shank, R. Yen, and C. Hirlimann, *Phys. Rev. Lett.* **50**, 454 (1983).
33. D. E. Aspnes, G. K. Celler, J. M. Poate, G. A. Rozgonyi, and T. T. Sheng, in "Laser and

Electron Beam Processing of Electronic Materials" (C. L. Anderson, G. K. Celler, and G. A. Rozgonyi, eds.), pp. 414–420. Electrochem. Soc., Princeton, New Jersey, 1980.
34. C. W. White, J. Narayan, and R. T. Young, *Science (Washington, D.C.)* **204**, 461 (1979).
35. J. Narayan, *in* "Laser and Electron Beam Processing of Electronic Materials" (C. L. Anderson, G. K. Celler, and G. A. Rozgonyi, eds.), pp. 294–332. Electrochem. Soc., Princeton, New Jersey, 1980.
36. J. L. Benton, L. C. Kimerling, G. L. Miller, D. A. H. Robinson, and G. K. Celler, *in* "Laser–Solid Interactions and Laser Processing—1978" (S. D. Ferris, H. J. Leamy, and J. M. Poate, eds.), p. 543. Am. Inst. Phys., New York, 1979.
37. L. C. Kimerling and J. L. Benton, *in* "Laser and Electron Beam Processing of Materials" (C. W. White and P. S. Peercy, eds.), pp. 385–396. Academic Press, New York, 1980.
38. N. M. Johnson, R. B. Gold, A. Lietoila, and J. F. Gibbons, *in* "Laser–Solid Interactions and Laser Processing—1978" (S. D. Ferris, H. J. Leamy, and J. M. Poate, eds.), AIP Conference Proceedings, No. 50, pp. 550–555. Am. Inst. Phys., New York, 1979.
39. N. M. Johnson, *in* "Laser and Electron-Beam Interactions with Solids" (B. R. Appleton and G. K. Celler, eds.), pp. 343–347. North-Holland Publ., Amsterdam, 1982, and references therein.
40. J. L. Benton, C. J. Doherty, S. D. Ferris, L. C. Kimerling, H. J. Leamy, and G. K. Celler, *in* "Laser and Electron Beam Processing of Materials" (C. W. White and P. S. Peercy, eds.), pp. 430–434. Academic Press, New York, 1980.
41. C. Hill, A. L. Butler, and J. A. Daly, *in* "Laser and Electron-Beam Interactions with Solids" (B. R. Appleton and G. K. Celler, eds.), pp. 579–584. North-Holland Publ., Amsterdam, 1982.
42. C. W. White, *in* "Laser and Electron-Beam Interactions with Solids" (B. R. Appleton and G. K. Celler, eds.), pp. 109–120. North-Holland Publ., Amsterdam, 1982.
43. C. W. White, S. R. Wilson, B. R. Appleton, and F. W. Young, Jr., *J. Appl. Phys.* **51**, 738 (1980).
44. P. Baeri, G. Foti, J. M. Poate, S. U. Campisano, and A. G. Cullis, *Appl. Phys. Lett.* **38**, 800 (1981).
45. P. L. Liu, R. Yen, N. Bloembergen, and R. T. Hodgson, *Appl. Phys. Lett.* **34**, 864 (1979).
46. R. Tsu, R. T. Hodgson, T. Y. Tan, and J. E. Baglin, *Phys. Rev. Lett.* **42**, 1356 (1979).
47. R. Yen, J. M. Liu, H. Kurz, and N. Bloembergen, *in* "Laser and Electron-Beam Interactions with Solids" (B. R. Appleton and G. K. Celler, eds.), pp. 37–42. North-Holland Publ., Amsterdam, 1982.
48. A. G. Cullis, H. C. Webber, and N. G. Chew, *in* "Laser and Electron-Beam Interactions with Solids" (B. R. Appleton and G. K. Celler, eds.), pp. 131–140. North-Holland Publ., Amsterdam, 1982.
49. A. G. Cullis, H. C. Webber, N. G. Chew, J. M. Poate, and P. Baeri, *Phys. Rev. Lett.* **49**, 219 (1982).
50. P. Baeri, G. Foti, J. M. Poate, and A. G. Cullis, *Phys. Rev. Lett.* **45**, 2036 (1980).
51. J. C. C. Fan, H. J. Zeiger, R. P. Gale, and R. L. Chapman, *Appl. Phys. Lett.* **36**, 158 (1980).
52. R. B. Gold, J. F. Gibbons, T. J. Magee, J. Peng, R. Ormond, V. R. Deline, and C. A. Evans, *in* "Laser and Electron Beam Processing of Materials" (C. W. White and P. S. Peercy, eds.), pp. 221–226. Academic Press, New York, 1980.
53. D. Bensahel and G. Auvert, *in* "Laser–Solid Interactions and Transient Thermal Processing of Materials" (J. Narayan, W. L. Brown, and R. A. Lemons, eds.), pp. 165–176. North-Holland Publ., Amsterdam, 1983.
54. H. J. Zeiger, J. C. C. Fan, B. J. Palm, R. P. Gale, and R. L. Chapman, *in* "Laser and

Electron Beam Processing of Materials" (C. W. White and P. S. Peercy, eds.), pp. 234–240. Academic Press, New York, 1980.
55. H. J. Leamy, W. L. Brown, G. K. Celler, G. Foti, G. H. Gilmer, and J. C. C. Fan, in "Laser and Electron-Beam Solid Interaction and Material Processing" (J. F. Gibbons, L. D. Hess, and T. W. Sigmon, eds.), pp. 89–96. North-Holland Publ., Amsterdam, 1981.
56. M. O. Thompson, G. J. Galvin, J. W. Mayer, P. S. Peercy, J. M. Poate, D. C. Jacobson, A. G. Cullis, and N. G. Chew, *Phys. Rev. Lett.* **52**, 2360 (1984).
57. W. L. Brown, in "Energy Beam–Solid Interactions and Transient Thermal Processing" (J. C. C. Fan and N. M. Johnson, eds.), pp. 9–23. North-Holland Publ., Amsterdam, 1984).
58. G. H. Gilmer and H. J. Leamy, in "Laser and Electron Beam Processing of Materials" (C. W. White and P. S. Peercy, eds.), pp. 227–233. Academic Press, New York, 1980.
59. H. J. Leamy, W. L. Brown, G. K. Celler, G. Foti, G. H. Gilmer, and J. C. C. Fan, *Appl. Phys. Lett.* **38**, 137 (1981).
60. L. Csepregi, E. F. Kennedy, J. W. Mayer, and T. W. Sigmon, *J. Appl. Phys.* **49**, 3906 (1978).
61. G. K. Celler, J. M. Poate, G. A. Rozgonyi, and T. T. Sheng, *J. Appl. Phys.* **50**, 7264 (1979).
62. A. Gat, J. F. Gibbons, T. J. Magee, J. Peng, V. R. Deline, P. Williams, and C. A. Evans, Jr., *Appl. Phys. Lett.* **32**, 276 (1978).
63. J. S. Williams, W. L. Brown, H. J. Leamy, J. M. Poate, J. W. Rodgers, D. Rousseau, G. A. Rozgonyi, J. A. Shellnutt, and T. T. Sheng, *Appl. Phys. Lett.* **33**, 542 (1978).
64. R. B. Fair and J. C. Tsai, *J. Electrochem. Soc.* **122**, 1689 (1975).
65. J. Narayan, O. W. Holland, and B. R. Appleton, *J. Vac. Sci. Technol., B* **1**, 871 (1983).
66. A. Lietoila, J. F. Gibbons, J. L. Regolini, T. W. Sigmon, T. J. Magee, J. Peng, and J. D. Hong, in "Laser and Electron Beam Processing of Electronic Materials" (C. L. Anderson, G. K. Celler, and G. A. Rozgonyi, eds.), pp. 350–360. Electrochem. Soc., Princeton, New Jersey, 1980.
67. H. Baumgart, F. Phillipp, and H. J. Leamy, in "Laser and Electron Beam-Interactions with Solids" (B. R. Appleton and G. K. Celler, eds.), pp. 355–360. North-Holland Publ., Amsterdam, 1982.
68. M. Mizuta, N. H. Sheng, and J. L. Merz, *J. Appl. Phys.* **52**, 6437 (1981).
69. R. L. Cohen, J. S. Williams, L. C. Feldman, and K. W. West, *Appl. Phys. Lett.* **33**, 751 (1978).
70. H. A. Bomke, H. L. Berkowitz, M. Harmatz, S. Kronenberg, and R. Lux, *Appl. Phys. Lett.* **33**, 955 (1978).
71. J. T. Lue, *Appl. Phys. Lett.* **36**, 73 (1980).
72. L. Correra and L. Peduli, *Appl. Phys. Lett.* **37**, 55 (1980).
73. T. Arai, T. Igarashi, and T. Hiramoto, *Proc. Conf. Jpn. Appl. Phys. Soc.* **28**, 557 (1981).
74. G. L. Olson, S. A. Kokorowski, J. A. Roth, and L. D. Hess, in "Laser–Solid Interactions and Transient Thermal Processing of Materials" (J. Narayan, W. L. Brown, and R. A. Lemons, eds.), pp. 141–154. North-Holland Publ., Amsterdam, 1983.
75. B. G. Bagley and H. S. Chen, in "Laser–Solid Interactions and Laser Processing—1978" (S. D. Ferris, H. J. Leamy, and J. M. Poate, eds.), AIP Conference Proceedings, No. 50, pp. 97–101. Am. Inst. Phys., New York, 1979.
76. F. Spaepen and D. Turnbull, in "Laser Annealing of Semiconductors" (J. M. Poate and J. W. Mayer, eds.), pp. 15–42. Academic Press, New York, 1982.
77. G. L. Olson, J. A. Roth, L. D. Hess, and J. Narayan, in "Energy Beam—Solid Interactions and Transient Thermal Processing" (J. C. C. Fan and N. M. Johnson, eds.), pp. 375–382. North-Holland Publ., Amsterdam, 1984; G. L. Olson, J. A. Roth, L. D. Hess, and J. Narayan, *Proc. US–Jpn. Semin. Solid Phase Epitaxy Interface Kinet.*, Oiso, Jpn. 1983.

78. T. O. Sedgwick, *J. Electrochem. Soc.* **130**, 484 (1983).
79. T. E. Seidel, *IEEE Electron Device Lett.* **EDL-4**, 353 (1983).
80. Heatpulse 210T, AG Associates, Palo Alto, California.
81. Radiant Impulse Processor-Model 180, Tamarack Scientific Co., Inc., Anaheim, California.
82. IA-200, Varian, Gloucester, Massachusetts; C. J. Russo, D. F. Downey, S. C. Holden, and R. T. Fulks, *Int. Conf. Ion Implantation Equip. Tech., 4th, Berchtesgaden, FRG, 1982.*
83. T. E. Seidel, D. J. Lischner, C. S. Pai, and S. S. Lau, *Proc.—Electrochem. Soc.* **84-7**, 184 (1984).
84. D. J. Lischner and G. K. Celler, in "Laser and Electron-Beam Interactions with Solids" (B. R. Appleton and G. K. Celler, eds.), p. 759–764. North-Holland Publ., Amsterdam, 1982.
85. D. J. Lischner, and A. G. Shatas, personal communication (1984).
86. D. J. Lischner, M. J. Kelley, L. Lancaster, and L. Adda, unpublished data (1983).
87. G. G. Bentini and L. Correra, *J. Appl. Phys.* **54**, 2057 (1983).
88. H. E. Cline, *J. Appl. Phys.* **56**, 2683 (1983).
89. H. D. Chiou, *Ext. Abst. Proc.—Electrochem. Soc.* **83-2**, 538 (1983).
90. C. J. Doerschel and F. G. Kirscht, *Phys. Status Solidi A* **64**, K85 (1981).
91. H. Cline and T. Anthony, *J. Appl. Phys.* **47**, 2332 (1976).
92. G. K. Celler, McD. Robinson, L. E. Trimble, and D. J. Lischner, *Appl. Phys. Lett.* **43**, 868 (1983).
93. W. G. Hawkins and D. K. Biegelsen, *Appl. Phys. Lett.* **42**, 358 (1983).
94. G. K. Celler, K. A. Jackson, L. E. Trimble, McD. Robinson, and D. J. Lischner, in "Energy Beam–Solid Interactions and Transient Thermal Processing" (J. C. C. Fan and N. M. Johnson, eds.), pp. 409–414. North-Holland Publ., Amsterdam, 1984.
95. T. E. Seidel, in "VLSI Technology" (S. M. Sze, ed.), p. 242. McGraw-Hill, New York, 1983.
96. Operational Manual for 210M, AG Associates, Palo Alto, California, 1982.
97. H. Baumgart, G. K. Celler, D. J. Lischner, McD. Robinson, and T. T. Sheng, in "Laser–Solid Interactions and Transient Thermal Processing of Materials" (J. Narayan, W. L. Brown, and R. A. Lemons, eds.), pp. 349–354. North-Holland Publ., Amsterdam, 1983.
98. R. Kwor, D. W. Kwong, and B. Y. Tsaur, *Ext. Abst. Proc.—Electrochem. Soc.* **84-1**, No. 114 (1984).
99. S. R. Wilson, R. B. Gregory, W. M. Paulson, and H. T. Diehl, *IEEE Trans. Nucl. Sci.* **NS-30**, 1734 (1983).
100. R. T. Hodgson, V. Deline, S. M. Mader, F. F. Morehead, and J. Gelpey, in "Energy Beam–Solid Interactions and Transient Thermal Processing" (J. C. C. Fan and N. M. Johnson, eds.), pp. 253–257. North-Holland Publ., Amsterdam, 1984.
101. W. K. Hofker, *Philips Res. Rep., Suppl.* No. 8 (1975).
102. J. B. Lasky, *J. Appl. Phys.* **54**, 6009 (1983).
103. T. E. Seidel, R. W. Knoell, F. A. Stevie, G. Poli, and B. Schwartz, *Proc.—Electrochem. Soc.* **84-7**, 201 (1984).
104. M. Y. Tsai and B. G. Streetman, *J. Appl. Phys.* **50**, 183 (1979).
105. T. M. Liu and W. G. Oldham, *IEEE Electron Device Lett.* **EDL-4**, 59 (1983).
106. W. Maszara, C. Carter, D. K. Sadana, J. Liu, V. Ozguz, J. J. Wortman, and G. A. Rozgonyi, in "Energy Beam–Solid Interactions and Transient Thermal Processing" (J. C. C. Fan and N. M. Johnson, eds.), pp. 285–291. North-Holland Publ., Amsterdam, 1984.
107. D. K. Sadana, E. Myers, J. Liu, T. Finstad, J. J. Wortman, W. K. Chu, and G. A. Rozgonyi, in "Energy Beam–Solid Interactions and Transient Thermal Processing" (J. C. C. Fan and N. M. Johnson, eds.), pp. 303–308. North-Holland Publ., Amsterdam, 1984.

108. J. Narayan, *J. Appl. Phys.* **53,** 8607 (1982).
109. J. L. Benton, G. K. Celler, D. C. Jacobson, L. C. Kimerling, D. J. Lischner, G. L. Miller, and McD. Robinson, *in* "Laser and Electron-Beam Interactions with Solids" (B. R. Appleton and G. K. Celler, eds.), pp. 765–778. North-Holland Publ., Amsterdam, 1982.
110. R. B. Fair, J. J. Wortman, and J. Liu, *J. Electrochem. Soc.* **131,** 2387 (1984).
111. T. O. Sedgwick, S. A. Cohen, G. S. Oehrlein, V. R. Delne, R. Kalish, and S. Shatas, *Proc.—Electrochem. Soc.* **84-7,** 192 (1984).
112. T. E. Seidel and D. J. Lischner, *Electrochem. Soc. Meet., Cincinnati, Ohio* (1984).
113. T. O. Sedgwick, S. A. Cohen, G. S. Oehrlein, V. R. Deline, R. Kalish, and S. Shatas, *Electrochem. Soc. Meet., Cincinnati, Ohio* (1984).
114. D. K. Sadana, S. C. Shatas, and A. Gat, *Conf. Ser.—Inst. Phys,* No. 67, p. 143 (1983).
115. D. Peters, S. Y. Chiang, P. Carey, and K. Chem, *Proc.—Electrochem. Soc.* **84-7,** 211 (1984).
116. B. Y. Tsaur and C. H. Anderson, Jr., *J. Electrochem. Soc., Rev. News, Washington, D.C.* Pap. 314 (1983).
117. S. Salmi, E. Landi, and P. Negrini, *IEEE Electron Device Lett.* to be published.
118. K. K. Ng, G. K. Celler, E. I. Povilonis, R. C. Frye, H. J. Leamy, and S. M. Sze, *IEEE Electron Device Lett.* **EDL-2,** 316 (1981).
119. H. Baumgart, H. J. Leamy, L. E. Trimble, C. J. Doherty, and G. K. Celler, *in* "Grain Boundaries in Semiconductors" (H. J. Leamy, G. E. Pike, and C. H. Seager, eds.), p. 311. North-Holland Publ., Amsterdam, 1982.
120. E. W. Maby, H. A. Atwater, A. L. Keigler, and N. M. Johnson, *Appl. Phys. Lett.* **43,** 482 (1983).
121. G. K. Celler, K. K. Ng, L. E. Trimble, and E. I. Povilonis, *in* "Comparison of Thin Film Transistors and SOI Technologies" (H. W. Lam and M. J. Thompson, eds.). pp. 127–132. North-Holland Publ., Amsterdam, 1984.
122. K. K. Ng, G. K. Celler, E. I. Povilonis, L. E. Trimble, and S. M. Sze, *in* "Energy Beam–Solid Interactions and Transient Thermal Processing" (J. C. C. Fan and N. M. Johnson, eds.), pp. 559–565. North-Holland Publ., Amsterdam, 1984.
123. K. K. Ng, G. W. Taylor, G. K. Celler, L. E. Trimble, R. J. Bayruns, and E. I. Povilonis, *Proc. IEEE IDEM* pp. 356–359. (1983).
124. M. Delfino and T. A. Reifsteck, *IEEE Electron Device Lett.* **EDL-3,** 116 (1982).
125. R. T. Fulks, R. A. Powell, and W. T. Stacy, *IEEE Electron Device Lett.* **EDL-3,** 179 (1982).
126. S. Shatas, A. Gat, and R. Ramani, personal communication; see also *Proc. Workshop Refract. Met. Silicides VLSI-II, Univ. California, Berkeley* (1984).
127. S. Murarka, "Silicides for VLSI Applications." Academic Press, New York, 1983.
128. R. A. Powell, R. Chew, C. Thridandam, R. T. Fulks, I. A. Blech, and J.-D. T. Pan, *IEEE Electron Device Lett.* **EDL-4,** 380 (1983).
129. J. M. Poate, K. N. Tu, and J. W. Mayer, "Thin Films—Interdiffusion and Reactions," p. 16. Wiley, New York, 1978.
130. P. S. Burggraaf, *Semicond. Int.* (Dec. 1983) p. 69, see also p. 73.
131. J. J. Towner and E. P. van de Ven, *Appl. Phys. Lett.* **44,** 198 (1984).
132. H. J. Leamy, *in* "Laser and Electron-Beam Interactions with Solids" (B. R. Appleton and G. K. Celler, eds.), pp. 459–470. North-Holland Publ., Amsterdam, 1982.
133. G. K. Celler, *J. Cryst. Growth* **63,** 429 (1983).
134. J. A. Knapp and S. T. Picraux, *J. Cryst. Growth* **63,** 445 (1983).
135. J. C. C. Fan, B.-Y. Tsaur, and M. W. Geis, *J. Cryst. Growth* **63,** 453 (1983).
136. McD. Robinson, D. J. Lischner, and G. K. Celler, *J. Cryst. Growth* **63,** 484 (1983).
137. H. J. Leamy, C. C. Chang, H. Baumgart, R. A. Lemons, and J. Cheng, *Mater. Lett.* **1,** 33 (1982).

138. N. M. Johnson, D. K. Biegelsen, H. C. Tuan, M. D. Moyer, and L. E. Fennell, *IEEE Electron Device Lett.* **EDL-3,** 369 (1982).
139. H. W. Lam, R. F. Pinizzotto, and A. F. Tasch, Jr., *J. Electrochem. Soc.* **128,** 1981 (1981).
140. G. K. Celler, L. E. Trimble, McD. Robinson, K. K. Ng, and H. J. Leamy, *Electron. Mater. Conf.*, Fort Collins, Colo., Abstr. J-8 (1982).
141. G. K. Celler, L. E. Trimble, K. K. Ng, H. J. Leamy, and H. Baumgart, *Appl. Phys. Lett.* **40,** 1043 (1982).
142. K. E. Bean and W. R. Runyan, *J. Electrochem. Soc.* **124,** 5C (1977).
143. G. K. Celler, McD. Robinson, and D. J. Lischner, *Appl. Phys. Lett.* **42,** 99 (1983).
144. G. K. Celler, McD. Robinson, D. J. Lischner, and T. T. Sheng, *in* "Laser–Solid Interactions and Transient Thermal Processing of Materials" (J. Narayan, W. L. Brown, and R. A. Lemons, eds.), pp. 575–580. North-Holland Publ., Amsterdam, 1983.
145. G. K. Celler, McD. Robinson, D. J. Lischner, and L. E. Trimble, *J. Electrochem. Soc.*, **132,** 211 (1985).
146. T. E. Seidel, *in* "VLSI Technology" (S. M. Sze, ed.), p. 255. McGraw-Hill, New York, 1983.
147. A. G. Cullis, T. E. Seidel, and R. L. Meek, *J. Appl. Phys.* **49,** 5188 (1978).
148. C. W. Pearce and V. J. Zaleckas, *J. Electrochem. Soc.* **126,** 1436 (1979).

Reactive Ion-Beam Etching and Plasma Deposition Techniques Using Electron Cyclotron Resonance Plasmas

SEITARO MATSUO

ATSUGI ELECTRICAL COMMUNICATION LABORATORY
NIPPON TELEGRAPH AND TELEPHONE PUBLIC CORPORATION
ATSUGI-SHI, JAPAN

I.	Introduction	75
II.	Reactive Ion-Beam Etching	76
	1. Introduction	76
	2. Broad-Beam ECR Ion Source	77
	3. Shielded Single-Grid Ion Extraction	80
	4. Reactive Ion-Beam Etching System	84
	5. Etching Characteristics	85
	6. Pattern Formation Process	90
III.	Plasma Deposition	95
	7. Introduction	95
	8. ECR Plasma Deposition Apparatus	95
	9. Divergent Magnetic Field Plasma Extraction	97
	10. Deposition Characteristics	102
	11. Material Supply by Sputtering	109
	12. Applications	113
	References	116

I. Introduction

Plasma technology has been successfully applied to semiconductor device fabrication, especially to plasma etching and plasma deposition processes. As a plasma generation method, rf discharge is most generally utilized because a stable and uniform plasma is easily obtained with relatively large area, and, further, the plasma is suitable for low-temperature processes because of its relatively small heating effect. The gas pressures used range from 10^{-2} to 1 Torr, and the plasma density is of the order of $10^{10}/cm^3$. Therefore, the ionization ratio is from 10^{-6} to 10^{-4}, and the value is very small. In these plasmas, neutral molecules and radicals, which constitute the greater part of the plasma, affect the etching or deposition reactions significantly, although the role of ions is often more essential.

On the other hand, electron cyclotron resonance (ECR) plasma can easily be generated at much lower gas pressures of 10^{-5}–10^{-3} Torr. By using an apparatus modified for semiconductor processes, as will be described, the plasma density is still left at the same order of magnitude as that of the rf plasma. Therefore, the ionization ratio becomes 10^{-3}–10^{-1}, larger by about three figures than that of the rf plasma. Furthermore, stable plasma generation is possible with reactive gases, as in the case of the rf discharge, since no electrode configuration is needed.

The ECR plasma generation method was originally studied in nuclear fusion plasma research[1-4] for the purpose of producing a high-temperature plasma. The plasma parameters required for this purpose are quite different from those of the plasma suitable for semiconductor processes. Plasma technology for fusion aims at the control of the plasma itself, confined by magnetic fields such as a mirrorlike configuration, to approach a limit of high temperature and high density. On the other hand, for semiconductor processes a moderate plasma generation and supply of reactants to the specimen with little heating are required. Thus, apparatus construction for semiconductor processes can be much simpler than for fusion.

In the following sections the applications of ECR plasmas to reactive ion-beam etching and plasma deposition are described, utilizing the distinctive features of the plasma such as high ionization ratio and low gas pressure.

II. Reactive Ion-Beam Etching

1. Introduction

Plasma etching technologies have been used widely in semiconductor device fabrications. In these applications, the characteristics of selective etching are very important, because the materials such as the etch mask should be preserved as much as possible. Carbon tetrafluoride (CF_4) plasma effectively etches the materials that form volatile fluorides, such as Si, Mo, W, and Ti, and is applied to patterning poly-Si, Si_3N_4, and other materials on SiO_2 layers.[5-7] However, this method brings some deterioration in pattern transfer accuracy due to the undercutting effect caused by the isotropic nature of the reactive radicals, fluorine atoms, generated by rf glow discharge. On the other hand, reactive sputter etching (reactive ion etching), the method of sputter etching using reactive gases,[8] has been applied to selectively etch

SiO_2, Si_3N_4, and other materials on Si substrate or on poly-Si layers with fluorocarbons.[9,10] In this method, the mask patterns are transferred to the underlying substance with high accuracy, since the etching process is based on the reactivity of incident ions such as CF_n^+ with kinetic energy.[10] Furthermore, reactive sputter etching with incident ions combining fluorocarbons with low reactive halogens ($CBrF_3$, $CF_4 + I_2$, etc.) shows etching characteristics with a selectivity similar to that of CF_4 plasma etching but without undercutting.[11]

By choosing a suitable condition, reactive sputter etching gives high-accuracy pattern transfer from mask to underlying substance for various etching selectivities. These advantageous features result from the effect of reactive ions incident on the specimen surface. However, the gas-phase reactivity must be suppressed by delicately controlling the gas constituents. Therefore, the method of reactive ion-beam etching, utilizing reactive ions extracted with energies precisely controlled and separated from the gas plasma at much lower gas pressures, is expected to provide a new, useful technology. The condition of low gas pressure inevitably decreases the gas-phase reactivity regardless of the kind of reactive gas used.

A method of reactive ion-beam etching has been examined that involves the introduction of reactive gases such as CF_4, C_2F_6, and Cl_2 into a Kaufman-type ion source.[10,12-14] However, this method is somewhat impractical, because the thermionic cathode in the discharge chamber is easily attacked by the reactive gas plasma, resulting in cathode wire breakdown or instability in thermionic emission.

The features of ECR plasma, low-gas-pressure discharge and no electrode configuration, are quite suitable for the ion sources for reactive ion-beam etching. Electron cyclotron resonance plasmas have already been applied as ion sources[15-17] for ion injection into accelerators and for ion implantation. However, the ion sources are of a narrow-beam type and are not suitable for etching.

The following section describes a broad-beam ECR ion source newly developed for reactive ion-beam etching, and further describes the etching characteristics of various materials, such as SiO_2 and Al. Reaction mechanism and pattern profiles fabricated with masks are also discussed.

2. Broad-Beam ECR Ion Source

Figure 1 illustrates an experimental apparatus using the ECR ion source for reactive ion-beam etching.[18] Microwave power (frequency, 2.45 GHz) is introduced into the plasma chamber through a rectangular waveguide and a window made of a fused quartz plate. The plasma chamber is 20 cm in diameter and 20 cm in height (inside dimensions) and operates as a microwave cavity resonator in TE_{113} mode. Magnet coils are arranged around the periphery of the plasma chamber for the microwave ECR condition:

$$\omega_c = \omega, \qquad \omega_c = eB/m, \tag{2.1}$$

where ω_c is the electron cyclotron frequency, ω, the microwave frequency; e, the absolute value of electron charge; B, the magnetic flux density; and m, the electron mass. This condition gives a magnetic flux density of 875 G for a microwave frequency of 2.45 GHz.

The ion extraction system, employing a 15-cm-diameter broad beam, is composed of two stainless steel grids with aligned multiple holes 2 mm in diameter, similar to the configuration in a Kaufman-type ion source. The upper grid makes electrical contact with the plasma chamber, and the lower grid is at ground potential. The ion extraction voltage of 0–2000 V is applied to the whole of the plasma chamber. The plasma chamber and the magnet coils are water cooled. The vacuum system consists of an oil diffusion pump (1200 liters/sec) and a mechanical rotary pump (500 liters/min). Etching gas

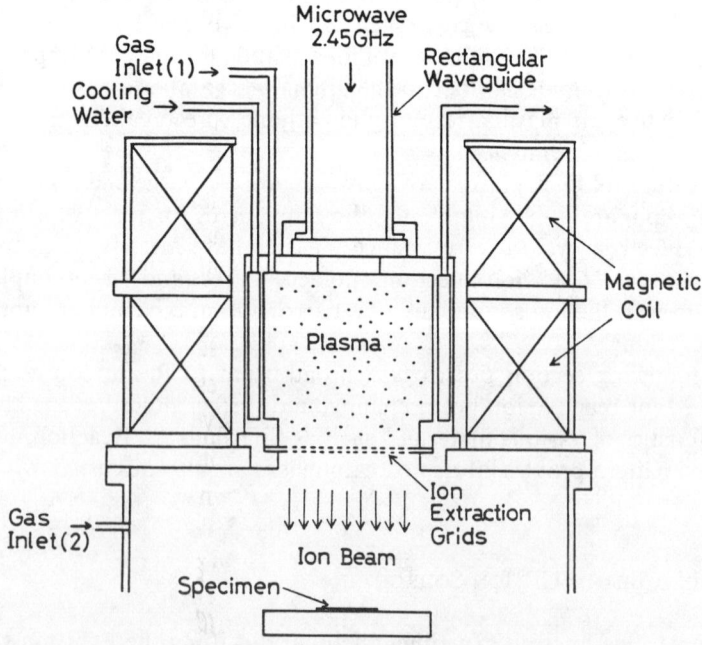

Fig. 1. Broad-beam ECR ion source for reactive ion-beam etching. Ion-beam diameter, 15 cm. A dual-grid system is used in this apparatus for ion extraction. (From Matsuo and Adachi.[18])

is introduced through the main gas inlet (1) or the auxiliary gas inlet (2) (Fig. 1).

The intensity of the magnetic field in the plasma chamber gradually weakens from the ECR region at the top toward the ion extraction grids. This magnetic field configuration serves to improve the intensity of the extracted ion current and the uniformity of the ion current density, because the plasma is transported at an accelerating rate along the divergent magnetic field and is spread over the grids. The effect of the interaction between ECR plasma and divergent magnetic field is described in more detail in Subsection 9. The ion-current density is monitored by a Faraday cup attached to a shutter plate that is used also as a total ion-current monitor.

The microwave ECR discharge is initiated and sustained at gas pressures above about 3×10^{-5} Torr, nearly independent of the gas introduced. In usual etchings, gas pressures of 10^{-4}–10^{-3} Torr are employed. Figure 2 shows the ion extraction characteristics of the ion source as a function of microwave power input, changing the grid spacing. The ion current increases with increasing microwave power and with decreasing grid spacing. An ion current density of 0.5–1 mA/cm² is obtained at the grid spacing of 1 mm by supplying microwave power of 100–300 W at an ion extraction voltage of 1000 V. Generally, the ion current extracted depends on ion extraction voltage and grid spacing, roughly according to Langmuir–Child's law:[19]

$$J = kV^{3/2}/d^2, \qquad (2.2)$$

where J is the current extracted, k, a constant depending on the charge-to-mass ratio of the ions, V, the extraction voltage, and d, the spacing between

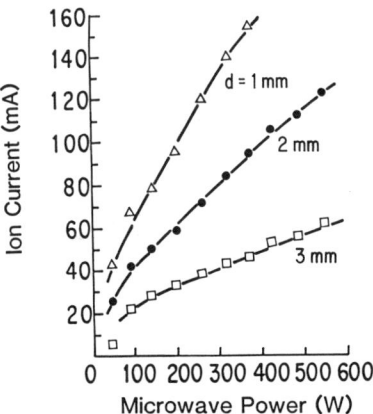

Fig. 2. Ion extraction characteristics as a function of microwave power, with the spacing of dual grids varied. Ion-beam diameter, 15 cm. C_4F_8, 3×10^{-4} Torr, 1 kV.

the grids. When the spacing is set at smaller than 1 mm, the ion current may be increased further. However, reliable operation and a uniform beam are not obtained, because the grid thermal deformation makes it difficult to ensure uniform spacing. Uniformity can be controlled to a certain degree by controlling the magnet coil current. In an extracted ion beam 15 cm in diameter, the area with uniformity within 10% is about 13 cm in diameter.

3. Shielded Single-Grid Ion Extraction

An ion-beam etching technique for semiconductor device fabrication must minimize surface damage caused by ion bombardment, while maintaining the etching characteristics such as etch rate, selectivity, and pattern accuracy. To realize such a condition, a high-current, moderately low-energy ion beam must be extracted. As an approach to reducing ion energy, an ion extraction system composed of a fine-mesh single grid has been applied to reactive ion-beam etching, modifying a Kaufman-type ion source.[20] If the extraction

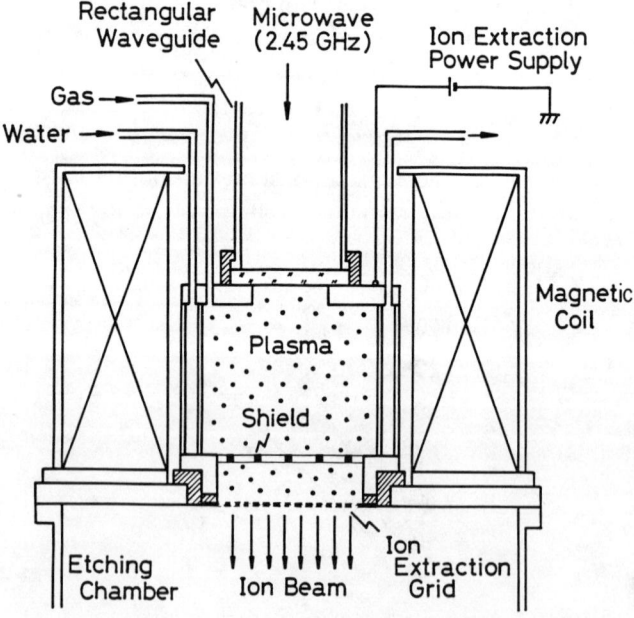

Fig. 3. Broad-beam ECR ion source with shielded single-grid ion extraction system. Ion-beam diameter, 15 cm. Single grid with multiple holes 2 mm in diameter is made of stainless steel or carbon. (From Ono et al.[21])

voltage is fairly low, the thickness of the ion sheath generated by the voltage becomes smaller than the spacing. In such a case, the upper grid (screen grid, Fig. 1), can be omitted to improve the ion current extraction; that is, only the lower (acceleration) grid is used. In fact, an ion-current density of about 1 mA/cm^2 is obtained at a voltage of about 100 V.[20] A fine mesh is used as the grid since the ion sheath thickness is very small at such a low voltage, and the thickness must be comparable to or larger than the grid aperture dimension. However, a fine-mesh grid is not so reliable, because sputtering damage and overheating result from ion impingement on the grid. Furthermore, because the ion extraction voltage is limited to low ion energies, it is difficult to ensure the desired etching characteristics. When relatively high voltages are employed, say higher than about 200 V, the electric field resulting from the grid potential influences the whole plasma in the discharge chamber and causes plasma instability and sparkover phenomena between the grid and discharge chamber wall.

To extend the advantages of the single-grid method to a wider range of ion extraction voltages, a shielded single-grid technique, which can employ a grid with dimensions similar to those in a conventional dual-grid system, has been developed.[21] Figure 3 shows the ECR ion source with the shielded single-grid ion extraction system. The plasma chamber is separated into two regions, the plasma generation region, and the plasma transport region by a shield electrode made of a very rough grating (20 mm in pitch, 2 mm in

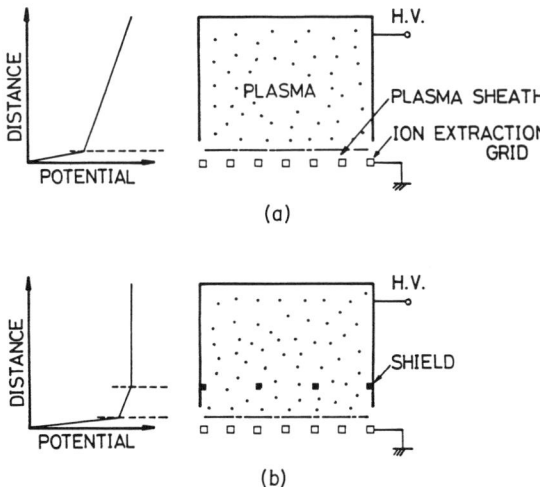

Fig. 4. Illustration of the influences of grid potential on plasma for (a) nonshielded single-grid ion extraction and (b) shielded single-grid ion extraction.

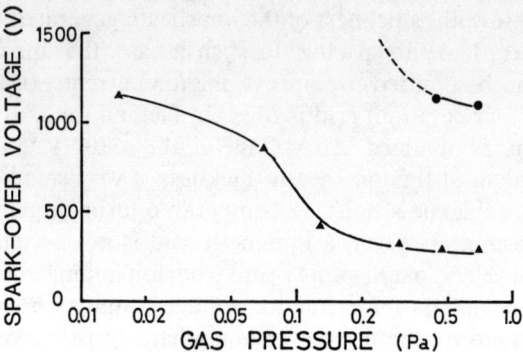

Fig. 5. Effect of the shield electrode is shown in sparkover voltage characteristics as a function of C_2F_6 gas pressure, for the distances of 30 mm (shielded, ●) and 100 mm (nonshielded, ▲) from grid to shield. Microwave power, 200 W. (From Ono et al.[21])

thickness), which also functions as a microwave reflector. The generation region is 17 cm in diameter and 20 cm in height in inside dimensions and operates as a microwave cavity resonator. The transport region is 15 cm in diameter and 3 cm in height.

A single grid with multiple holes 2 mm in diameter, made of stainless steel or carbon, is arranged at the end of the transport region, to extract a 15-mm-diameter beam. The grid thickness is 0.5 mm for stainless steel or

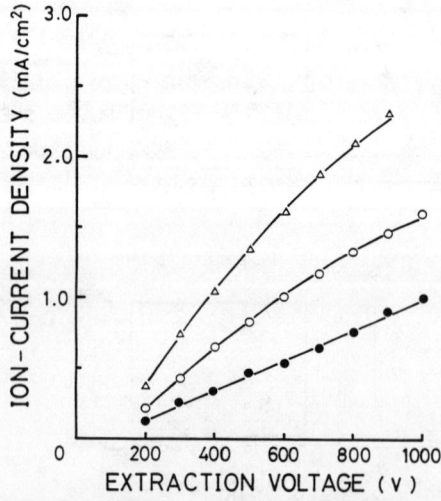

Fig. 6. Ion extraction characteristics as a function of extraction voltage. Multiple holes, 2 mm in diameter. Thickness: 0.5 mm for (△) single- and (●) dual-grid and 1.5 mm for single-grid carbon (○). Spacing in dual grid SUS, 1.5 mm. Microwave power, 200 W; C_2F_6 gas pressure, 0.28 Pa. (From Ono et al.[21])

1.5 mm for carbon. The carbon grid is useful to prevent metallic contamination of the specimen caused by sputtered particles from the grid. In this configuration the electric field generated by the grid is confined in the plasma transport region 3 cm in thickness and does not influence the plasma in the generation region. Figure 4 illustrates the concept of shielded single-grid ion extraction, compared with that with no shield electrode.

Figure 5 shows the effect of the shield electrode on the sparkover voltage characteristics as a function of the distance between the grid and the shield electrode. Here the position of the grid and the gas pressure are varied. When the distance is 10 cm, corresponding to the distance in a conventional ion source, the sparkover occurs at a voltage of 300–500 V at gas pressures of about 10^{-3} Torr. Therefore, stable ion extraction is possible only at below 200–300 V. On the other hand, when the distance is 3 cm, as in the present shielded single grid, an ion beam is stably extracted up to 1100 V. From this result it is clear that the sparkover voltage is basically determined by Paschen's law governing the dependence of discharge-initiation voltage on electrode separation and gas pressure, although the plasma already exists in the region between the grid and the shield electrode.

Figure 6 shows the ion extraction characteristics using the shielded single grid. The ion-current density of over 2 mA/cm^2 is obtained at a voltage of 800 V, with a stainless steel grid that is 2.5 times larger than that using the dual grid with the same aperture dimensions. Figure 7 shows the distributions of the ion current density for the stainless steel single grid. The uniformity of 5% is obtained in the middle area of 10 cm in diameter. When a

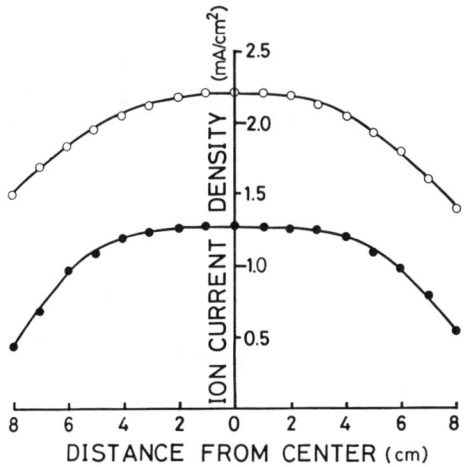

Fig. 7. Distribution of ion current density. Microwave power, 300 W; C$_2$F$_6$, 0.28 Pa; extraction voltage: ○, 1000 V; ●, 600 V.

carbon single grid is used, ion-current density decreases somewhat due to the increase in the grid thickness. However, with the carbon grid, sufficient current over 1 mA/cm² is obtained without metallic contamination, and further reliable operation is ensured because of its very high resistance to heat.

4. REACTIVE ION-BEAM ETCHING SYSTEM

Figure 8 shows a photograph of the reactive ion-beam etching system developed using the broad-beam ECR ion source.[21] Silicon wafers, 4 in. in diameter, are loaded and unloaded with a mechanical wafer chucking system. They are etched sequentially with automated cassette-to-cassette operation.

A silicon wafer to be etched is placed on an etching table, surrounded with a carbon plate. A wafer-cooling unit is arranged under the wafer. The carbon plate prevents wafer metallic contamination. The problem of wafer positive charge buildup caused by ion-beam irradiation is settled by the neutralization effect of the secondary electrons emitted from the carbon

Fig. 8. A reactive ion-beam etching system using a broad-beam ECR ion source. Wafers are etched sequentially with automated cassette-to-cassette operation. (From Ono et al.[21])

plate, which is grounded. In fact, when the entire etching table is put into a floating state, etch rates decrease considerably because of the positive charge buildup.

In the application to semiconductor device fabrication, it is important to maintain wafer temperatures sufficiently low during etching, since, usually, the etching mask resist patterns are poor in heat resistance. Generally, a wafer placed on a cooled table in vacuum cannot be effectively cooled, because the existence of a microscopic gap between them causes thermal conduction to be very poor. To improve the thermal contact between the wafer and the water-cooled table, a wafer-cooling technique utilizing an electrostatic force is employed in the system. An electrode plate covered with a dielectric film is positioned under the wafer, and the wafer itself is used as the second electrode, utilizing the fact that wafer during etching is automatically grounded by the secondary electron neutralization effect. The voltage applied to the first electrode does not affect the wafer potential. This wafer-cooling method is simple and reliable.

5. ETCHING CHARACTERISTICS

a. SiO_2 *Etching*

As the etching gas, the fluorocarbon C_4F_8 was mainly investigated, from among CF_4, C_2F_6, and C_4F_8. These gases possess similar characteristics with regard to SiO_2–Si selectivity, ranging from 5 to 20, whereas they exhibit quite different characteristics in conventional plasma etching. For instance, CF_4 is most popularly used for selective Si etching to SiO_2, with reverse selectivity, in plasma etching. In reactive sputter etching, these gases exhibit intermediate characteristics. The differences in respective etching methods are related to the gas pressure used and the degree of ion bombardment.

Figure 9 shows the relationship between ion-current density and etch rates for various materials. The ion-current density was controlled by the microwave power input. High selectivities of SiO_2 etching to Si and resists (AZ 1350J, PMMA) are seen. The selectivity of SiO_2 to Si and AZ 1350J reach about 20 and 5, respectively. Etching characteristics with high selectivity are stably obtained with high reproducibility, whereas polymeric compound deposition is apt to occur in conventional reactive sputter etching.

b. Al *Etching*

As the etching gas, chlorosilicon, $SiCl_4$, was investigated from among CCl_4, BCl_3, and $SiCl_4$, from the viewpoint of the selectivities of Al to

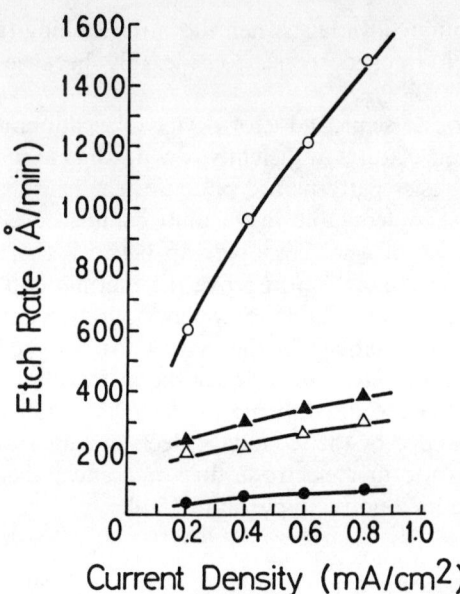

Fig. 9. Relation between ion current density and etch rates of SiO_2, Si, and resists. Ion extraction voltage, 1000 V; C_4F_8 gas pressure, 8×10^{-4} Torr (0.11 Pa); ○, SiO_2; ▲, PMMA; △, AZ1350 J; ●, Si. (From Matsuo and Adachi.[18])

SiO_2 and resist. A gas containing the element carbon such as CCl_4 tends to enhance SiO_2 etching, since the ion energy is larger than that in conventional plasma etching or reactive sputter etching. Figure 10 shows the relationship between ion extraction voltage and etch rates for various materials. The etch rate ratios of Al to SiO_2 and resist (AZ 1350J) are about 6 and 3.5, respectively, at an ion extraction voltage of 1000 V. At ion extraction voltages lower than 500 V, the etch rate of Al decreases considerably. It has been found that, by adding Cl_2 to $SiCl_4$, etch rate, selectivity, and their dependence on ion extraction voltage can also be controlled. This is because etch rates are determined by the difference between the removal reaction process by Cl elements and the deposition reaction process by Si elements.

c. *Reaction Mechanism*

Bombardment of energetic ions plays an essential role in the etching reaction. The typical form of the reaction induced by ion bombardment is interpreted as follows.[10]

In SiO_2 etching with fluorocarbons, ion bombardment causes dissociation both of the incident CF_n^+ ion and of the lattice of the bombarded material surface, resulting in dissociated CF_n radicals adsorbed on the surface. Then

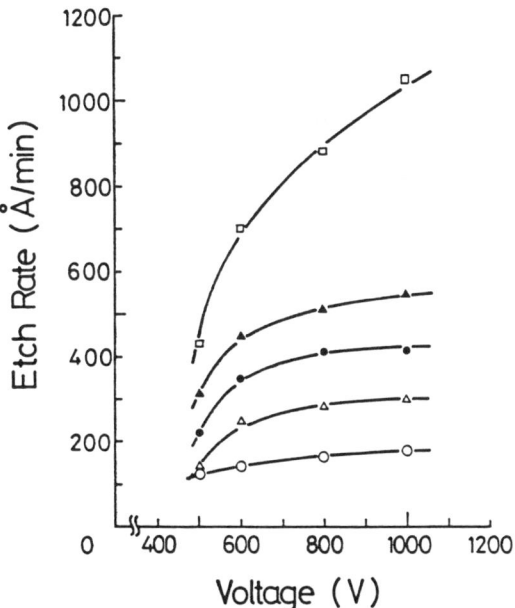

Fig. 10. Relationship between ion extraction voltage and etch rates of Al, SiO_2, Si, and resists. $SiCl_4$ gas pressure, 8×10^{-4} Torr (0.11 Pa); □, Al; ▲, PMMA; ●, Si, △, AZ1350 J; ○, SiO_2. (From Matsuo and Adachi.[18])

the most stable compounds among the various combinations of the atoms are formed. When the formed compounds are all volatile, the etching of the material proceeds effectively. Therefore, in the case of SiO_2, the following reaction occurs:

$$CF_n + SiO_2 \rightarrow SiF_m + CO, CO_2. \tag{2.3}$$

The compounds formed are all volatile, and the etching proceeds effectively. Similar reactions occur for other oxides such as Ta_2O_5 and TiO_2.

On the other hand, for Si, carbon atoms are to remain on the surface by the reducing reaction:

$$CF_n + Si \rightarrow SiF_m + C. \tag{2.4}$$

When the carbon surface layer is once generated, the decomposition and the recombination of CF_n species are merely repeated on the surface:

$$CF_n + C \rightarrow CF_m + C. \tag{2.5}$$

That is, Si etching hardly proceeds. For metals such as Ta and Ti an analogous relation also holds. There have been similar discussions of the reaction mechanism,[22-25] reporting that ion bombardment enhances formation of volatile molecules through surface damage (damage-induced chemical reaction) and that ion bombardment enhances the release of quasi-volatile molecules from the surface (chemical sputtering).

Another effect of ion bombardment is physical sputtering. The removal of carbon atoms from the Si surface by this effect causes the etching of Si to a certain extent. Accordingly, to improve the etching selectivity of SiO_2 to Si, the ratio of C to F about CF_n must be larger than that about the fluorocarbon CF_m being released quasi-chemically from the Si surface. The deposition reactions from the species containing inhibitors such as carbon atoms are generally important in the control of etching selectivities.

Figure 11 shows the dependence of the SiO_2 etch rate on the ion extraction voltage (ion energy). The etch rate is normalized to the value at the ion current density of 0.2 mA/cm^2. The etch rate of SiO_2 can be expressed by the following equation:[18]

$$v = v_0 \exp(-E_0/E), \qquad (2.6)$$

where E is the ion energy and E_0 is a constant corresponding to the activation energy. The constant v_0 is determined by the ion-current density and ion composition. This equation was introduced by the analogy between the roles of the ion energy E in the ion-induced reaction and the thermal energy of kT in the thermal reaction, which can be expressed by the Arrhenius reac-

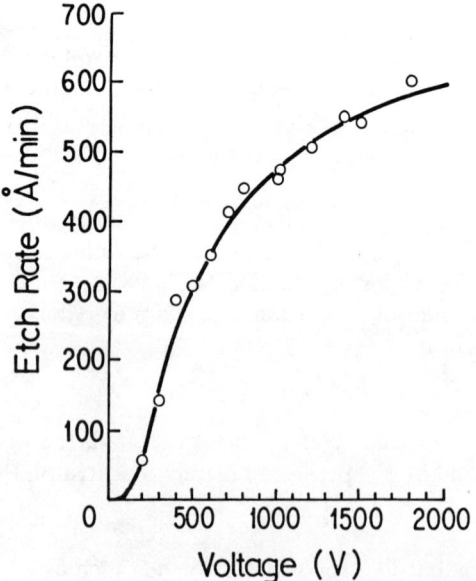

Fig. 11. Dependence of SiO_2 etch rates on ion extraction voltage (ion energy). C_4F_8 gas pressure, 0.2 mA/cm^2; $v = v_0 \exp(-E_0/E)$ where $E_0 = 450$ V and $v_0 = 750$ Å/min. (From Matsuo and Adachi.[18])

tion rate formula:

$$v = v_0 \exp(-E_a/kT),\quad (2.7)$$

where E_a is the activation energy, k is Boltzmann's constant, and T is the temperature. The solid line in Fig. 11 shows the calculated result when it is assumed that $v_0 = 750$ Å/min and $E_0 = 450$ V. The calculated result coincides with the experimental result. This suggests that ion bombardment creates an effect equivalent to that caused by a thermal spot containing a few hundred atoms. The number of the atoms is roughly estimated by the equation:

$$N \simeq E_0/E_a,\quad (2.8)$$

because the ion energy E is dispersed into the atoms contained in the thermal spot with a thermal energy kT. The number of molecules etched per incident ion (etching yield) is estimated to be about two at a voltage of 1500 V. This implies that the adsorbed CF_n radicals bombarded by ions and the energetic neutrals generated through charge exchange between ions and gas molecules also contribute to the etching reaction. On the other hand, for the other materials such as Si, the accumulating effect (deposition reaction) of inhibitors such as carbon atoms and the process by which reaction products are removed must be considered simultaneously.

Figure 12 shows the dependence of etch rates on the angle of ion incidence for SiO_2 and Si. The etch rate of SiO_2 decreases with increasing angle of

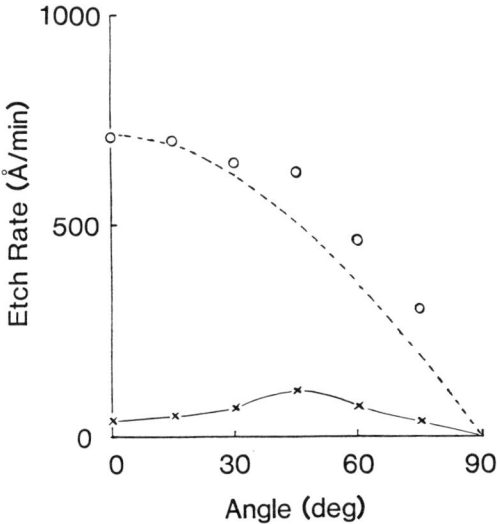

Fig. 12. Dependence of SiO_2 and Si etch rates on the angle of ion incidence. Broken line exhibits cosine rule. C_4F_8 gas pressure, 0.3 mA/cm^2; ○, SiO_2; ×, Si.

ion incidence and can be expressed roughly by the cosine rule. This result means that etching proceeds in proportion to the amount of incident ions, supporting the reaction model discussed above. On the other hand, the etch rate of Si has a maximum value at an angle of about 45°, similar to the case of physical sputter etching, although the etch rate itself is low. The dependence of etch rates on the angle of ion incidence is important in understanding how the mask pattern profiles are transferred to the substance to be etched in the application to fine pattern fabrications. The cosine-rule dependence for SiO_2 etching gives rise to pattern transfer with high accuracy, because that relation implies projective pattern transfer.

6. Pattern Formation Process

a. Analytical Treatment

The pattern formation process with a mask can be analyzed,[26] based on the dependence of etch rates on the angle of ion incidence $v(\theta)$. When the ion incidence is normal to the substrate, the angle of incidence becomes equal to the inclination angle at the surface under consideration. Figure 13 illustrates the formation of surface profile by etching with a mask. The surface profile $z = f(x, t)$, where t is time, satisfies the following equations:

$$dz/dt + v(\theta)/\cos\theta = 0, \tag{2.9}$$

$$\tan\theta = dz/dx. \tag{2.10}$$

In the pattern formation process, the edge of the mask has a greater influence on the profile generated into the substrate than any other part of the mask. For this reason it is assumed that the initial profile is expressed by straight lines, as shown in Fig. 13. Based on these assumptions, we suggest that the profile formed in the substrate is constructed with the characteristic curve, which is determined only by $v(\theta)$, and its tangent lines, which are determined by the initial profile and mask conditions. The characteristic curve is expressed in polar coordinates as

$$\phi = \theta + \tan^{-1} v'(\theta)/v(\theta), \tag{2.11}$$

$$r = t\sqrt{v(\theta)^2 + v'(\theta)^2}, \tag{2.12}$$

or in the orthogonal coordinates as

$$x/t = v'(\theta)\cos\theta + v(\theta)\sin\theta, \tag{2.13}$$

$$z/t = v'(\theta)\sin\theta - v(\theta)\cos\theta. \tag{2.14}$$

The geometrical meaning of the characteristic curve is as follows: The initial profile has a distinctive feature with respect to the singularity of the origin in Fig. 13. The origin in the initial profile can be considered to contain every value of θ, ranging from zero to a certain value. On the other hand, a straight line that intersects the origin with an inclination angle θ moves by distance $v(\theta)t$ from the origin during time t. Therefore, the profile formed by etching is expected to be given by the envelope of the family of the etched straight lines, given by

$$z = x \tan \theta - v(\theta)t/\cos \theta. \tag{2.15}$$

In fact, the characteristic curve Eqs. (2.13) and (2.14) are easily obtained.

The features of various etching methods in pattern profile formation are described by assuming the form of the function $v(\theta)$. To discuss qualitatively the profiles formed by various etchings, the dependences of etch rates on ion incidence $v(\theta)$ are assumed as the following form with parameters A, B, and C:

$$v(\theta) = A + B \cos \theta + C \sin^2 2\theta, \tag{2.16}$$

where A stands for the degree of isotropic feature; B, the degree of directional feature; and C, the degree of physical sputtering feature that gives the maximum etch rate at an inclined surface. When the deposition effects influence the etching process, the parameter A or B can be put into a negative value. Further, a negative value for the parameter C is also possible to represent the removal of adsorbed reactants on the surface by sputtering.

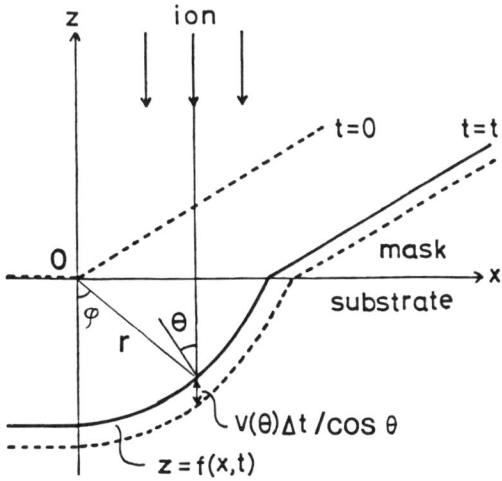

Fig. 13. Pattern profile formation with mask. (From Matsuo.[26])

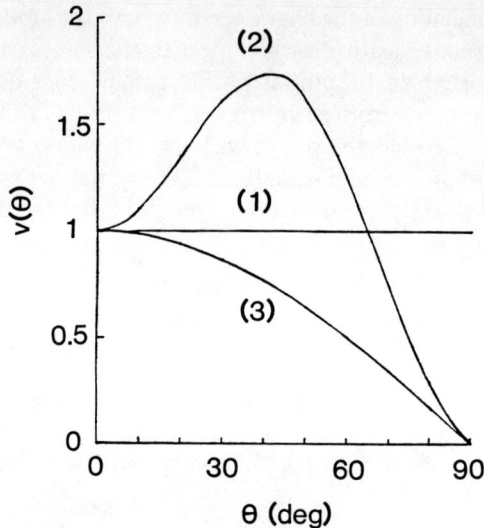

Fig. 14. Dependence of etch rates on the angle of ion incidence, assuming Eq. (6.8) with parameters: (1) $A = 1$, $B = 0$, $C = 0$; (2) $A = 0$, $B = 1$, $C = 1$; (3) $A = 0$, $B = 1$, $C = 0$.

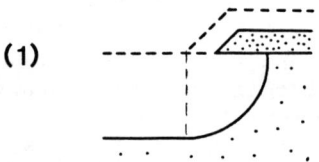

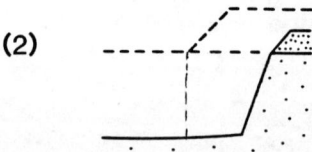

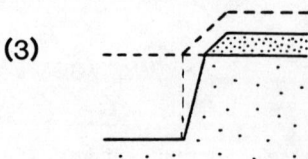

Fig. 15. Illustration of pattern profiles obtained from the functions shown in Fig. 14. (From Matsuo.[26a])

The functions for typical cases are plotted in Fig. 14 by assuming the following parameters:

(1) $A = 1,$ $B = 0,$ $C = 0$ (isotropic),
(2) $A = 0,$ $B = 1,$ $C = 1$ (sputtering), (2.17)
(3) $A = 0,$ $B = 1,$ $C = 0$ (directional).

The function $v(\theta)$ is normalized to $v(0) = 1$. Curves (1), (2), and (3) in Fig. 14 correspond to those of isotropic plasma etching, physical sputter etching, and ideal directional etching, respectively.

The results of the profiles analytically obtained are shown in Fig. 15. In (1) undercutting occurs, as is often observed in plasma etching, and in (2) a considerable lateral shift of the pattern edge occurs, although the undercutting does not occur at all. In (3) a high-accuracy pattern transfer is performed. As mentioned in Section II.5, the function $v(\theta)$ in reactive ion-beam etching can be approximately expressed by case (3).

b. *Fine-Pattern Fabrications*

Figure 16 shows an SEM photograph of a SiO_2 pattern on Si substrate with a PMMA resist mask remaining after an etch. A high-accuracy pattern is obtained with nearly perpendicular sidewalls.

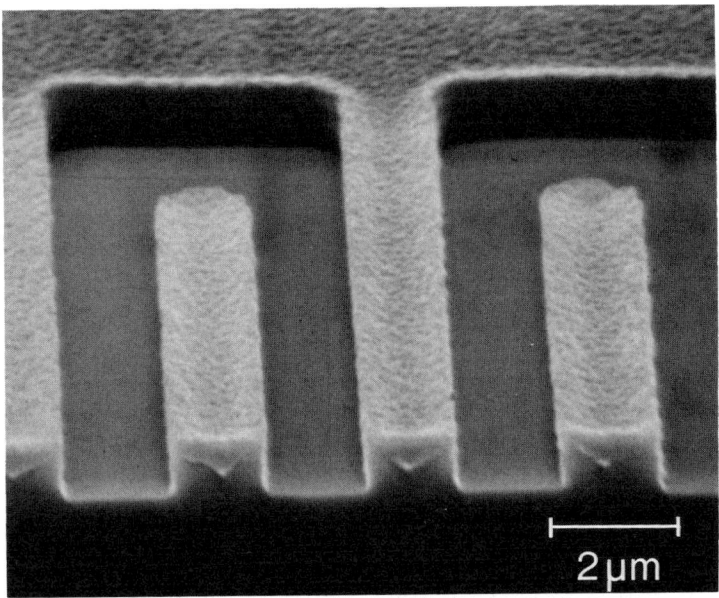

Fig. 16. SEM photograph of SiO_2 pattern on Si substrate with PMMA resist remaining after an etch. SiO_2 thickness, 0.6 μm. (From Matsuo and Adachi.[18])

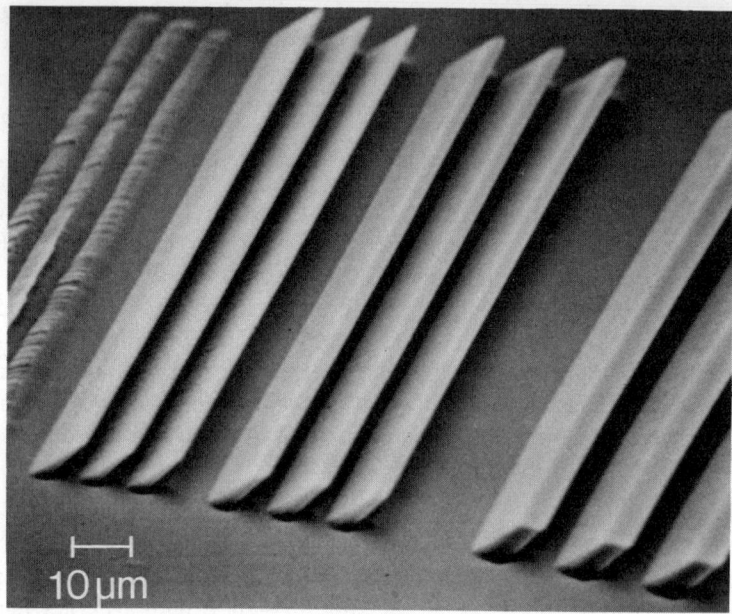

Fig. 17. SEM photograph of pattern engraved into fused quartz substance inclined by 45° with Ti mask. Mask pattern is transferred projectively in direction of ion incidence. (From Matsuo and Adachi.[18])

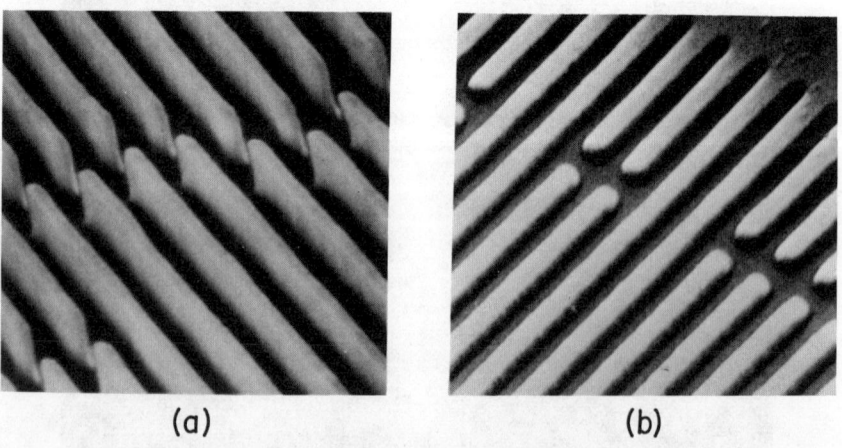

Fig. 18. SEM photographs of SiO_2 and Al patterns obtained by reactive ion-beam etching. Submicrometer patterns are realized with high accuracy. (a) SiO_2: line width, 0.4 μm; thickness, 0.8 μm. (b) Al: line width, 0.5 μm; thickness, 0.3 μm. (From Ono et al.[21])

Figure 17 shows a pattern engraved into a fused quartz substrate inclined by 45° with a metal (Ti) mask. A peculiar profile that cannot be obtained in conventional reactive sputter etching can be realized because of the independence of the specimen–table configuration and the direction of ion incidence. The pattern profile controllability of reactive ion-beam etching is very useful for various applications.

Figure 18 shows SEM photographs of a SiO_2 pattern on a Si substrate with 0.4-μm lines and spaces and 0.8-μm thickness, and an Al pattern on a SiO_2 layer with 0.5-μm lines and spaces and 0.3-μm thickness. High-accuracy patterns are indeed obtained for submicrometer patterns.

III. Plasma Deposition

7. INTRODUCTION

In semiconductor device fabrication processes, the plasma chemical vapor deposition (CVD) technique, which employs plasma reactions by rf discharge at low temperature, has become an important research subject.[27-29] A deposition technique using microwave discharge and plasma transport at a low gas pressure with a parallel magnetic field has also been reported.[30] However, in both these techniques, the specimen substrate must still be heated to a temperature of from 250 to 350°C. Furthermore, the quality of the deposited film is inadequate, possibly because raw material gases, such as SiH_4, do not decompose sufficiently, and the deposition reaction on the specimen surface is not complete. These conditions might allow hydrogen and poor molecular bonds to remain in the film.

The newly developed ECR plasma deposition apparatus allows deposition of high-quality thin films at room temperatures without the need for thermal reactions. This is made possible by enhancing the plasma excitation efficiency and by the acceleration effect of ions, using the ECR plasma, with plasma extraction by a divergent magnetic field.

8. ECR PLASMA DEPOSITION APPARATUS

Figure 19 illustrates the ECR plasma deposition apparatus.[31] Microwave power is introduced into the plasma chamber through a rectangular waveguide and a window made of a fused quartz plate. Microwave frequency is 2.45 GHz, and output power is delivered at a 50-Hz duty cycle, at the convenience of a power supply. The plasma chamber is 20 cm in diameter and 20 cm in height in inside dimensions and operates as a microwave cavity

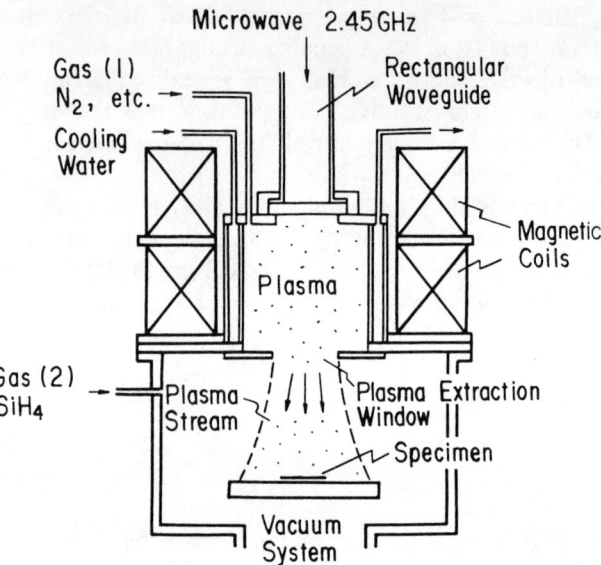

Fig. 19. ECR plasma deposition apparatus. Deposition area, 20 cm in diameter. Gas pressure, 10^{-4}–10^{-3} Torr (about 0.01–0.1 Pa). Substrate, without heating. (From Matsuo and Kiuchi.[31])

Fig. 20. ECR plasma deposition apparatus. Plasma chamber is arranged inside the magnetic coils and connected to microwave power supply through rectangular waveguide.

resonator (TE_{113}). Magnetic coils are arranged around the periphery of the chamber for ECR plasma excitation. The circular motion frequency—electron cyclotron frequency—is controlled by the magnetic coils so as to coincide with the microwave frequency (magnetic flux density, 875 GHz) in a proper region inside of the chamber. The design is similar to that of the broad-beam ECR ion source described in Section II. The ECR condition enables the plasma to absorb the microwave energy effectively. Thus, highly activated plasma is easily obtained at low gas pressures of 10^{-5}–10^{-3} Torr.

In this apparatus ions are extracted in the form of a plasma stream from the plasma chamber to the specimen chamber, along a divergent magnetic field, and the film is deposited on the specimen substrate. Reactive deposition gases are introduced through two inlet systems, one into the plasma chamber and the other into the specimen chamber. Figure 20 shows a photograph of the apparatus. The plasma chamber and the magnetic coils are water-cooled. The vacuum system consists of an oil diffusion pump (2400 liters/sec) and a mechanical rotary pump (500 liters/min).

9. Divergent Magnetic Field Plasma Extraction

A divergent magnetic field method has been developed for ion extraction in the form of a plasma stream from the plasma chamber to the specimen chamber. The intensity of the magnetic field in the specimen chamber is gradually weakened from the plasma chamber to the specimen table, as shown in Fig. 21. High-energy electrons in circular motion peculiar to ECR

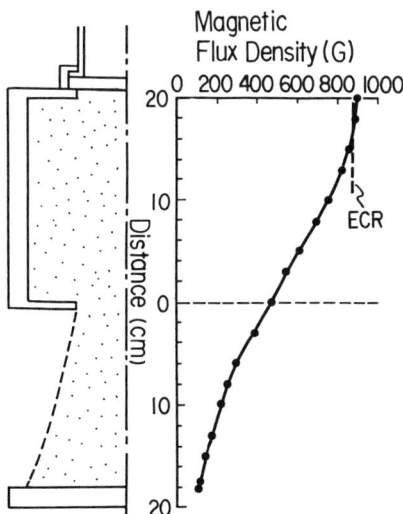

Fig. 21. Distribution of magnetic field intensity (magnetic flux density) from the top of plasma chamber to the specimen table. (From Matsuo and Kiuchi.[31])

plasma are accelerated by the interaction between their magnetic moments and the magnetic field gradient. The accelerated electrons bring about a negative potential toward the specimen table, which is electrically isolated from the plasma chamber. Therefore, a static electric field, which accelerates ions and decelerates electrons, is generated along the plasma stream between the plasma chamber and the specimen table so as to satisfy the neutralization condition. Ion extraction, transport, and bombardment of the specimen surface with moderate-energy ions are thus enhanced during deposition.

Under these conditions electrons and ions have the same acceleration along the divergent magnetic field, as follows:

$$F_i/M = F_e/m, \tag{3.1}$$

where M and m are the respective masses of the ion and the electron. The quantities F_i and F_e are the forces on the ion and the electron, respectively. These forces are expressed by the following equations:

$$F_i = eE, \tag{3.2}$$

$$F_e = -\mu \, dB/dz - eE, \tag{3.3}$$

where μ is the magnetic moment of the electron in circular motion and E is the electric field generated in the plasma stream. The magnetic moment μ is an adiabatic invariant, and it is given by the kinetic energy of the electron in circular motion W, as follows:

$$\mu = W/B. \tag{3.4}$$

From these relations the electric field E can be obtained as

$$E = (W_0/eB_0)[(-dB/dz)/(1 + m/M)], \tag{3.5}$$

The potential ϕ is obtained, by integration with an approximation, $(1 + m/M) \approx 1$, as follows:

$$\phi = -W_0/e(1 - B/B_0), \tag{3.6}$$

where W_0 and B_0 are the electron energy and the magnetic flux density in the plasma chamber, respectively. This equation states that the ion energy is given approximately by the product of the electron energy in the plasma chamber and the ratio of the decreased magnetic field intensity to the initial intensity. Ions are thus accelerated and transported toward the specimen table, and electrons lose the energy of circular motion by the same amount. As a result, deposition reactions induced by ions are enhanced, and heating effects caused by electrons are reduced. The divergent magnetic field plays a role in converting the electron energy of circular motion into the ion energy along the magnetic field. Therefore, the divergent magnetic field method is particularly effective when it is combined with ECR plasma generation. An electric potential generation in an ECR plasma, related to a magnetic field distribution, has been previously observed and investigated, though in a much higher energy range, in the field of plasma research.[32]

Figure 22 shows a photograph of the plasma stream, being extracted from the plasma chamber by the divergent magnetic field method. The plasma extraction window is 10 cm in diameter, and the plasma stream at the specimen table (i.e., the deposition area) is 20 cm in diameter.

The negative potential generated by the divergent magnetic field was measured using a plane probe from the floating potential, which had a larger area than the plasma stream cross section. The result is shown in Fig. 23 as a function of the distance from the plasma extraction window. The negative potential increases, corresponding to the decrease in the magnetic field intensity. The energy of the accelerated ion through the plasma stream, from the plasma extraction window to the specimen table, is of the order of 10–15 eV. Figure 24 shows the dependence of the negative potential at the specimen table on the gas pressure. The gas pressure was controlled by changing the gas flow rate. The negative potential increases rapidly as the gas pressure decreases, owing to the increase in the electron mean free path

Fig. 22. Photograph of plasma stream, extracted from the plasma chamber.

and the electron energy in circular motion. Thus, the ion energy can be easily controlled in the range from 5 to 30 eV by changing the gas pressure, besides the microwave power input.

The plasma potential through the plasma stream was measured directly using an emissive probe method,[33] in order to distinguish from each other the respective effects of the electric field in the plasma stream and the electric field due to the ion sheath in the vicinity of the specimen surface generated by the thermal motion of electrons. The emissive probe method utilizes the fact that the usual potential difference between the plasma and the probe surface (ion sheath region) does not occur when electron exchange between the probe and the plasma is made free, as in the case of a thermionic filament probe. The results are shown in Fig. 25, where the substrate potential is chosen as zero. The potential difference due to the ion sheath, of the order of 0.3 mm in thickness, is about 10 V. The ion energy incident to the specimen surface is given by the sum of the energy gained in the divergent magnetic field and that due to the ion sheath potential difference, and is about 20–30 eV. The ions in such an energy range are expected to enhance deposition reactions and to improve the film quality, but not to cause surface damage.

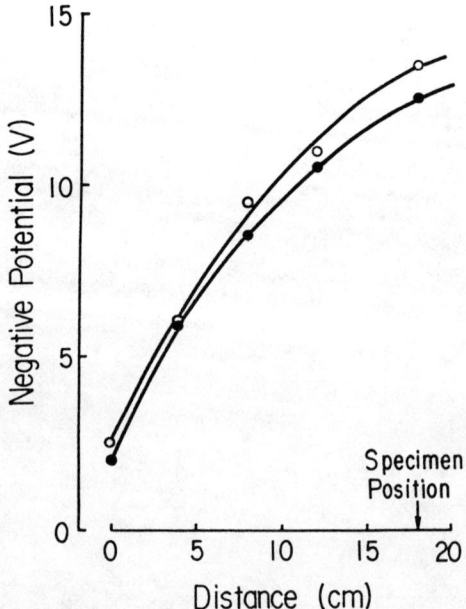

Fig. 23. Negative potential generation by divergent magnetic field as a function of distance from the plasma extraction window. N_2: ○, 15 cm^3/min, 1.5×10^{-4} Torr; ●, 20 cm^3/min, 2×10^{-4} Torr; power = 100 W. (From Matsuo and Kiuchi.[31])

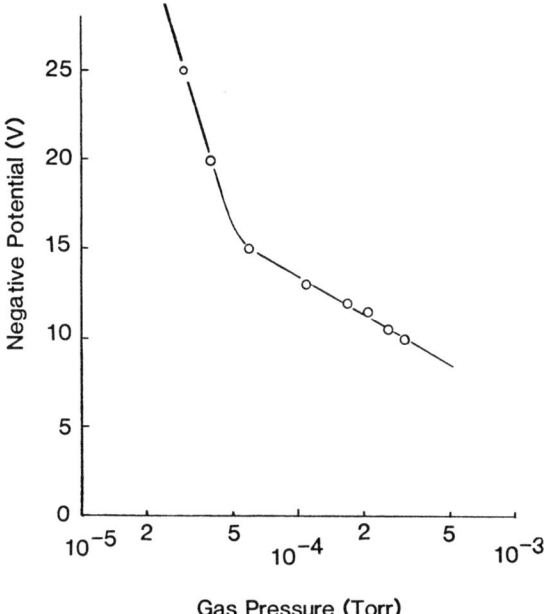

Fig. 24. Relationship between gas pressure and negative potential at the specimen table. N_2, power = 100 W.

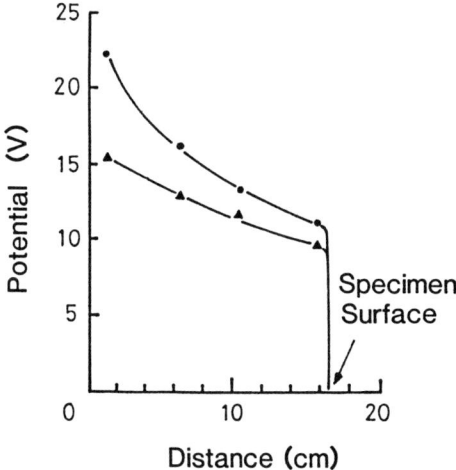

Fig. 25. Plasma potential variation along plasma stream, measured with emissive probe method. Substrate potential is chosen as zero. There is a potential drop due to the ion sheath at the surface. Ar: ●, 4×10^{-4} Torr; ▲, 1.1×10^{-3} Torr; power = 100 W.

The electric field in the plasma stream plays an important role by very effectively transporting the ions generated in the plasma chamber toward the specimen, in contrast to conventional methods in which ions are utilized only by the transport they provide through their thermal diffusion. In fact, high ion current density of 3–5 mA/cm^2 is easily obtained at the specimen table position, measured with a negatively biased plane probe.

10. DEPOSITION CHARACTERISTICS

All the experiments on film deposition were carried out without substrate heating. The wafer temperature rise during deposition is shown in Fig. 26. The specimen temperature was in the range 50–150°C, owing to some heating effect by the plasma. The wafer temperature can easily be kept below 50°C by employing a simple wafer-cooling scheme. In the preceding temperature range, deposition characteristics for silicon nitride and silicon dioxide hardly depend on the temperature. Deposition uniformity is within 5% in the 10-cm-diameter middle area.

a. Si_3N_4 *Deposition*

For silicon nitride (Si_3N_4) film deposition, nitrogen (N_2) and silane (SiH_4) gases are introduced into the plasma chamber and the specimen chamber, respectively. Figure 27 shows Si_3N_4 deposition characteristics as a function of microwave power, when the introduced gas flow rates are N_2, 10 cm^3/min, and SiH_4, 10 cm^3/min. The deposition rate increases from about 200 to 300 Å/min, and the refractive index (wavelength, 6328 Å) gradually decreases

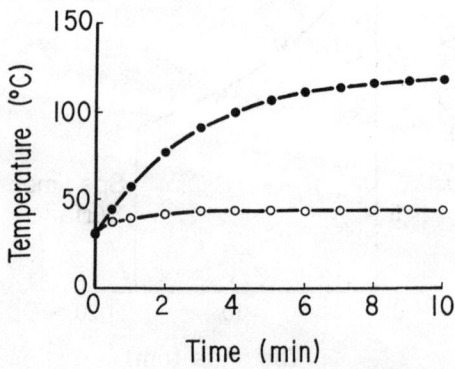

Fig. 26. Wafer temperature rise during deposition. ●, without cooling; ○, with cooling; power = 100 W.

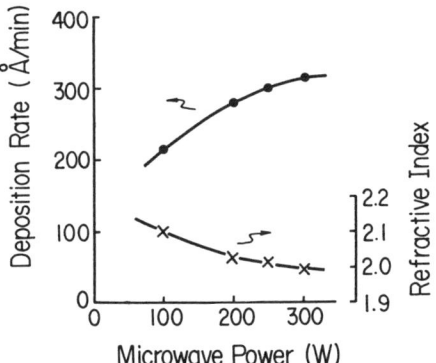

Fig. 27. Si_3N_4 deposition characteristics. Deposition rates and refractive indices are shown as functions of microwave power. Gas pressure, 2×10^{-4} Torr; SiH_4, 10 cm^3/min; N_2, 10 cm^3/min. (From Matsuo and Kiuchi.[31])

from about 2.1 to 2.0, with increasing microwave power from 100 to 300 W. The deposition rate is high. This means that the introduced gases are effectively transported and react to form a film, even at low microwave power. Figure 28 also shows the deposition characteristics when the deposition rate is increased by increasing the introduced gas flow rates. The deposition rate increases up to about 700 Å/min, and the refractive index markedly decreases with microwave power to 150 W, and then becomes almost constant. This means that microwave power larger than 150 W is required for

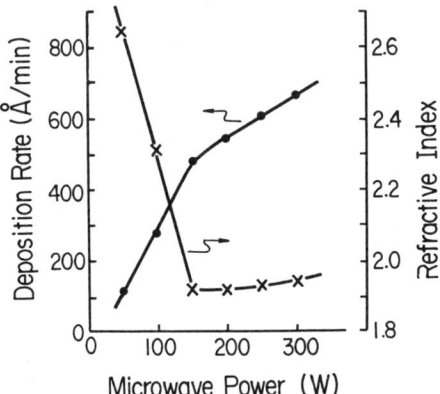

Fig. 28. Si_3N_4 deposition characteristics. Gas pressure, 5×10^{-4} Torr; SiH_4, 20 cm^3/min; N_2, 30 cm^3/min.

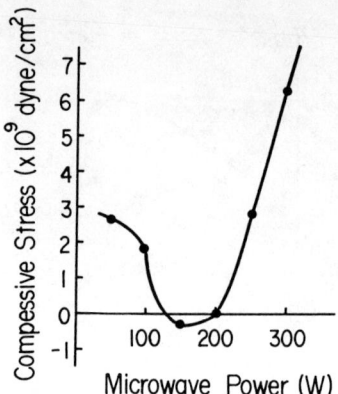

Fig. 29. Internal stress of deposited Si_3N_4 films. SiH_4, 20 cm³/min; N_2, 30 cm³/min.

sufficiently complete reactions to deposit Si_3N_4 film in this condition. Figure 29 gives the internal stress of the films shown in Fig. 28, also as a function of microwave power. The internal stress was measured from the bowing of the silicon substrate caused by film deposition. The stress is mainly com-

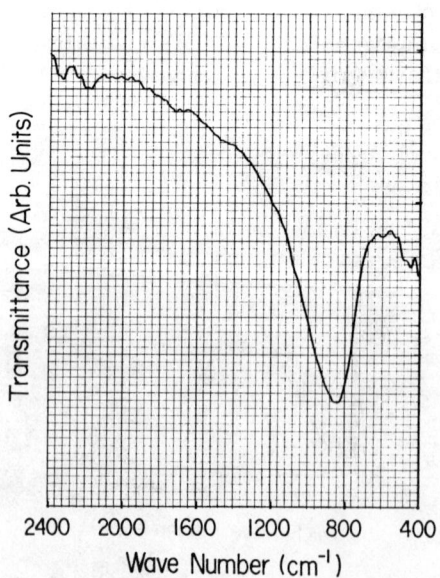

Fig. 30. Infrared absorption spectrum for deposited Si_3N_4 film. (From Matsuo and Kiuchi.[33a] This figure was originally presented at the Fall 1982 Meeting of The Electrochemical Society, Inc., held in Detroit, Michigan.)

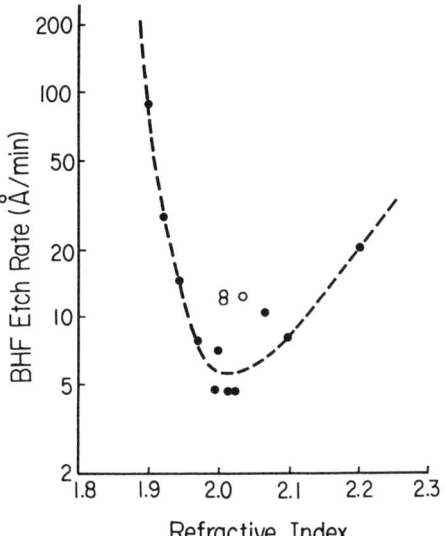

Fig. 31. (●) Si_3N_4 film etch rates with BHF solution. BHF, 50% HF:40% NH_4F = 15:85, 20°C. Flow rates: SiH_4, 10 cm³/min; N_2, 10–20 cm³/min, without heating. (○) Etch rate for CVD at 800°C shown for comparison. (From Matsuo and Kiuchi.[31])

pressive but becomes tensile to some extent at a power of about 150 W. This tendency seems to be correlated to the variation of refractive index in Fig. 28. The internal stress of the Si_3N_4 film can thus be controlled to about zero. The film stress controllability is advantageous for various applications.

Figure 30 shows the infrared absorption spectrum for Si_3N_4 film deposited at microwave power of 150 W and gas flow rates of N_2, 10 cm³/min and SiH_4, 10 cm³/min. The Si–N bond peak is clearly observed at the wave number of 845/cm, while the Si–H bond peak at about 2100/cm is hardly observed. The amount of hydrogen in the film seems very small.

The etch rates of the Si_3N_4 films with a buffered HF solution (BHF, 50% HF:50% NH_4F = 15:85, 20°C) were further examined for film quality evaluation. These rates are shown in Fig. 31 as a function of the film refractive index, which was changed by controlling the ratio of the introduced gas flow rates of N_2 and SiH_4. The etch rate reaches the minimum value at a refractive index of about 2.0. The value there is lower than 10 Å/min, which is comparable to values for high-temperature (800°C) CVD films, in spite of the deposition at a low temperature without substrate heating.

b. SiO_2 *Deposition*

Silicon dioxide (SiO_2) can also be deposited by introducing oxygen (O_2) and silane (SiH_4) gases into the plasma and specimen chamber, respectively.

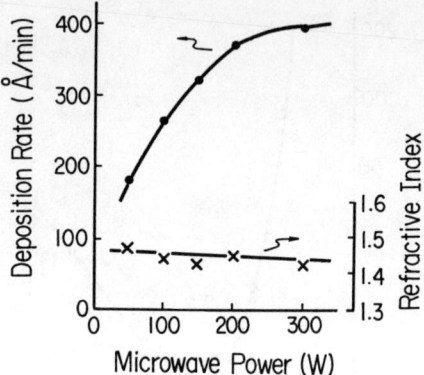

Fig. 32. SiO_2 deposition characteristics. Deposition rates and refractive indices are shown as functions of microwave power. Gas pressure, 2×10^{-4} Torr; SiH_4, 10 cm³/min; O_2, 10 cm³/min. (From Matsuo and Kiuchi.[31])

Figure 32 shows the SiO_2 deposition characteristics as a function of microwave power, when the introduced gas flow rates are O_2, 10 cm³/min, and SiH_4, 10 cm³/min. The refractive index is almost constant at values from 1.46 to 1.48 in a wide range of microwave power. The deposition rate increases from about 200 to 400 Å/min with increasing microwave power from 50 to 300 W. Figure 33 also shows SiO_2 deposition characteristics when depo-

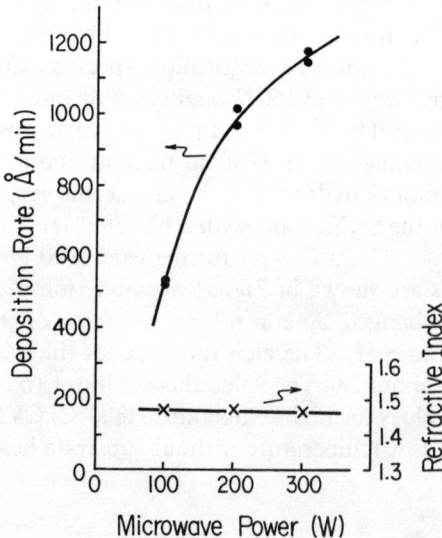

Fig. 33. SiO_2 deposition characteristics. Gas pressure, 6×10^{-4} Torr; SiH_4, 30 cm³/min; O_2, 30 cm³/min.

sition rate is increased by increasing the introduced gas flow rates. A high deposition rate, over 1000 Å/min is obtained, and the refractive index is almost constant in this condition, which differs from the tendency in the Si_3N_4 deposition shown in Fig. 28. The internal stress of the SiO_2 film is hardly affected by the deposition condition, giving values of $2-3 \times 10^9$ dynes/cm^2 (compressive).

Figure 34 shows the infrared absorption spectrum for SiO_2 film deposited at a microwave power of 100 W and with gas flow rates of O_2, 10 cm^3/min, and SiH_4, 10 cm^3/min. The Si–O bond peak is clearly observed at the wavenumber of 1065 cm^{-1}, and a Si–H bond peak is hardly observed.

The etch rates of the SiO_2 films with buffered HF solution were examined and compared to SiO_2 films prepared by a thermal oxidation method (wet, 1000°C). The result is shown in Fig. 35. Although the refractive index of the deposited film ($n = 1.48$) is somewhat larger than that of the thermally oxidized film ($n = 1.46$), the etch rates of the films almost coincide with each other at various solution temperatures. This result indicates that the SiO_2 films deposited without substrate heating by the ECR plasma method are dense, of high quality, and a match for the thermally oxidized films.

By introducing an inert gas such as argon (Ar) into the plasma chamber instead of oxygen or nitrogen, silicon film can be deposited. The deposition rate is similar to that of the silicon nitride. Therefore, by introducing a

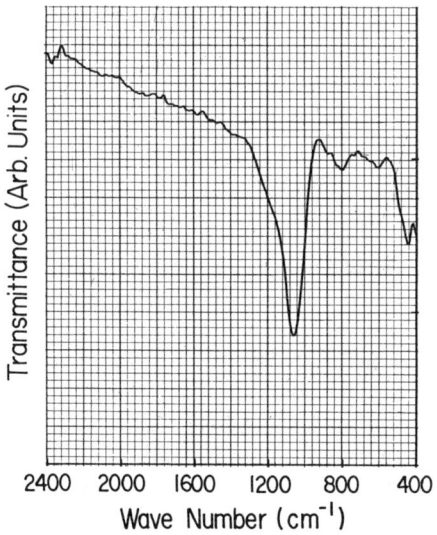

Fig. 34. Infrared absorption spectrum for deposited SiO_2 film. (From Matsuo and Kiuchi.[33a] This figure was originally presented at the Fall 1982 Meeting of The Electrochemical Society, Inc., held in Detroit, Michigan.

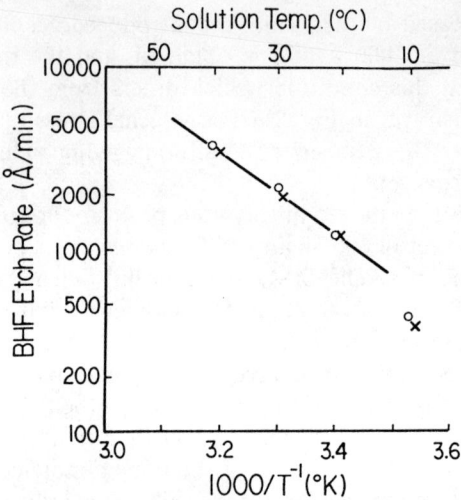

Fig. 35. SiO$_2$ film etch rates with BHF solution. BHF, 50% HF:40% NH$_4$F = 15:85, 20°C. ○, ECR plasma; ×, thermal. (From Matsuo and Kiuchi.[31])

mixture of O$_2$, N$_2$, and Ar into the plasma chamber, films of various composition can easily be obtained.

As mentioned earlier, high-quality film deposition of materials, such as silicon nitride and silicon dioxide, at room temperature is possible. The deposition reaction does not require the assistance of a thermal reaction, because both the highly activated plasma and ion bombardment effect with moderate energy sufficiently enhance the film deposition reactions.

c. *Ion Incidence Effects*

To clarify the contribution of energetic ions to the deposition process, the influence of the existence or absence of ion incidence was examined. Figure 36 shows the distributions of the deposition rates and the BHF etch rates when the substrate was partially masked with a 10-mm gap during a Si$_3$N$_4$ deposition, parameterized by the substrate heating temperature. The deposition rates at the shadowed area are much lower than those at the plasma-irradiated area. Further, the BHF etch rates of the film deposited at the shadowed area are much larger in spite of low deposition rates. The effect of the substrate heating at 300°C contributes to improve, to a certain degree, the film quality deposited at the shadowed area, but the film quality is still much inferior to that of the plasma-irradiated area. The deposition

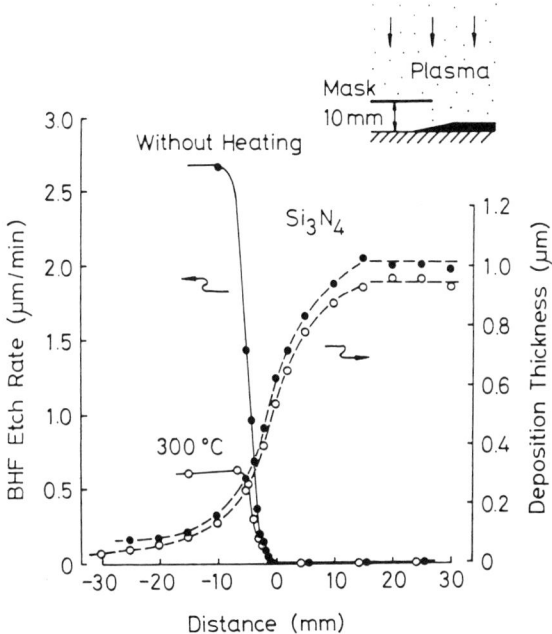

Fig. 36. Evaluation of Si_3N_4 films deposited partially masked with a gap of 10 mm with heating at 300°C (○) and without heating (●). Flow rates: SiH_4, 20 cm³/min; N_2, 30 cm³/min. Power: 200 W. Deposition time: 20 min. The inset shows the process schematically.

characteristics at the shadowed area seem similar to those of conventional rf plasma CVD, including the temperature dependence.

On the other hand, for films deposited at the plasma-irradiated area, change in the BHF etch rates by substrate heating is hardly observed. These results indicate that low-energy ion bombardment at a low gas pressure, combined with a highly activated plasma, even at a low temperature, considerably enhances deposition reactions, such as hydrogen atom release and molecular bonding.

11. Material Supply by Sputtering

To extend the advantages of the ECR plasma method to metallic compound deposition, an ECR plasma deposition apparatus has been developed to add a material supply by sputtering.[34] It realizes low-temperature deposition for various compound films by combining plasma extraction with a divergent magnetic field and the raw material supplied by sputtering.

Figure 37 shows the ECR plasma deposition apparatus. A sputtering target with a shield electrode is placed at the plasma extraction window around the extracted plasma stream. The target is indirectly (radiatively) cooled by water-cooling the shield electrode. Sputtering gas (Ar) and material gas (such as O_2) are introduced into the plasma chamber and the specimen chamber. Figure 38 illustrates the deposition mechanism. A dc voltage is supplied to the target so that sputtering will occur with ions from the plasma stream. Sputtered particles are transported with the plasma stream to the specimen substrate. The introduced material gas is ionized and transported in the same way. In this target configuration, a stable operation is realized without abnormal discharges such as the sparkover phenomenon, because the target surface is arranged approximately parallel to the magnetic field.

Figure 39 shows the target current characteristics as a function of the target voltage. High target currents, over 600 mA, are obtained in spite of low gas pressures because of the use of highly ionized ECR plasma. In the

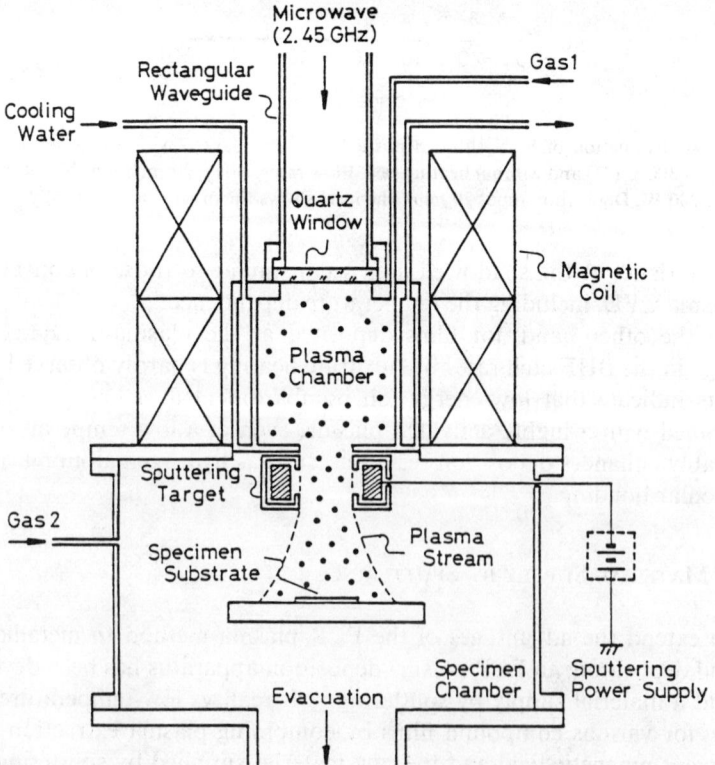

Fig. 37. ECR plasma deposition apparatus with material supply by sputtering. (From Ono et al.[34] This figure was originally presented at the Spring 1984 Meeting of The Electrochemical Society, Inc., held in Cincinnati, Ohio.)

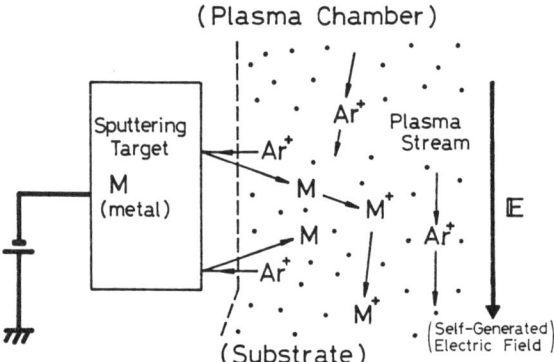

Fig. 38. Illustration of deposition mechanism with sputtering material supply. (From Ono et al.[34] This figure was originally presented at the Spring 1984 Meeting of The Electrochemical Society, Inc., held in Cincinnati, Ohio.)

range of voltages from 300 to 1000 V, which are used in practice, the target current is almost constant, determined by the microwave power. Therefore, target currents and target voltages suitable for deposition can be chosen and controlled at pressures of 10^{-5}–10^{-3} Torr.

The deposition experiments were carried out using a tantalum (Ta) target and an aluminum (Al) target, without substrate heating, to deposit tantalum and aluminum oxide films (Ta_2O_5, Al_2O_3). For deposition, Ar gas was introduced for sputtering into the plasma chamber, and O_2 gas was introduced into the specimen chamber.

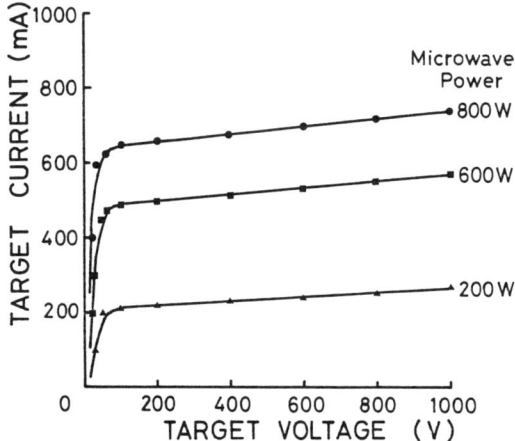

Fig. 39. Target current characteristics as a function of target voltage, with microwave power varied. Ar gas pressure, 8×10^{-2} Pa (6×10^{-4} Torr). (From Ono et al.[34] This figure was originally presented at the Spring 1984 Meeting of The Electrochemical Society, Inc., held in Cincinnati, Ohio.)

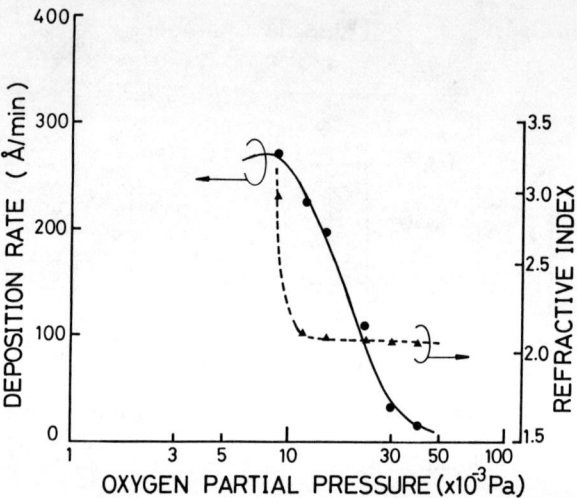

Fig. 40. Ta_2O_5 deposition characteristics. Deposition rates and refractive indices are shown as functions of O_2 partial pressure. Total gas pressure, 8×10^{-2} Pa (6×10^{-4} Torr). (From Ono et al.[34] This figure was originally presented at the Spring 1984 Meeting of The Electrochemical Society, Inc., held in Cincinnati, Ohio.)

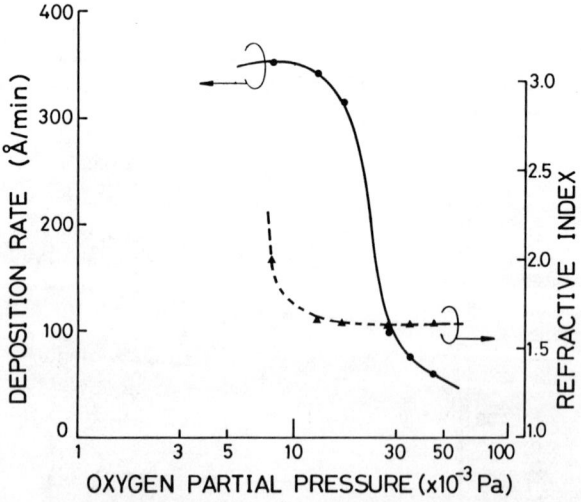

Fig. 41. Al_2O_3 deposition characteristics. Deposition rates and refractive indices are shown as functions of O_2 partial pressure. Total gas pressure, 8×10^{-2} Pa (6×10^{-4} Torr). (From Ono et al.[34] This figure was originally presented at the Spring 1984 Meeting of The Electrochemical Society, Inc., held in Cincinnati, Ohio.)

Figure 40 shows the Ta_2O_5 deposition characteristics. Oxygen partial pressures were varied under a total gas pressure of 6×10^{-4} Torr. The refractive index of deposited film decreases with increasing oxygen partial pressure and saturates at an index of about 2.1. The deposition rate decreases at higher oxygen partial pressures. In conventional sputtering, fully oxidized films are obtained only in the oxygen partial pressure range for which the deposition rates become extremely low. That is, high-quality oxide film deposition and high-rate deposition are difficult to perform simultaneously. In this ECR method, high deposition rates, over 200 Å/min, are obtained even when fully oxidized film is deposited.

Figure 41 shows the Al_2O_3 deposition characteristics. Deposition characteristics are similar to those of Ta_2O_5 in Fig. 40. Aluminum oxide films with a refractive index of 1.65 are obtained under the deposition rate of 350 Å/min. High-rate deposition is possible for aluminum oxide with this method using an aluminum target, although, conventionally, aluminum sputtering is severely affected by the oxygen atmosphere.

The ECR plasma deposition method employing a sputtering material supply can be applied, combined with introducing various gases such as O_2, N_2, CH_4, and SiH_4, to deposition of various metals, their oxides, nitrides, carbides, and silicides.

12. APPLICATIONS

As discussed, high-quality film deposition of various materials is possible by introducing chemical vapors and/or by using sputtering material supply at a low temperature without substrate heating. ECR plasma deposition techniques can be applied to various substrates, including those with poor heat resistance, utilizing the advantages of the low-temperature process.

a. *Surface Planarization for LSI Fabrications*[35]

For higher packing density in LSI a multilevel structure is attractive. A multilevel structure, however, is often accompanied by a severe surface step coverage problem. A planarization technology for the multilevel structure is, therefore, expected to be very useful.

Figure 42 shows the basic procedure for making planar structures for multilevel metallization using the ECR technique. First, the Al layer is etched with a resist mask using a dry etching technique (step A). Second, silicon dioxide is deposited using the ECR plasma deposition technique,

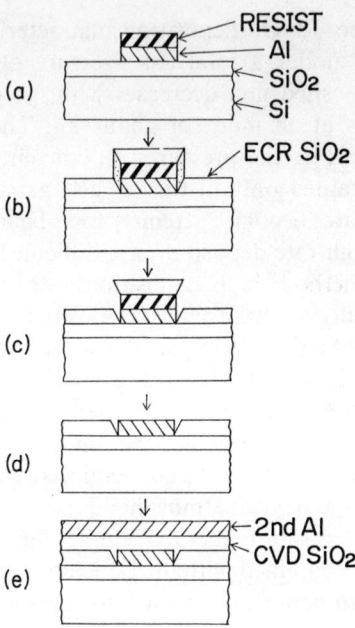

Fig. 42. Basic process steps for making planar interconnections. (a) Reactive ion etching, (b) ECR plasma deposition, (c) slight etching, (d) lift-off, (e) interlevel insulator deposition and second Al layer formation. (From Ehara et al.[35] Reprinted by permission of the publisher, The Electrochemical Society, Inc.)

utilizing the feature of low-temperature deposition (step B). Third, sidewall-deposited film is etched away with a wet etchant (step C). Fourth, silicon dioxide film on the resist is removed through a lift-off process by resist removing (step D). Finally, an interlevel insulator, CVD SiO_2, is deposited, and then the next Al layer is formed. The groove generated in step C is filled up through interlevel insulator deposition (step E).

Step C in Fig. 42 utilizes the fact that the film deposited on the sidewall has a much higher etch rate than that deposited on the flat surface in a HF solution. This phenomenon results because the sidewall receives hardly any ion bombardment and the film quality is poor compared to that on the flat surface. This feature is undesirable from the viewpoint of step coverage but is very useful for application to the lift-off process.

Figure 43 shows an SEM photograph of a cross-sectional view of the planarized structure. The buried Al line is 2 μm wide, and the CVD SiO_2 used as the interlevel insulator is 0.8 μm thick. A second Al layer is formed on the planarized structure. This planarization technique offers a great advantage for LSI fabrication, because it leads to a good yield and fine Al interconnections are easily obtained.

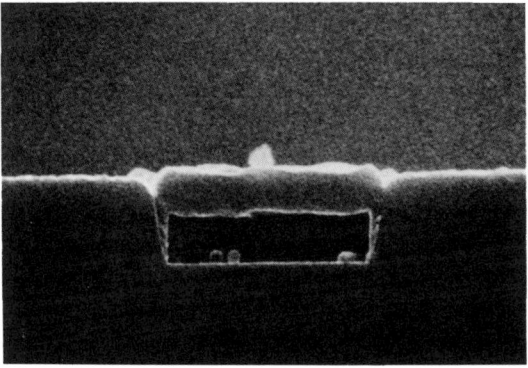

Fig. 43. SEM photograph of buried Al line, cross-sectional view. A part of the Al is slightly etched away so the structure can be clearly observed. (From Ehara et al.[35] Reprinted by permission of the publisher, The Electrochemical Society, Inc.)

b. Si_3N_4–InP *MIS diode*[36]

The operation of InP MISFETs has become an interesting subject in recent years. To investigate the MIS device, the ECR plasma deposition technique has been applied to deposit Si_3N_4 insulating films on InP substrates. Before film deposition, plasma cleaning of the substrate surface is

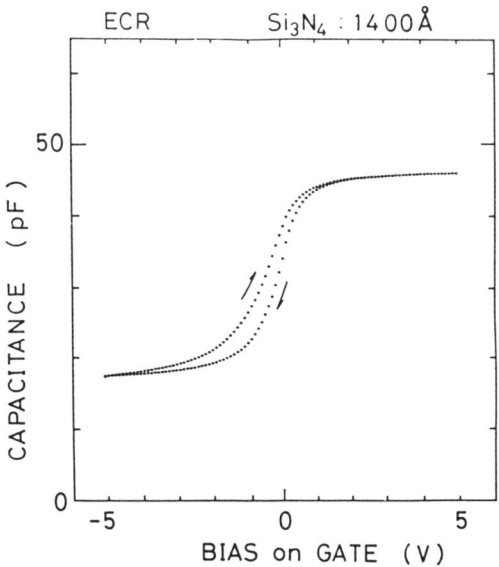

Fig. 44. Typical C–V characteristics for InP MIS diode with ECR plasma deposited Si_3N_4 film. Frequency, 1 MHz. $C_{ox} = 46.2$ pF.

performed using N_2 gas for 1 min. This cleaning is effective and sensitive for surface state densities. To fabricate MIS diodes, the insulators were deposited on bulk n-type substrates in InP with carrier densities of $5 \times 10^{15}/cm^3$. Alloyed Au–Ge–Ni/Au was used as a back contact to the samples and circular gate electrodes of Au (375 μm in diameter) were formed on the insulators.

Figure 44 shows a typical $C-V$ curve for the MIS diodes measured at a frequency of 1 MHz. Hysteresis is observed with a width of about 0.5 V. Surface state density N_{ss} was estimated from the $C-V$ curves using Terman's method and was in the range from 10^{11} to 10^{12} cm^{-2}/eV near the conduction-band edge. This result means that the film deposited by the ECR method is as good as the film carefully prepared by an anodic oxidation process that is not expected to cause surface damage. In other words, effects of both the highly activated plasma and the ion bombardment with energies of about 20 eV markedly contribute to enhance deposition reactions and improve film quality, but appear not to induce surface damage, even for heat-sensitive compound semiconductors.

REFERENCES

1. T. Consoli and R. B. Hall, *Nucl. Fusion* **3**, 237 (1963).
2. R. Bardet, T. Consoli, and R. Geller, *Nucl. Fusion* **5**, 7 (1965).
3. H. Ikegami, H. Ikeji, M. Hosokawa, S. Tanaka, and K. Takayama, *Phys. Rev. Lett.* **19**, 778 (1967).
4. Y. Sakamoto, *Jpn. J. Appl. Phys.* **16**, 1993 (1977).
5. H. Abe, Y. Sonobe, and T. Enomoto, *Jpn. J. Appl. Phys.* **12**, 154 (1973).
6. Y. Horiike and M. Shibagaki, *Proc. Conf. Solid State Devices* **7**, 13 (1976).
7. R. L. Bersin, *Solid State Technol.* **19**, May 31 (1976).
8. N. Hosokawa, R. Matsuzaki, and T. Asamaki, *Proc. Int. Vac. Congr., 6th, Kyoto* p. 435 (1974).
9. S. Matsuo and Y. Takehara, *Jpn. J. Appl. Phys.* **16**, 175 (1977).
10. S. Matsuo, *J. Vac. Sci. Technol.* **17**, 587 (1980).
11. S. Matsuo, *Appl. Phys. Lett.* **36**, 768 (1980).
12. Y. Horiike, M. Shibagaki, and K. Kadono, *Jpn. J. Appl. Phys.* **18**, 2309 (1979).
13. S. Matsui, T. Yamato, H. Aritome, and S. Namba, *Jpn. J. Appl. Phys.* **19**, L126 (1980).
14. D. F. Downey, W. R. Bottoms, and P. R. Hanley, *Solid State Technol.* **24**, 121 (1981).
15. R. Geller, *Appl. Phys. Lett.* **16**, 401 (1970).
16. Y. Okamoto and H. Tamagawa, *Rev. Sci. Instrum.* **43**, 1193 (1972).
17. N. Sakudo, K. Tokiguchi, H. Koike, and I. Kanomata, *Rev. Sci. Instrum.* **48**, 762 (1977).
18. S. Matsuo and Y. Adachi, *Jpn. J. Appl. Phys.* **21**, L4 (1982).
19. C. D. Child, *Phys. Rev.* **32**, 492 (1911).
20. J. M. E. Harper, J. J. Cuomo, P. A. Leary, G. M. Summa, H. R. Kaufman, and F. J. Bresnock, *J. Electrochem. Soc.* **128**, 1077 (1981).
21. T. Ono, Y. Adachi, and S. Matsuo, *Proc. Int. Ion Eng. Congr.—ISIAT '83 IPAT '83, Kyoto* p. 753 (1983).
22. J. L. Mauer, J. S. Logan, L. B. Zielinski, and G. S. Schwartz, *J. Vac. Sci. Technol.* **15**, 1734 (1978).

23. D. L. Flamm and V. M. Donnelly, *Plasma Chem. Plasma Process.* **1**, 317 (1981).
24. J. W. Coburn and H. F. Winters, *J. Appl. Phys.* **50**, 3189 (1979).
25. H. F. Winters, J. W. Coburn, and T. J. Chuang, *J. Vac. Sci. Technol.*, B **1**, 469 (1983).
26. S. Matsuo, *Jpn. J. Appl. Phys.* **15**, 1253 (1976).
26a. S. Matsuo, *Proc. Int. Ion Eng. Congr.—ISIAT '83 IPAT '83, Kyoto*, p. 1597 (1983).
27. A. K. Sinha, H. J. Levinstein, T. E. Smith, G. Quintana, and S. E. Haszko, *J. Electrochem. Soc.* **125**, 601 (1978).
28. W. A. Lanford and M. J. Rand, *J. Appl. Phys.* **49**, 2473 (1978).
29. A. C. Adams, F. B. Alexander, C. D. Capio, and R. E. Smith, *J. Electrochem. Soc.* **128**, 1545 (1981).
30. T. Tsuchimoto, *J. Vac. Sci. Technol.* **15**, 70 (1978).
31. S. Matsuo and M. Kiuchi, *Jpn. J. Appl. Phys.* **22**, L210 (1983).
32. R. Geller, N. Hopfgarten, B. Jacquot, and C. Jacquot, *J. Plasma Phys.* **12**, 467 (1974).
33. R. F. Kemp and J. M. Sellen, Jr., *Rev. Sci. Instrum.* **37**, 455 (1966).
33a. S. Matsuo and M. Kiuchi, *Proc. Int. Symp. VLSI Sci. Technol., 1st, Detroit*, p. 79 (1982).
34. T. Ono, C. Takahashi, and S. Matsuo, *Proc.—Electrochem. Soc.* **PV84-7**, 373 (1984).
35. K. Ehara, T. Morimoto, S. Muramoto, and S. Matsuo, *J. Electrochem. Soc.* **131**, 419 (1984).
36. M. Minakata, E. Yamaguchi, and S. Matsuo, to be published.

Physics of VLSI Processing and Process Simulation

W. FICHTNER

AT&T BELL LABORATORIES
MURRAY HILL, NEW JERSEY

I.	Introduction	119
II.	Physics of Processing and Process Simulation	122
	1. Epitaxy	124
	2. Ion Implantation	137
	3. Diffusion	168
	4. Lithography	226
	5. Etching and Deposition	314
III.	Conclusions and the Future	324
	1. Multidimensional Oxidation	324
	2. Defect Formation and Annealing	325
	References	325

I. Introduction

The progress in microelectronics during the past two decades has been tremendous. The first silicon integrated circuits (ICs) were demonstrated in 1959 by R. N. Noyce at Fairchild[1] and J. Kilby at Texas Instruments.[2] While Kilby proposed to use wire bonding to connect different areas of a chip, Noyce suggested the use of diffused resistors, a much more general idea that is still used today. In his patent application Noyce defined an IC as "a body of semiconductor ... containing adjacent P-type and N-type region with a junction therebetween extending to said surface ... two contacts upon opposite sides of said junctions, an insulating layer consisting essentially of oxide and said surface ... and an electrical connection to one of said contacts comprising a conductor adherent to said layer ...." The first metal–oxide–semiconductor field-effect transistor (MOSFET) was demonstrated in 1960 by D. Kahng and M. M. Atalla.[3,4] About a decade later, the MOSFET had become the basic building block and fundamental device structure for complex ICs.

Figure 1 shows the evolution of component complexity per chip since the invention of the IC.[5] The density of components per chip has increased exponentially in this time. At the time this paper is written, we have reached the complexity level of 10^6 components per chip.[6] The points in the figure indicate the most complex circuit types introduced at the time indicated.

The active chip area has remained more or less constant (about 1 square centimeter) during all this time. The increase in complexity of silicon ICs has been achieved by drastically reducing the lateral and vertical dimensions of the individual building blocks. The inset in Fig. 1 shows the reduction of minimum feature size for the same time period. This feature size has also decreased exponentially, as is indicated in the figure.

The reduction in active device dimensions to micrometer submicrometer geometries has resulted in an intimate coupling of process conditions, device behavior, and circuit performance to a degree unknown a few years ago. It becomes more and more difficult to develop new processes because of the inherent complexity of IC fabrication.

The use of computer-aided design tools has proved to be invaluable in the development of new technologies and in IC design. A modern IC process contains several hundred individual steps. Computer simulations have emerged as a very elegant way to aid the process and device engineer in their task of finding an optimum process.

Traditionally, new technologies have been developed guided by an experi-

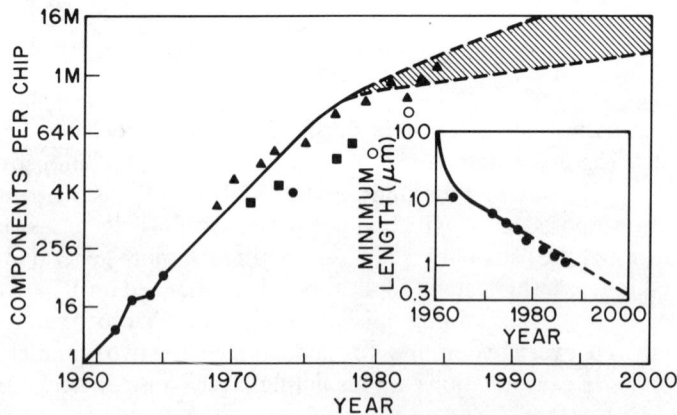

Fig. 1. Increase in chip component complexity for different technologies since the invention of the IC in 1959. The inset shows the reduction in the smallest feature size for the same period. ▲, MOS memory; ■, MOS logic; ●, bipolar logic; ○, bipolar memory. (From G. E. Moore,[5] updated. © 1979 IEEE.)

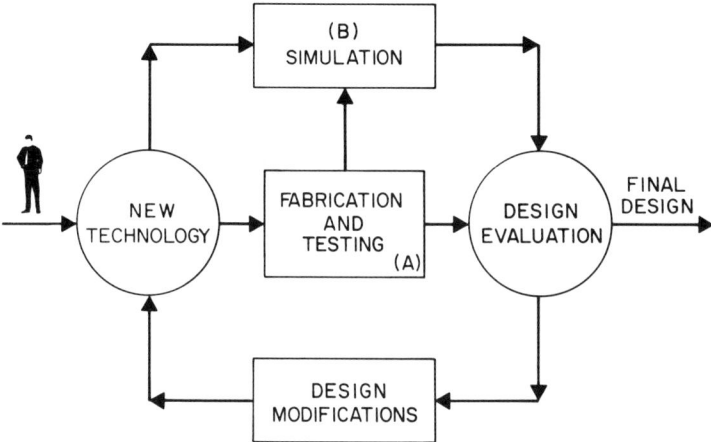

Fig. 2. Simulation and experiment are alternative routes to technology development and CAD. Route A schematically represents the experimental approach. Route B is the software approach; simulation stands for process, device and circuit simulation. (From Fichtner et al.[7] © 1984 IEEE.)

mental "trial-and-error" approach. Starting with an existing process, certain steps in the process are changed together with structural dimensions. This *modified* process is then fabricated by processing several lots (1 lot = 20–25 wafers). Completed test structures are subsequently evaluated to investigate whether the original goals have been met. In Fig. 2, route A schematically describes this approach, which might require many iterations to optimize the new process.[7] Based on selected parameters of this new process, new circuits can be designed.

The application of software tools in the development of new processes and novel device structures has become a worthwhile albeit challenging alternative to the experimental route (see path B in Fig. 2). Fabricating one lot in a modern process can cost considerably more than $10,000, consuming weeks or even months. The use of accurate simulation tools in the proper computing environment allows for comparatively cheap "computer experiments."

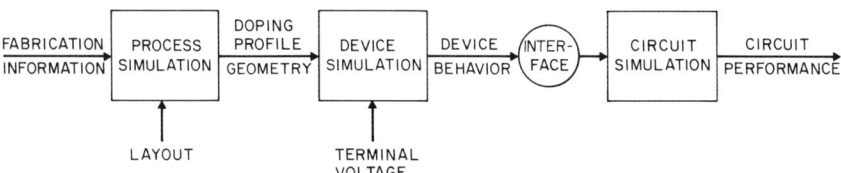

Fig. 3. Coupling between the various simulation fields.

Figure 3 shows schematically the various software aids for IC technology development and their coupling.

Process simulation deals with all aspects of IC fabrication. With the proper input parameters (processing steps, layout geometry), process simulation determines the details of the resulting device structure, including the boundaries of the different materials of the structure and the distribution of impurity ions within the structure.

The output of the process simulation, together with applied terminal voltages and currents, is the input to a device simulation program that determines the electrical characteristics of the device. Furthermore, important insight can be gained by properly analyzing the behavior of internal variables such as potential and carrier densities.

The current–voltage characteristics of the device are then used as the essential data to construct a compact model of the device. Such a model can be used as input to a circuit simulation program to determine the electrical characteristics of a circuit comprised of multiple interconnected devices.

This chapter was written to serve as an introduction to the field of process simulation. I have given special emphasis to the foundations of this rather new field of research, which merges several disciplines, including chemistry, computer science, mathematics, and physics. At this time, all major industrial semiconductor companies and universities have at least one group pursuing research in this field.

Section II of this chapter presents a thorough discussion of the physics of the various process steps and process simulation. All the process steps of importance are covered in detail.

In Section III we draw conclusions and look into the future for further needs in process simulation.

II. Physics of Processing and Process Simulation

Modern IC fabrication consists of many individual processing steps. To illustrate this, Fig. 4 outlines schematically the process flow of an n-channel metal–oxide–semiconductor (NMOS) process with 1-μm design rules.[8,9] Table I is a summary of the wafer process. The starting substrates are doped uniformly with boron. The alignment marks used in the lithography are defined by steps following those in the figure.[7] After the field oxide (F_{ox}) is grown over the entire wafer, a boron implant through this oxide sets the threshold of the enhancement-mode devices and forms the "chain-stop" regions [Fig. 4a]. At this stage, the first lithography step defines the active transistor areas. This process allows for the additional option of another boron implant at this stage in order to optimize devices with gate length smaller then 1 μm (Fig. 4b). Steps 6 and 7 in Table I define the depletion-load devices and set the depletion-load threshold voltage (Fig. 4c). Subsequently, the gate oxide (G_{ox}) is grown (Fig. 4d), the polysilicon is deposited, and the gates are defined. These gates act as self-aligned masks for the high-dose source–drain implant (Fig. 4e). The rest of the process is standard, with intermediate insulator deposition, contact formation, metallization, and cap insulator deposition (Fig. 4f).

Modern processes like this one allow the classification of IC fabrication steps into three areas: (1) thermal processing and doping, (2) pattern definition (lithography), and (3) pattern transfer (etching and deposition). Each of these areas contains several subcategories, as shown in Table II.

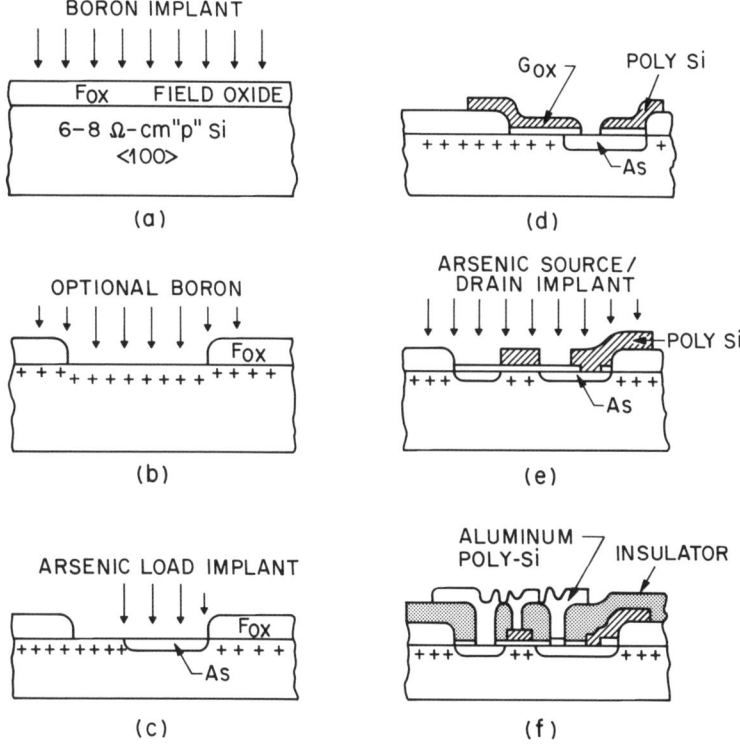

Fig. 4. An advanced NMOS process. For details, see text and Watts et al.[8] (a) Sheet implant through field oxide. (b) Active area definition. (c) Depletion-level lithography with depletion implant. (d) Gate definition. (e) Source–drain formation. (f) Metal lithography. (From Fichtner et al.[7] © 1984 IEEE.)

TABLE I
Wafer Process[a]

1. Field oxide
2. Ion implant B
3. Active area lithography
4. Active area definition
5. Ion implant, boron (optional)
6. Depletion-level lithography
7. Ion implant As
8. Grow gate oxide
9. Polysilicon definition
10. First poly lithography
11. Define polysilicon
12. Deposit second polysilicon
13. Second polysilicon lithography
14. Define polysilicon
15. Source–drain implant As
16. Anneal implant
17. Contact lithography
18. Form contacts
19. Metal deposition
20. Metal lithography
21. Caps processing

[a] The substrate is 6–8-Ω-cm p type.

TABLE II
IC Process Steps

Thermal processing and doping	Pattern definition	Pattern transfer
Epitaxy	Optical lithography	Wet chemical etching
Ion implantation	Electron-beam lithography	Ion milling
Predisposition	Ion-beam lithography	Reactive ion etching
Annealing	X-ray lithography	CVD
Drive-In		Evaporation
Oxidation		Sputtering

1. Epitaxy

The distribution of impurities in epitaxially grown silicon layers plays an important role in the operation of integrated circuits. In very large-scale integrated (VLSI) devices it is quite common to use epitaxial layers on highly doped substrates (e.g., to reduce the alpha particle sensitivity in dynamic RAM circuits or to avoid latch-up in complementary MOS (CMOS) gates).

The impurity redistribution that occurs during epitaxial growth and subsequent processing is dependent on processing temperatures and times, the diffusivities of the impurities present, the corresponding evaporation rates, and the segregation rates of the dopants in the solid and gaseous phases.

Any time a lightly doped region is exposed to a heavily doped region, transport of dopant will occur through both the solid and gaseous phases. Therefore, the distribution of impurities near the interface between the epitaxial film and the underlying substrate is in general nonuniform. During the initial stages of growth, a certain time period (a few minutes) is required before the steady-state doping level is established. A special case of this transient problem is the autodoping phenomenon, which occurs when lightly doped films are grown on top of heavily doped substrates.

The existence of dopant in an epitaxial layer caused by out-diffusion from the substrate or autodoping has been the subject of interest from the earliest work on epitaxial deposition. Thomas et al.[10] and Grossman[11] were the first authors to explain this extra dopant by a mechanism that involves dopant transfer from the front surface of the wafer, mixing of this dopant with the gas phase, and reincorporation of part of this dopant into the wafer. Solid-state diffusion was ignored in that model and application of this model near the interface to the substrate was questionable. Rice,[12] Grove et al.,[13]

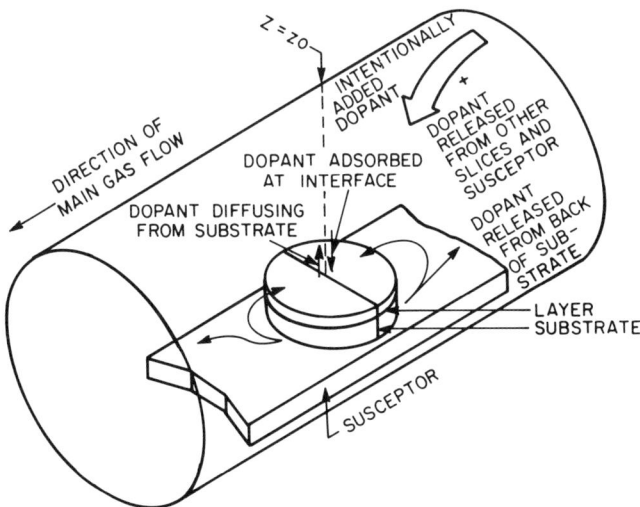

Fig. 5. Schematic representation of an epitaxial reactor showing sources of dopant. (From Langer and Goldstein.[18] Reprinted by permission of the publisher, The Electrochemical Society, Inc.)

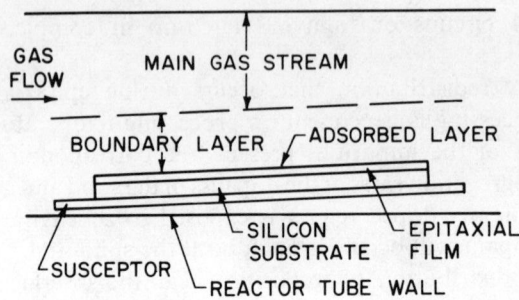

Fig. 6. Schematic section of the horizontal reactor tube along the length of the susceptor. (From Reif and Dutton.[24] Reprinted by permission of the publisher, The Electrochemical Society, Inc.)

and Abe et al.[14] assumed that solid-state diffusion was the only mechanism for dopant transport from the substrate into the layer and ignored the influence of the ambient gas phase. Joyce et al.,[15] Shepherd,[16] and Skelly and Adams[17] showed that dopant from the backside of the wafer could also act as an additional source of added impurity to the growing layer. This phenomenon is called *back-side autodoping*.

The first computational model of the epitaxial process incorporating all of these effects was published by Langer and Goldstein[18] in 1974. Their model was solved numerically using a program called CASPER (computer-aided semiconductor processing and epitaxial redistribution).

The original Langer–Goldstein model has been incorporated in the second version of the process-modeling program, the Stanford University Process Engineering Models program (SUPREM),[19,20] neglecting front-side and back-side autodoping.

Meindl et al.[21] and Reif et al.[22–25] put forward a new computational model. This model considers gas-phase dynamics and the physicochemical processes occurring in the gas phase and at the growing epitaxial surface.

Figure 5 shows a section of an epitaxial reactor with the wafer on a heated susceptor. The solution of the problem involves the treatment of three different regions, which can be seen in Fig. 6, which is a cross section of the reactor.

a. *The Bulk Gas Phase*

In most epitaxial reactors, gas-phase depletion of dopant species in the main gas stream is negligible. Therefore, it can be assumed that all the dopant is uniformly distributed. Furthermore, the partial pressure of the dopant species in the main gas stream region is more or less independent of position, and nearly equal to the partial pressure of dopant species at the input to the reactor. Reif et al.[26,27] have shown that the time constant associated with changes in dopant concentration in the main gas stream is much smaller than the time constant of the overall doping process, which allows the assumption that

$$P_{Dop}(t) \cong P^0_{Dop}(t), \tag{2.1}$$

where P_{Dop} is the partial pressure of the dopant species in the main stream gas region, and P^0_{Dop} is the corresponding value at the input.

b. *The Gas-Phase Boundary Layer*

The time associated with dopant transport through the boundary layer by diffusion is much shorter than the overall doping process. Therefore, it can be assumed that the flux of dopant species leaving the boundary layer by adsorption on the silicon surface, F_s, follows instantaneously any variation of the dopant flux entering the boundary layer from the main gas stream.

Figure 7 shows an enlarged view of the boundary layer regions with the main gas stream at the top and the solid silicon below. Mass balance of dopant species within an element of thickness Δz and unit length and width leads to the expression

$$-\partial F_z/\partial z \cong \partial C_z/\partial t, \tag{2.2}$$

where F_z is the dopant flux in z direction and C_z is the dopant species concentration per unit volume in the boundary layer.

Equation (2.2) is obtained from a continuity equation assuming that no reaction occurs in the respective volume element (i.e., neither formation nor consumption) and that transport occurs only in the z direction. It can be further simplified by noting that the dopant species in the boundary layer is always in a steady-state condition $\partial C_z/\partial t = 0$, which leads to

$$\frac{\partial F_z}{\partial t} \cong 0. \tag{2.3}$$

The flux F_z is independent of position regardless of whether the overall doping process is at steady state or undergoes a transient.

The dopant flux leaving the boundary layer because of surface adsorption follows instantaneously the variations in the dopant flux entering the boundary layer,

$$F_s(t) \cong F_z(z, t). \tag{2.4}$$

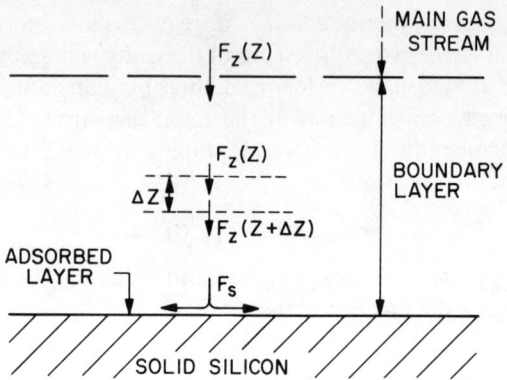

Fig. 7. Enlarged view of the boundary layer region. (From Reif and Dutton.[24] Reprinted by permission of the publisher, The Electrochemical Society, Inc.)

The two fluxes in Eq. (2.4) can be expressed in terms of deposition parameters. F_z can be written as

$$F_z(z, t) = k_m(P_{Dop}^0(t) - P_{Dop}^*) \qquad (2.5)$$

in which k_m is the boundary layer mass transport coefficient of dopant species, and P_{Dop}^* is the dopant partial pressure above the gas–solid interface.

c. *The Adsorbed Layer*

The different processes can be divided into a sequence of steps that occur sequentially. In the following we assume that arsine (AsH_3) is the dopant species considered, and silicon growth is from silane (SiH_4) in a hydrogen

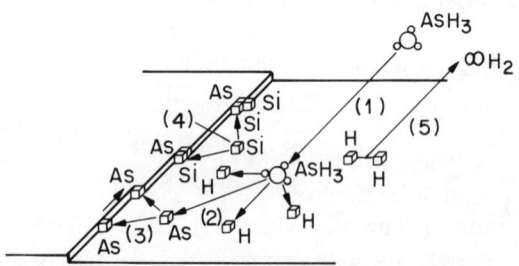

Fig. 8. Sequence of steps occurring on the surface: (step 1) adsorption of the As-containing compound, (step 2) surface chemical dissociation, (step 3) surface diffusion and site incorporation, (step 4) "burying" of As by subsequently arriving Si atoms, (step 5) desorption of hydrogen. (From Reif et al.[22] Reprinted by permission of the publisher, The Electrochemical Society, Inc.)

ambient in an atmospheric pressure reactor. A similar approach has been taken by Faktor and Garrett[28] and Shaw[29] discussing crystal growth from the vapor.

The different mechanisms at the growing surface are classified using the terrace–ledge–kink model for the crystal surface. Three different sites are assumed: terrace (adsorption) sites, ledge (step) sites, and kink sites.

Figure 8 shows the sequence of steps occurring at the surface.

Step 1. When an arsine molecule in the gas phase gets close to the silicon surface, it can undergo an adsorption process. This is characterized by the reaction

$$\text{AsH}_3(g) + s \rightleftarrows \text{AsH}_3\text{-}s \tag{2.6a}$$

with reaction constant

$$K_1 = \frac{\theta_{\text{Dop}} N_s}{P^*_{\text{Dop}}(\theta C_s)} = \frac{k_F}{k_R} \tag{2.6b}$$

Step 2. The adsorbed arsine molecule decomposes chemically into elemental arsenic.

Step 3. The elemental arsenic diffuses on the surface until it finds an incorporation site and attaches to it.

Step 4. The newly incorporated arsenic atom is quickly covered by subsequently arriving silicon atoms and is ready to diffuse into or out of the bulk silicon.

Step 5. Desorption of hydrogen takes place.

Steps 2–4 can be summarized by the reaction

$$\text{AsH}_3\text{-}s \underset{k_R}{\overset{k_F}{\rightleftarrows}} \text{As(ss)} + s + \tfrac{3}{2}\text{H}_2(g), \tag{2.7a}$$

with reaction constant

$$K_2 = \frac{(k_H C)}{(\theta C_s)} P_{H_2}^{3/2}. \tag{2.7b}$$

In Eqs. (2.6) and (2.7), s represents a vacant adsorption site on the surface, ss an incorporation site, $\text{AsH}_3\text{-}s$ represents an arsine molecule occupying an adsorption site, K_1 and K_2 are equilibrium constants, θ_{Dop} is the fraction of adsorption sites occupied by arsine, C_s is the surface density of adsorption sites per unit area, k_F and k_R are the forward and reverse reaction rate constants for arsine adsorption, θ is the vacant fraction of adsorption sites, k_H is Henry's law constant, and P_{H_2} is the hydrogen partial pressure.

From Eq. (2.6), it follows that

$$F_s = k_F(\theta C_s) P^*_{\text{Dop}} - k_R(\theta C_s), \tag{2.8}$$

which can be rewritten as

$$F_s = k_F(\theta C_s)\left[P^*_{\text{Dop}} - \frac{k_R}{k_F}\frac{(\theta_{\text{Dop}} C_s)}{(\theta C_s)}\right] \tag{2.9}$$

and with Eq. (2.7)

$$F_s = k_F(\theta C_s)\left(P^*_{\text{Dop}} - \frac{k_H P_{H_2}^{3/2} C}{K_1 K_2}\right) \tag{2.10}$$

or finally

$$F_s = k_f(P^*_{\text{Dop}} - C/K_p), \tag{2.11}$$

where $k_f = k_F(\theta C_s)$ is a kinetic constant associated with arsine adsorption, $K_p = K_1 K_2/k_H$ is a thermodynamic constant relating dopant species in the gas phase and the solid silicon, and $P_{H_2} \simeq 1$ atm.

Equations (2.5) and (2.11) can be used to obtain the final expression relating F_s and the input partial pressure P^0_{Dop},

$$F_s = k_{\text{mf}}(P^0_{\text{Dop}} - C/K_p) \tag{2.12}$$

with

$$k_{\text{mf}} = (1/k_m + 1/k_f)^{-1}.$$

The redistribution of impurities within the solid is controlled by diffusion, and the computational procedure involves a solution of Fick's second law throughout the solid

$$\frac{\partial C(z,t)}{\partial t} = D\frac{\partial^2 C}{\partial z^2} \quad \text{for} \quad z_f < z < \infty, \tag{2.13}$$

where C is the dopant concentration in the silicon, D is the corresponding diffusion coefficient, and z and t are the spatial and time variables.

The solution of the diffusion equation, Eq. (2.13), must satisfy the following initial and boundary conditions:

$$C(z, 0) = f_1(z), \tag{2.14}$$

$$D\frac{\partial C}{\partial z} = 0 \quad \text{as} \quad z \to \infty, \tag{2.15}$$

$$D\frac{\partial C}{\partial t} = f_2(t) \quad \text{for} \quad z = z_f. \tag{2.16}$$

The function $f_1(z)$ stands for distribution of impurities in the substrate before the deposition starts and $f_2(t)$ indicates that the diffusive flux at the surface is a function of time during deposition. Equation (2.15) states that deep in the substrate the impurity flux is zero.

An expression for $f_2(t)$ can be derived from the detailed understanding of the different mechanisms considered in steps 1–4. From mass balance considerations, we obtain the relation

$$F_s - gC(z_f) + D\left.\frac{\partial C}{\partial z}\right|_{z=z_f} = \frac{d(\theta_{Dop}C_s)}{dt} \qquad (2.17)$$

with the epitaxial growth rate g. The second term represents the rate at which the adsorbed layer decreases its concentration of dopant species because of coverage by silicon atoms. The third term accounts for diffusive fluxes between the adsorbed layer and the bulk silicon. The right-hand side represents the rate of change of the concentration of dopant species per unit area in the adsorbed layer, and goes to zero when the total doping process reaches steady state.

Substituting Eq. (2.12) into Eq. (2.17) leads to

$$k_{mf}\left[P^0_{Dop} - \frac{C(z_f)}{k_p}\right] - gC(z_f) + D\left.\frac{\partial C}{\partial z}\right|_{z=z_f} = K\frac{\partial C(z_f)}{\partial t}, \qquad (2.18)$$

where we have used Eq. (2.7) with $K \equiv k_H(\theta C_s)P_{H_2}^{3/2}/K_2$.

Equation (2.18) can be rearranged to give the final expression for $f_2(t)$

$$D\left.\frac{\partial C}{\partial z}\right|_{z=z_f} = f_2(t) = -k_{mf}\left[P^0_{Dop} - \frac{C(z_f)}{k_p}\right] + gC(z_f) - K\frac{\partial C(z_f)}{\partial t}. \qquad (2.19)$$

From Eqs. (2.19) and (2.17), it is evident that any abrupt variation in P^0_{Dop} is not followed instantaneously by $C(z_f)$ because some finite time is needed to adjust the population of dopant species in the absorbed layer θC_s to the new condition.

Using Eq. (2.19) in Eq. (2.16) allows a solution of Fick's second law by numerical means. Reif and Dutton proceed in a similar way to Langer and Goldstein in the numerical solution. The epitaxial doping model is contained in Eq. (2.19), which relates $C(z_f)$ to P^0_{Dop}. Reif and Dutton have implemented Eq. (2.14) into SUPREM with the further simplification that only steady-state conditions are considered. This assumption results in the equation

$$0 = k_{mf}[P^0_{Dop} - (C/K_p)] + gC \qquad (2.20)$$

in which C is the uniform epitaxial doping level and P^0_{Dop} and g are assumed to be time dependent. Equation (2.20) can be used to determine the doping level that results if the total deposition time is much longer than the system time constant and if the gas flow conditions do not change during the time of the deposition. Rewriting Eq. (2.20) as

$$P^0_{Dop}/C = 1/K_p + (g/k_{mf}) \qquad (2.21)$$

indicates that a plot of P^0_{Dop}/C versus g that is generated from experiments under steady-state conditions should yield a straight line from which values of k_{mf} and K_p can be extracted.

Figure 9 shows an example of a P^0_{Dop}/C versus g plot generated from experiments. The hydrogen, silane, and arsine flows remained unchanged during the deposition cycle. Thick layers (7–9 μm) were grown to neglect any influence arising from the initial transient period. The arsine partial pressure was varied to produce doping levels in the 10^{15}–10^{17}/cm^3 range, and the silane partial pressure was adjusted to yield growth rates between 0.07 and 0.42 μm/min. By fitting Eq. (2.21) to the experiments, the following values are obtained: $K_p = 1.05 \times 10^{26}$/cm^3 and $k_{mf} = 4.85 \times 10^{19}$/cm^2.

Equation (2.20) has been incorporated into SUPREM by decoupling it from the solution to the diffusion equation, Eq. (2.13). Figure 10 shows the discretization space at two different time steps during the calculation. In Fig. 10a the solid line indicates the interface between the gas phase and the solid phase. The bulk silicon region is subdivided into finite cells with broken lines showing the cell boundaries. In each cell the dopant concentration is assumed to be uniform. The solution proceeds in the following way:

(1) At $t = t_0$ the doping profile in the silicon is either known from Eq. (2.14) or given by the calculation up to the time point t_0. A new cell z_{i-1} is

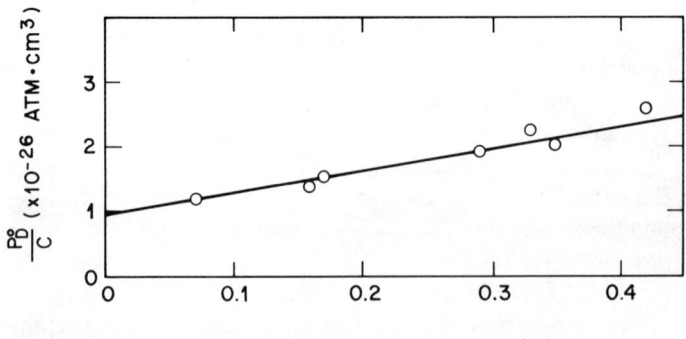

Fig. 9. Plot of P^0_D/C vs. silicon growth rate used to determine k_{mf} and K_p. —, Model; O, experiment. (From Reif and Dutton.[24] Reprinted by permission of the publisher, The Electrochemical Society, Inc.)

added (Fig. 10b), and the corresponding concentration C_{i-1} is calculated from Eq. (2.20), i.e.,

$$0 = k_{mf}\left[P^0_{Dop} - \frac{C_{i-1}}{K_p}\right] - gC_{i-1} - K\frac{dC_{i-1}}{dt}. \quad (2.22)$$

It is important to note that this procedure accounts only for dopant introduction into the new cell but ignores the simultaneous redistribution in the solid silicon.

(2) A solution of Fick's second law allows the calculations of the thermal redistribution of impurities during the time interval under consideration. This is illustrated in Fig. 10c, where the arrows represent diffusive fluxes across cell boundaries in a schematic way. No flux crosses the gas–solid interface, indicating the decoupling between deposition and redistribution.

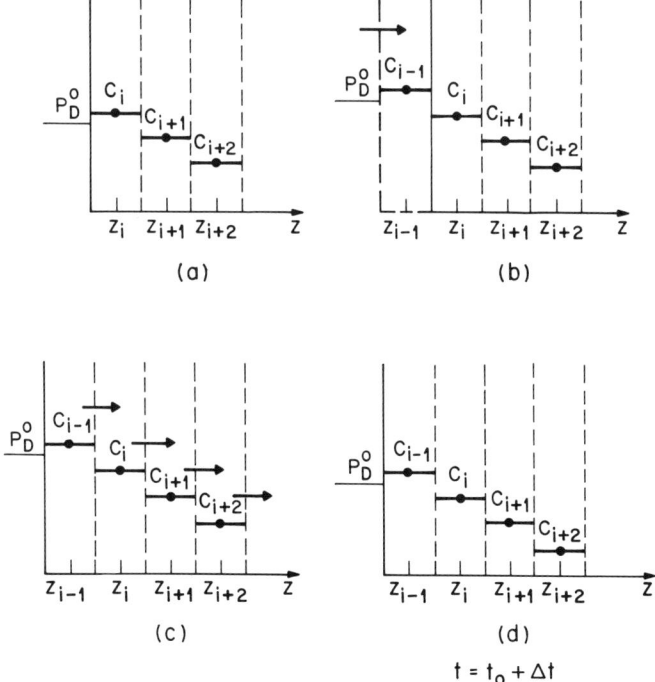

Fig. 10. Implementation of the numerical technique used to solve Fick's Second Law with the surface boundary condition dictated by the epitaxial deposition process. For explanation see text. (After Reif and Dutton.[24] Reprinted by permission of the publisher, The Electrochemical Society, Inc.)

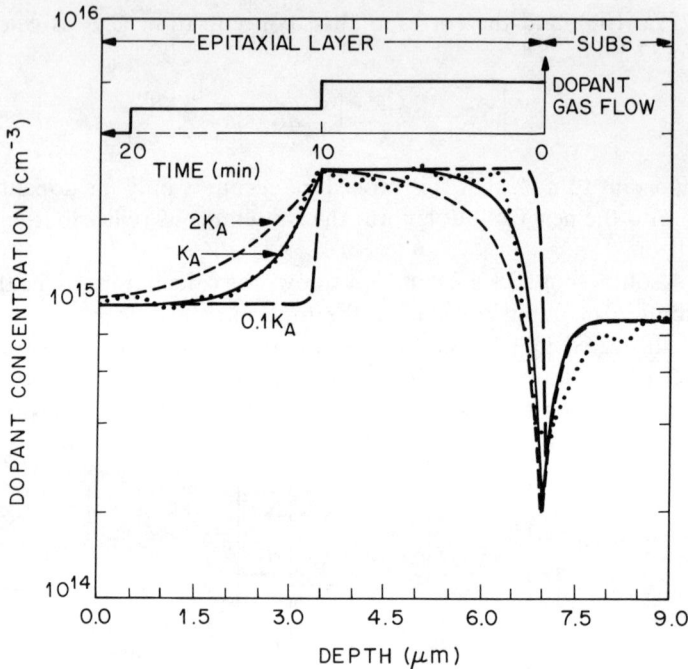

Fig. 11. Measured and simulated doping profiles used to determine K_A. Solid and dashed lines, simulated; dotted lines, measured. (From Reif and Dutton.[24] Reprinted by permission of the publisher, The Electrochemical Society, Inc.)

After Eq. (2.13) is solved, the concentration in each cell has been rearranged and one cycle has been completed (Fig. 10d). This two-step process is repeated until the final deposition time is reached.

This numerical procedure is finally used to determine the constant K in Eq. (2.20). A family of profiles is generated numerically by varying K. Next the simulated deposition conditions are carried out experimentally and the resulting profile is measured. The value of K is selected that best fits the experimental profile.

Figure 11 shows a typical result of an experiment with three theoretical results for three values of K. As indicated in the figure, the arsine flow was kept constant for 10 min, and then abruptly decreased to be left constant at a lower level for another 20 min. The corresponding arsine partial pressures were 7.5 and 2.4×10^{-11} atm, respectively. The growth rate of 0.35 μm/min was achieved with a silane partial pressure of 8.4×10^{-4} atm. The dotted line is the measured profile as determined by the spreading resistance technique, corrected using multilayer correction factors. The doping profile corresponding to $K = 5.7 \times 10^{-5}$ cm shows the best overall fit.

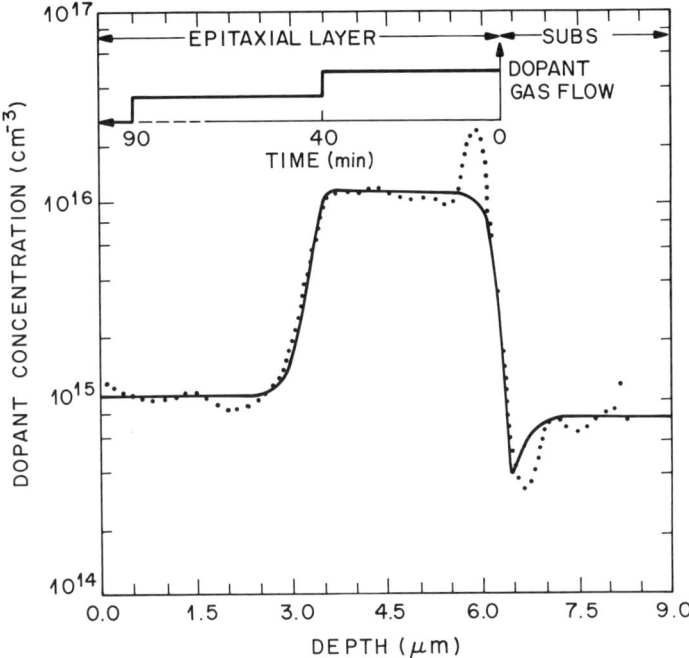

Fig. 12. Measured and simulated doping profiles resulting from the decreasing step change in arsine flow indicated in the inset. —, Simulated; ···, measured. (From Reif and Dutton.[24] Reprinted by permission of the publisher, The Electrochemical Society, Inc.)

The preceding model compares extremely well to experimental results. Figure 12 shows measured (dotted line) and simulated (full line) doping profiles resulting from a decreasing step change similar to the result in Fig. 11. The growth rate is 0.07 μm/min at $T = 1050°C$. The doping level charge is approximately one order of magnitude. The full transition range between the two doping levels is roughly 1 μm. The dip in the experimental profile is a measurement artifact.

d. *Analytical Expressions*

Sometimes we must obtain quick estimates on out-diffusion from a growing epitaxial layer when computer programs such as SUPREM are not available. Analytical solutions of Eq. (2.13), neglecting the full complexity of Eq. (2.20) have been published by Rice[30] and are given in standard textbooks on diffusion or heat transport.[31-33]

Consider initially the case of an undoped substrate and the epitaxial layer being put down with a doping concentration C_0. Choosing the constant concentration layer as origin of coordinates, then $z = 0$ defines the growing surface and the layer–substrate interface recedes with a constant velocity v. With the transformation $z \to z + vt$, Eq. (2.19) becomes

$$D \frac{\partial^2 C}{\partial z^2} = \frac{\partial C}{\partial t} + v \frac{\partial C}{\partial z} \qquad (2.23)$$

to be solved subject to the set of simplified boundary conditions

$$C = C_0 \quad \text{at} \quad z = 0 \quad \text{for all } t, \qquad (2.24a)$$
$$C = 0 \quad \text{as} \quad z \to \infty, \qquad (2.24b)$$
$$C = 0 \quad \text{for} \quad z > 0 \quad \text{at} \quad t = 0. \qquad (2.24c)$$

The solution becomes

$$C = \frac{C_0}{2}\left(\operatorname{erfc}\frac{z - vt}{2\sqrt{Dt}} + \exp\frac{vz}{D}\operatorname{erfc}\frac{z - vt}{2\sqrt{Dt}}\right) \qquad (2.25)$$

where erfc is the complementary error function defined by

$$\operatorname{erfc} y = \frac{2}{\sqrt{\pi}} \int_y^\infty \exp(-t^2)\, dt = 1 - \operatorname{erf} y \qquad (2.26)$$

with the error function erf y.[34] Simple rational approximations are available for the error function in the literature.

The second case is that of an undoped (or very low doped) layer being deposited on a heavily doped substrate. This case is more interesting since it occurs quite often in standard MOS or bipolar processing. Some of the dopant will out-diffuse into the growing layer. If it diffuses all the way through, it can reach the growth surface and evaporate. A net loss of dopant will occur if the evaporation rate exceeds the adsorption rate of dopant species from the gas boundary layer.

Equation (2.23) is subject to the boundary and initial conditions Eq. (2.24) with an additional rate equation at the surface

$$D \left.\frac{\partial C}{\partial z}\right|_{z=0} = (K + v)C(0, t), \qquad (2.27)$$

where K is the evaporation rate. The solution is

$$C = C_0 - \frac{C_0}{2}\left(\operatorname{erfc}\frac{z - vt}{2\sqrt{Dt}} + \frac{K + v}{K}\exp\frac{vt}{D}\operatorname{erfc}\frac{z + vt}{2\sqrt{Dt}}\right)$$
$$- C_0 \left(\frac{2K + v}{2K}\right)\exp\frac{(K + v)(z + vt)}{D}\operatorname{erfc}\frac{z + (2K + v)t}{2\sqrt{Dt}}. \qquad (2.28)$$

For $K \to 0$, Eq. (2.28) reduces to

$$C = \frac{C_0}{2}\left(1 + \operatorname{erf}\frac{z - vt}{2\sqrt{Dt}}\right), \qquad (2.29)$$

which means that if there is no loss at the surface because of evaporation, then the diffusing species does not "see" the surface.

2. Ion Implantation

The successful application of ion implantation in silicon processing depends strongly on the ability to predict and to control mechanical and electrical effects resulting from given implant conditions. The theory of penetration of charged particles through solids, pioneered by Bohr,[35-37] was developed to a sophisticated stage by Lindhard and co-workers[38-41] (the LSS theory) and Firsov.[42-45]

Several in-depth treatments of ion implantation are available in the literature[46-49] reviewing the basic foundations and concepts of ion implantation.

This section summarizes the major theoretical approaches to ion implanation in solids, with a particular emphasis on numerical methods.

a. Classical Scattering Theory

The following assumptions are usually made in the description of scattering events between particles:[50-54]

(1) Validity of the binary collision approximation: Collisions between atoms of reasonably high energy (keV and higher) result in a very close approach of the collision partners, which makes the probability for three- and more particle collisions extremely small. Collective effects become important only in the low-energy region (<1 keV).

(2) Validity of classical mechanics: Atomic collisions are hardly ever classical in the sense that all measurable characteristics are derivable with reasonable accuracy from the laws of classical mechanics. The applicability of classical mechanics is normally limited to specific quantities such as the total differential cross section $d\sigma(\theta)$, where θ is the scattering angle in the center-of-mass (CM) system.

(3) Excitation or ionization of electrons are only a source of energy loss, and do not influence the collision: This statement is justified if the energy transferred to the electrons is small compared to the exchange of kinetic energy between the atoms [see Eq. (2.34)], which is usually fulfilled in our

case. Therefore, the electronic energy loss enters as a superimposed energy absorption process.

(4) One of the two collision partners is initially at rest.

For elastic two-particle collisions of particles with masses M_1 and M_2, with M_2 initially at rest, we obtain for the angles and energies in the laboratory and CM system:

$$\tan \psi' = \frac{M_2 \sin \theta}{M_1 + M_2 \cos \theta}, \quad (2.30)$$

$$\psi'' = (\pi - \theta)/2, \quad (2.31)$$

$$E' = E - T, \quad (2.32)$$

$$E'' = T, \quad (2.33)$$

$$T = T_m \sin^2(\theta/2) = \gamma E \sin^2(\theta/2), \quad (2.34)$$

$$\gamma = 4M_1 M_2/(M_1 + M_2)^2. \quad (2.35)$$

Figure 13 describes the collision in both systems. The quantity E is the initial energy, T the energy transfer (recoil energy), ψ' and ψ'' are the scattering and recoil angles in the laboratory system, E' and E'' are the corresponding energies, and θ is the CM scattering angle.

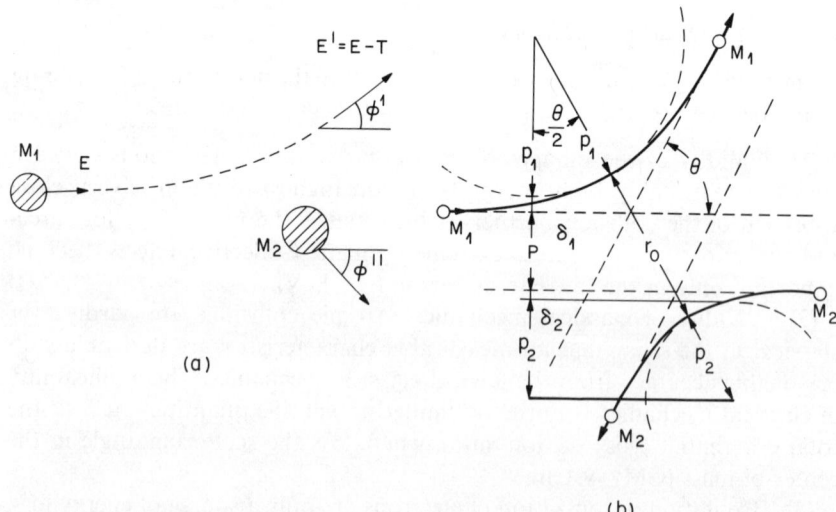

Fig. 13. Two-particle scattering in the (a) laboratory and (b) center-of-mass system. (From Biersack and Haggmark.[87] Copyright North-Holland Physics Publishing, Amsterdam, 1980.)

The scattering angle θ is calculated by integrating the equations of motion to yield

$$\theta = \pi - 2p \int_{r_0}^{\infty} \frac{dr}{r^2 [r - V(r)/E_r - p^2/r^2]^{1/2}}, \quad (2.36)$$

where p is the impact parameter (Fig. 13b), $V(r)$ is the interaction potential between the incident ions and the target atoms, and r_0 is the distance of closest approach, given by the zero of the square root in Eq. (2.36). E_r is the incident energy in the CM system

$$E_r = \frac{E}{1 + (M_2/M_1)}. \quad (2.37)$$

The total differential cross section $d\sigma(\theta)$ is obtained by inverting Eq. (2.36)

$$d\sigma(\theta) = -2\pi p(dp/d\theta)\, d\theta = -(p/\sin\theta)(dp/d\theta)\, d\omega, \quad (2.38)$$

where $d\omega = 2\pi \sin\theta\, d\theta$.

b. Nuclear Stopping and Scattering Cross Section

The form of the repulsive potential $V(r)$ is of critical importance in range calculation.[54] The potential used in essentially all calculations is a screened Coulombic Thomas–Fermi potential [39-41]

$$V(r) = (Z_1^2 Z_2^2 q^2/r)\phi(r/r_{Sep}) \quad (2.39)$$

with the interatomic separation r_{Sep} and the screening function ϕ.

The techniques for the determination of nuclear cross sections use reduced energy and length parameters[39,43]

$$\epsilon = \epsilon_1 E = \left(\frac{M_2}{M_1 + M_2} \frac{a}{Z_1 Z_2 q_2}\right) E, \quad (2.40)$$

$$\rho = \rho_1 z = N\pi q^2 \gamma z, \quad (2.41)$$

where N and z are the density and depth of the target and a is the screening radius

$$a = 0.8853 a_0/(Z_1^{2/3} + Z_2^{2/3})^{1/2}, \quad (2.42)$$

where a_0 is the Bohr radius $\hbar^2/m_0^2$ and m_0 is the electron mass.

The LSS theory introduces the parameter

$$t = (T/T_m)\epsilon^2 = T\epsilon^2/\gamma E \quad (2.43)$$

to approximate the cross section in Eq. (2.38) by a function depending only on one variable t instead of two variables ϵ and θ (or E and T). For screened Coulomb interaction, we obtain for the scattering cross section

$$d\sigma(E, T) = \pi a^2 (dt/2t^{3/2}) f(t^{1/2}). \tag{2.44}$$

The corresponding total reduced nuclear cross section is

$$S_n = \left. \frac{d\epsilon}{d\rho} \right|_{\text{nucl}} = \left. \frac{\epsilon_1}{\rho_1} \frac{dE}{dz} \right|_{\text{nucl}} = \frac{\epsilon_1}{\rho_1} N \int T \, d\sigma$$

$$= \frac{1}{\epsilon} \int_0^\epsilon f(t^{1/2}) \, dt. \tag{2.45}$$

The f function is known as the scattering function and depends on the form of $V(r)$. A suggested form is

$$f(t^{1/2}) = \lambda t^{(1/2)-m}/(1 + (2\lambda t^{1-m})^q)^{1/q}. \tag{2.46}$$

The constants $\lambda = 1.309$, $m = \frac{1}{3}$ and $q = \frac{2}{3}$ were originally determined to fit Thomas–Fermi interaction[55] and have been used in calculations up to the present.[56-58] The significance of Eq. (2.45) lies in the fact that one single functional expression describes the scattering of a two-fold variety of

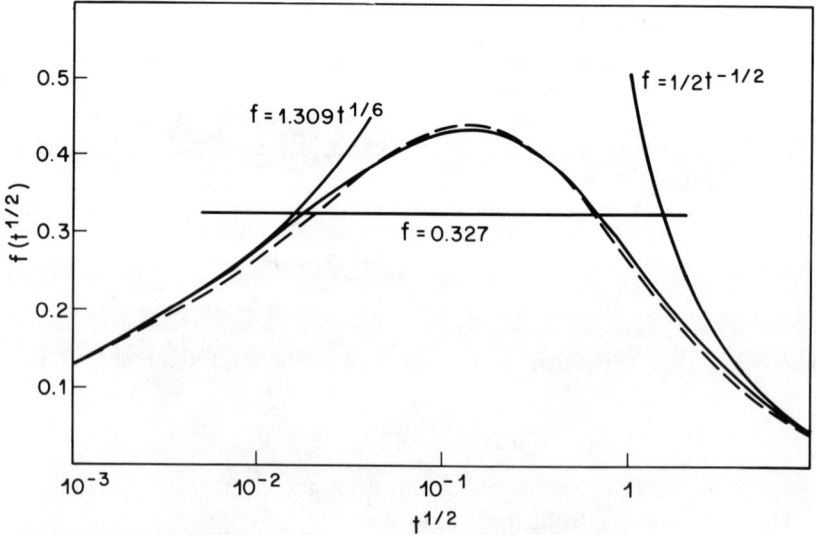

Fig. 14. Scattering function (Eq. 2.46) for a Thomas–Fermi potential (···). Full curves show different polynomial approximations. (From Winterbon et al.[55])

ion–target combinations. Figure 14 gives an indication of the accuracy of Eq. (2.45).

For normalized energies $\epsilon < 1.0$; however, the Thomas–Fermi potential overestimates the interaction at large interparticle separations (i.e., small t) and has been replaced by another set of constants[59] $\lambda = 2.54$, $m = 0.25$, and $q = 0.475$. These data, however, underestimate interactions at intermediate and high energies.

In contrast to the potentials of the Thomas–Fermi type, a realistic diatomic repulsive potential should drop to zero rather sharply for atomic separation greater than about 1 Å, thus reducing drastically atomic scattering at large impact parameters. Wilson et al.[60] have calculated interatomic potentials from first principles in the free-electron approximation for 14 diatomic interactions representing light particles incident on heavy targets ($M_1 \ll M_2$), self-irradiation ($M_1 = M_2$), and heavy particles on light targets ($M_1 \gg M_2$). The potentials were fitted to a Moliere-like form given by [see Eq. (2.39)]

$$\phi(x) = \sum_{i=1}^{3} C_i \exp\left(-\frac{b_i x}{a}\right), \quad (2.47)$$

where a is given by Eq. (2.42). The coefficients C_i are constrained by the condition

$$\sum_{i=1}^{3} C_i = 1$$

following Moliere.[61] Based on these results, a scattering function which

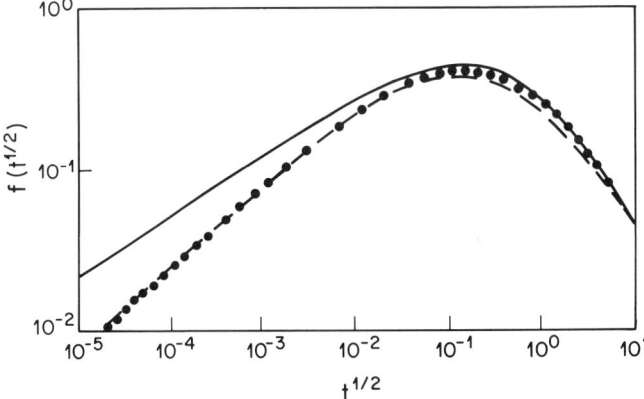

Fig. 15. Scattering function as a function of the reduced energy for the Thomas–Fermi potential (—), by Kalbitzer and Oetzmann[59] (---) and Wilson et al.[60] (···).

combines the favorable attributes of the Thomas–Fermi form at high energies and the Kalbitzer form at low energies has been obtained

$$f(t^{1/2}) = A \frac{X^{1+C} - 1 - (1 + C)\ln X}{X^{2+C} - 2X + X^{-C}}, \qquad (2.48)$$

where $X = Bt^{1/2}$, $A = 0.56258$, $B = 1.776$, and $C = 0.62680$. Equation (2.48) represents the actual results to within 10% error. This deviation is not considered serious, particularly when one considers the advantages of a closed-form expression. Figure 15 compares the various scattering functions. Littmark and Ziegler[62,63] have used a different fitting procedure to the exact data of Wilson et al.[60] They approximate the f function by a continuous power form

$$f(t^{1/2}) = \lambda_i t^{(1/2) - m_i}, \qquad (2.49)$$

where

$$m_i = 1 - \exp[-A - B(10 + \ln X_i)^C] \qquad (2.50)$$

$$\lambda_{i+1} = \lambda_i^{2(m_i - 1 - m_i)} \qquad (2.51)$$

with the coefficients

$A = 0.2089, \quad B = 3.235 \times 10^{-9}, \quad C = 8.58, \quad \lambda_1 = 7.575, \quad X_1 = 10^{-9}$.

By using this procedure, they have published stopping cross sections and range distributions for energetic ions in all elements.[63]

c. *Electronic Stopping*

Electronic collisions are basically different from nuclear collisions in the following aspects:

(1) The energy loss per collision is a function of the instantaneous velocity v_1 of the projectile rather than of its kinetic energy E_1

$$(-dE/dx)_{\text{elec}} \cong f(v_1) \qquad (2.52)$$

(2) The energy loss per collision is always very small in terms of the projectile's energy, since electrons have very small mass. Thus,

$$T \ll E_{\text{r}}, \qquad (2.53)$$

giving rise to a continuous slowing down similar to a friction-type stopping.

(3) The momentum transfer to electrons is small, resulting in very small angular deflections of the ion, hence yielding a more or less straight particle trajectory.

(4) Electronic losses are statistical in nature.

Section 4.b presents the theory of electronic stopping in more detail. For ion velocities $v > q^2 Z_1^{2/3}/\hbar;\ q^2 Z_2^{2/3}/\hbar$, the Bethe–Bloch[64] electronic stopping formulation is valid. It is commonly called the continuous slowing down approximation (CSDA) and can be written as

$$\left(\frac{dE}{dz}\right)_{\text{elec}}^{\text{high}} = \frac{4\pi Z_1^2 q^4 Z_2^2}{m_0 v^2} \ln\left(\frac{2m_0 v^2}{I}\right) \quad (2.54)$$

with the Bloch mean excitation energy I. The energy I depends on the valence number of the target and is accurately described by

$$I = \begin{cases} (12 + 7/Z_2)Z_2, & Z_2 < 13, \\ (9.76 + 58.5/Z_2^{1.19})Z_2, & Z_2 \geq 13. \end{cases} \quad (2.55)$$

For low projectile velocities the electron stopping power is given by the Lindhard–Scharff relation

$$(dE/dz)_{\text{elec}}^{\text{low}} = kE^p \quad (2.56a)$$

with

$$k = k_L = \frac{1.212 Z_1^{7/6} Z_2}{(Z_1^{2/3} + Z_2^{2/3})^{3/2} M_1^{1/2}} \quad (2.56b)$$

and $p \cong \frac{1}{2}$, which approximates well the Z_1 and Z_2 dependencies and can be adjusted to experimental values.

The total electronic cross section is modeled as

$$(dE/dz)_{\text{elec}} = \{[(dE/dz)_{\text{elec}}^{\text{low}}]^{-1} + [(dE/dz)_{\text{elec}}^{\text{high}}]^{-1}\}^{-1}. \quad (2.57)$$

Figure 16 compares the normalized nuclear and electronic energy loss as a function of reduced energy. The characteristic energies ϵ_1, ϵ_2, and ϵ_3 are strongly dependent on the parameters of the ion–target combination. Table III gives values for the important elements implanted into silicon.

For electronic stopping, a family of lines (one for each combination of projectile and target) is obtained. The majority of cases falls between the limits shown. The dot-dashed line represents the electronic stopping for $k = 0.15$. The horizontal line labeled S^0 represents the constant-stopping power approximation suggested by Nielsen.[65]

We see that nuclear stopping is more important at low energies, reaching a maximum around $\epsilon = 0.35$. Electronic stopping, however, increases linearly with velocity over a wide range, completely dominating at energies $\epsilon \gtrsim 3$. At even higher energies $(-d\epsilon/d\rho)_{\text{elec}}$ also passes through a maximum and then falls off as ϵ^{-1}.

For most of the heavier elements in silicon, nuclear stopping remains the dominant energy loss mechanism for ion energies up to several hundred kilo-electron-volts (i.e., $\epsilon \approx 3$). However, for the case of boron in silicon, we

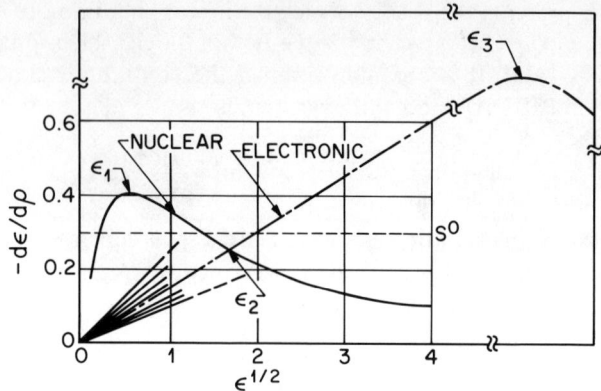

Fig. 16. Nuclear stopping power for Thomas–Fermi potential (solid line) and electronic stopping power (dash and dot line) for $k = 0.15$ in terms of the reduced variable ϵ and ρ, based on the LSS theory.[39] The family of curves for electronic stopping represents the majority of the usual projectile–target combinations. The values of the characteristic energies ϵ_1, ϵ_2, and ϵ_3 are given in Table III for various ions implanted in silicon. (From Mayer et al.[66])

find that the correction for electronic stopping becomes important at relatively low energies. At an energy of 10 keV ($\epsilon \approx 1.1$, $k \approx 0.22$), electronic stopping produces a 25% decrease in ρ as compared to the value calculated from nuclear stopping alone.[66]

The use of Thomas–Fermi statistical concepts in the original derivation of Eq. (2.56a) leads to an electronic stopping power that increases monotonically with increasing Z_1 (projectile) for a given value of Z_2 and ion velocity. However, experimental studies and theoretical calculations[67–71] using Hartree–Fock wave functions—rather than using a Thomas–Fermi electron distribution—have shown that electronic stopping has a marked periodic dependence on the atomic number of the incident ion. Figure 17

TABLE III

CHARACTERISTIC ENERGIES (keV) FOR DIFFERENT ELEMENTS IN Si[a]

Ion	ϵ_1	ϵ_2	ϵ_3
B	3	17	3×10^3
P	17	140	3×10^4
As	73	800	2×10^5
Sb	180	2000	6×10^5

[a] From Mayer et al.[66]

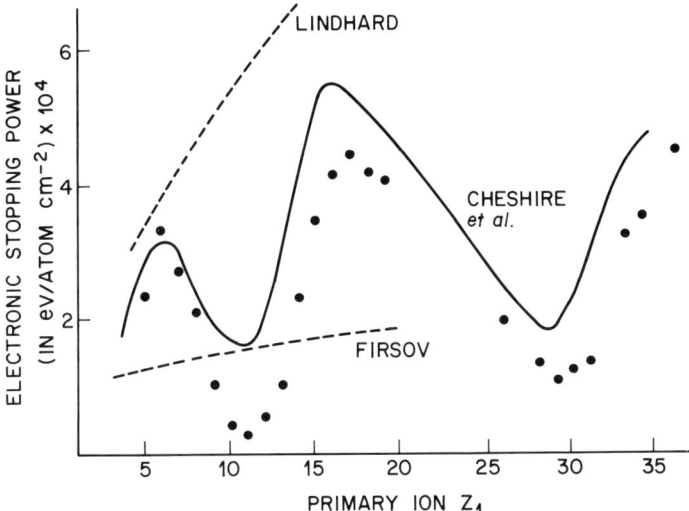

Fig. 17. Comparison of the measured electronic stopping power (from Eisen[72]) with theoretical results using Thomas–Fermi atomic models (from Firsov[44]) or Hartree–Fock calculations. (From Cheshire et al.[68] Copyright North-Holland Physics Publishing, Amsterdam, 1968.)

indicates the measured and predicted electronic stopping power for a range of Z_1.

Table IV presents some data[66] that illustrate the magnitude of this correction for various dopants in silicon, on the assumption that the Z_1 oscillations in silicon are comparable to those observed in carbon.

TABLE IV

Effect of Z_1 Oscillations in "Amorphous" Silicon

		$R_{corrected}/R_{LSS}$			
Ion	$(k^1/k)^a$	1 keV	10 keV	100 keV	1 MeV
^{11}B	1.20	0.98	0.95	0.89	0.85
	1.5^b	0.95	0.89	0.76	0.69
^{14}N	1.30	0.98	0.95	0.86	0.79
^{23}Na	0.84	0.98	0.95	0.86	0.79
^{27}Al	0.90	1.01	1.02	1.06	1.14
^{31}P	1.14	0.99	0.99	0.96	0.92
^{70}G	0.55	1.01	1.02	1.05	1.21
^{75}As	0.60	1.01	1.02	1.05	1.18

^{a} k^1/k is the ratio of the *observed* k-value in amorphous carbon targets to that predicted by LSS.

^{b} Taken from experimental data of Eisen.[72]

d. Classification of Theories

A classification of the important theories and calculational procedures is shown in Table V,[73] which also summarizes major developments in this field during the last 20 years.

The pioneering work of Lindhard and co-workers on the transport equation formalism has been refined by a large number of authors. Essentially all the available data on ion ranges in solids have been derived by the original LSS approach.

The procedure for solving the LSS equations[56,57] considers the changes in distribution probabilities after traveling a small distance δR. This procedure defines the integro-differential equation for the probability density function $P(R, E)$ as

$$\frac{\partial P(R, E)}{\partial R} = N \int_0^{T_m} [P(R, E - T, \cos\theta) - P(R, E)]\, d\sigma(T) - NS_l(E)\frac{\partial P(R, E)}{\partial E},$$

(2.58)

where E is the energy of the ion, N the target density, T the recoil energy in Eq. (2.34), θ the scattering angle, S_l the electronic energy cross section, and σ the energy-loss cross section.

A similar equation allows the determination of a probability density function of finding ions at any lateral displacement. Schiott[89] has solved

TABLE V

CLASSIFICATION OF THEORIES[a]

Method	Range					Damage	Reference
	R_p	ΔR_p	$\gamma, \beta \ldots$	ΔX	Multi-layer		
Transport theory							
First order	○	×	×	×	×	×	74
Second order	○	○	×	○	×	×	75,76
Third order	○	○	○	△	×	×	77
Integral	○	○	○	○	×	○	78–80
Equation	○	○	○	○	○	○	81–83
Intermediate methods							
Two-Step method	○	○	×	△	△	○	84,85
Semi-Monte Carlo method	○	○	×	×	○	○	86
Monte Carlo	○	○	○	○	○	○	87,88

[a] ○; calculated, △; possible, ×; impossible or inaccurate.

these equations by taking moments of the distributions and solving the recurrence relations

$$m\langle R^{m-1}(E)\rangle = N \int_0^{T_m} [\langle R^m(E)\rangle - \langle R^m(E - T, \cos\theta)\rangle] \, d\sigma$$

$$+ NS_l(E)\frac{\partial\langle R^m\rangle}{\partial E}, \quad (2.59)$$

where

$$\langle R^m(E)\rangle = \int_0^\infty dR \, R^m P(R, E). \quad (2.60)$$

The order of the moment is given by m. Both recurrence relations are solved by a fourth-order Runge–Kutta method.

Sigmund and Sanders[78] and Winterbon et al.[55] have presented an integro-differential equation analogous to Eq. (2.58) that governs the spatial distribution of energy deposited into atomic processes by ions moving through a solid. They have also introduced moments to the damage distribution. Because of the difficulty involved in including electronic stopping, their approach becomes impractical to obtain solutions for more than the first few moments. Range tables for a variety of different mass ratios have been published by Winterbon.[90]

A different approach has been taken by Brice,[91] who set up an equation based on the schedule of physical events occurring during the implant process. As a starting point, he determines the location of the incident ions within the target, and then from the interaction cross sections determines the amount of energy deposited at each location as the ions slow to a stop. Several approximations are involved, such as the assumption of a Gaussian distribution for the ion ranges. The effect of these approximations is small and decreases with increasing energy. This "direct" method is quite accurate at high energies, and reasonably accurate at lower energies.

At sufficiently high energies, energy transport by recoiling target atoms can be neglected and the damage energy distribution $Q(E, z)$ can be obtained from

$$Q(E, z) = N \int_0^E P(E, E', z) \left[\int_0^{T_m} q(T) \, d\sigma(E, T) \right] \frac{dR}{dE'} \, dE', \quad (2.61)$$

where $P(E, E', z)$ is the distribution of ions having energy E' at depth z and $q(T)$ is that portion of the recoil energy T that will ultimately be deposited into atomic processes (i.e., the damage energy of the recoiling target atom). The factor dR/dE' is a geometrical factor that takes into account the angular spreading of the penetrating ions. The basic physical concepts involved in writing Eq. (2.61) are illustrated in Fig. 18 for 100-keV boron ions incident on a silicon target. The narrow distributions, labeled with the ion energy E' in kilo-electron-volts, show $P(E, E', z)$ as a function of E' and z. The broad distribution in the figure is the depth distribution of deposited damage energy $Q(E, z)$. It is obtained by accumulating the energy deposited by the ions in the narrow distributions as they pass through a given depth. Brice has published a companion volume to Winterbon[90] that contains tables on primary ion ranges, damage ranges, etc., for a large variety of projectile–target combinations.[58]

Figure 19 shows contours of constant damage density for ^{11}B ions incident on an amorphous silicon target as function of incident energy E and depth z into the solid. A large collection of damage contours can be found in the original reference. Depth distribution profiles, such as the one in Fig. 19, can be constructed for any incident energy in this range by plotting the contour values as a function of the z values of the intercept of a horizontal line at the given energy with the contour.

All methods mentioned have the severe restriction that they require the assumption of a homogeneous target. Therefore, they cannot be used to simulate multilayered targets. Several attempts to study multilayered

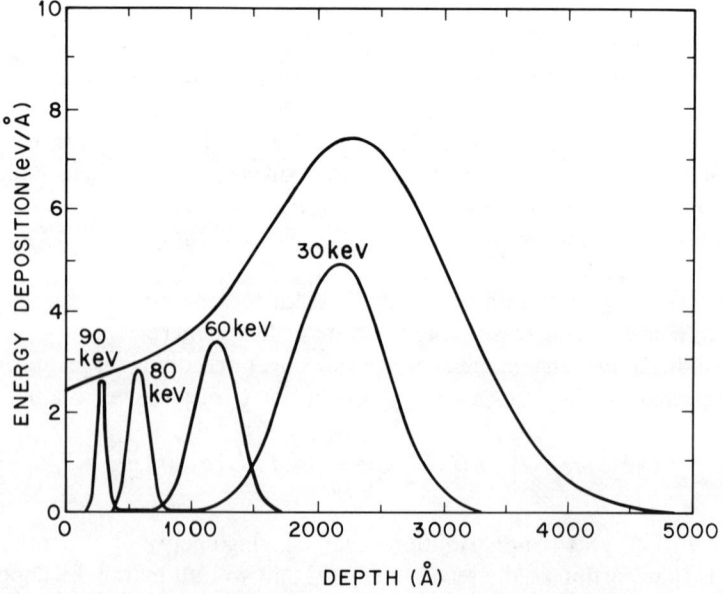

Fig. 18. Damage energy deposition distributions as a function of depth into target for 100-keV boron ions incident on silicon. Deposition rate distributions are shown after ion energy has been reduced to 90 keV, 80 keV, 60 keV, and 30 V in Gaussian approximations. Also shown is the final depth distribution of the deposited damage energy. (From Brice.[58])

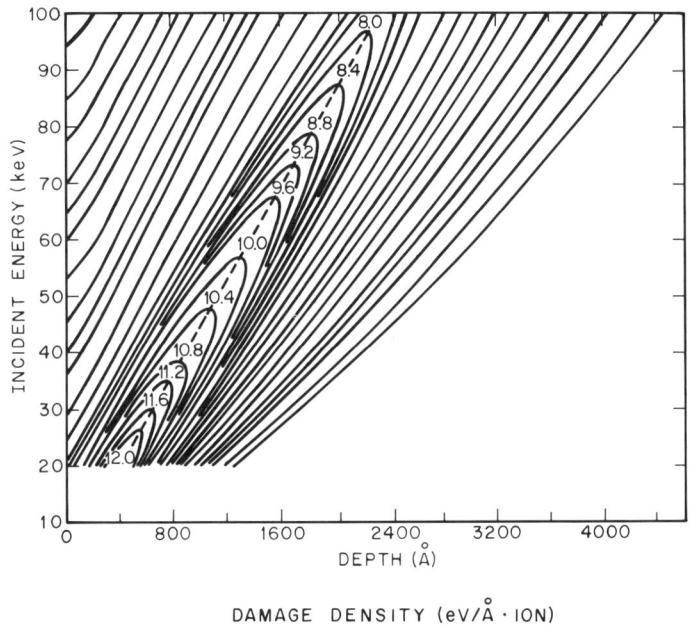
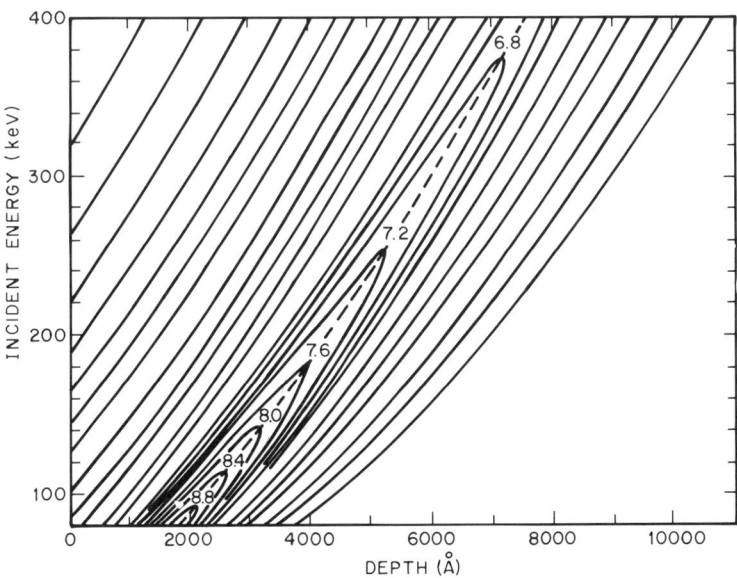

Fig. 19. Contours of constant damage density for ^{11}B ions incident on an amorphous silicon target. Incident energies are in the range 20–100 keV and 80–400 keV. (From Brice.[91])

structures have been reported, but they lack generality and/or are difficult to apply.

Only two methods have been developed so far that can be applied to range and damage analysis in solid arbitrary targets: the Boltzmann transport equation (BTE) method, pioneered by Smith and Gibbons[80] and successively refined by others,[81-83] and the Monte Carlo (MC) method.

Basically, both methods will give the same answer to the same problem. The major advantage of the MC model over the BTE approach lies in the fact that it is intrinsically a three-dimensional technique. In modern device processing, ions are implanted into finite areas—windows—of a wafer, which results in a lateral distribution of the ions under the mask edge. Although the BTE model could be generalized to more than one dimension, this has not yet been done.

A second advantage of the MC model arises in the case of implanting light ions into heavy substrates ($M_1/M_2 \ll 1$), such as in the case of ion-beam lithography. Many ions are backscattered toward the surface, which poses no problem in the MC model. In the BTE technique, however, these ions scatter back into regions where the solution is supposedly already known. Giles and Gibbons[83] have recently extended the Boltzman method to take this problem into account.

A third advantage arises in the simulation of ion implantation into crystalline media. No BTE results have been published accounting for lattice effects.

e. *Monte Carlo Calculations*

The basic idea behind Monte Carlo calculations is the simulation of the history of a projectile through its successive collisions with target atoms. The evaluation is based on the summation of these scattering events occurring in a large number N ($N > 1000$) of simulated particle trajectories within the target. By following N histories, distributions for the range parameters of primary and recoiled ions and the associated damage (given by the nuclear energy loss) can be obtained.

Each particle history begins with a given energy, position, and direction. The particle is assumed to change direction because of binary nuclear collisions and to move in straight paths between collisions. The energy of the particle is reduced as a result of nuclear [Eq. (2.34)] and electronic [Eq. (2.44)] energy losses. The ion will stop either when its energy drops below a certain value or when its position is outside the target (a reflected ion).

With the availability of high-speed digital computers, ion transport calculations based on the Monte Carlo method have been used by a variety of authors. Major differences between the various approaches have been in the treatment of elastic nuclear scattering, the representation of the target structure, and the formulation of the mean free path.

Three different models have been developed: (1) the local structure model,[87,92-98] (2) the dense gas approximation model,[99-104] and (3) the liquid structure model.[105] The major difference between the local structure and the liquid model is in the treatment of the nuclear collision. The liquid structure and the dense gas model differ in their choice of the mean free path formulation. In the gas model the mean free path is calculated using a random number, whereas in the liquid model it is approximated by the mean interatomic distance.

Oen, Robinson, and co-workers[92-97] treat the scattering process exactly by numerically evaluating the classical scattering integral in Eq. (2.36) for realistic atomic potentials by a four-point Gauss–Mehler quadrature. A comprehensive computer program called MARLOWE[98] is available that is based on this exact technique.

Other authors either invoke the momentum approximation extended to large angles[99,100] or use fitted truncated Coulomb potentials[101-104] to obtain analytical representations of the scattering integral.

Biersack and Haggmark[87] have developed an elegant analytical technique to evaluate Eq. (2.36). The method is applicable to a wide range of incident energies (0.1 keV to several MeV), depending on the masses involved. The lower limit is given by the binary collision approximation (see earlier), while the upper limit results from the neglect of relativistic effects.

Let us again consider the scattering problem in the center-of-mass (CM) system in Fig. 13b. Superimposed on the orbits of the two particles is the scattering triangle, determined by

$$\cos\frac{\theta}{2} = \frac{\rho + p + \delta}{\rho + r_0} \quad \text{with} \quad \begin{cases} \rho = \rho_1 + \rho_2, \\ \delta = \delta_1 + \delta_2. \end{cases} \quad (2.62)$$

The distance of closest approach r_0 is obtained from

$$1 - V(r_0)/E_r - (p/r_0)^2 = 0, \quad (2.63)$$

which can be solved by Newton's method in two to four iterations to an accuracy of better than 0.1%.

The radius of curvature ρ is obtained from the relation

$$\rho = \frac{2[E_r - V(r_0)]}{-V'(r_0)}, \quad (2.64)$$

where $V'(r_0)$ is the spatial derivative of the potential evaluated at r_0.

By expressing the various lengths in Eq. (2.62) in units of the screening length, Eq. (2.42), we obtain

$$\cos(\theta/2) = (B + R_c + \Delta)/(R_0 + R_c) \quad (2.65)$$

with the known normalized quantities $B = p/a$, $R_0 = r_0/a$, $R_c = \rho/a$ and the

TABLE VI

VALUES FOR CONSTANTS
IN EQ. (2.68) BASED ON
THE MOLIERE POTENTIAL

C_1	0.6743
C_2	0.009611
C_3	0.005175
C_4	10.00
C_5	6.314

unknown $\Delta = \delta/a$. The quantity Δ can be expressed as

$$\Delta = A(R_0 + B)/(1 + G) \tag{2.66}$$

with

$$A = 2\alpha\epsilon B^\beta \quad \text{and} \quad G = \alpha[(1 + A^2)^{1/2} - A]^{-1}, \tag{2.67}$$

where

$$\alpha = 1 + C_1 \epsilon^{-1/2}$$
$$\rho = (C_2 + \epsilon^{1/2})/(C_3 + \epsilon^{1/2}) \tag{2.68}$$
$$\alpha = (C_4 + \epsilon)/(C_5 + \epsilon)$$

and ϵ is the dimensionless reduced energy Eq. (2.40). The factors C_1 to C_5 are fitting parameters to be determined for the potential of interest. Table VI gives values for the five parameters for the Moliere potential obtained from least-squares fitting routines to exact values for $\sin^2 \theta/2$ in Eq. (2.34). A typical example showing the potential of Monte Carlo calculations is given in Fig. 20. Common elements (Sb, As, B, P) have been implanted at (a) 50 keV and (b) 100 keV into silicon. For all cases, 10^4 ion trajectories have been simulated. The correction factor in the expression for electronic stopping in Eq. (2.57), k/k_L, has been set to 1 for all elements except boron, where $k/k_L = 1.59$.

f. *BTE Calculations*

In range theory the calculation of the range distribution is regarded as a transport problem describing the motion of the ions during their slowing down to zero energy. A collection of a large number of particles with different velocities **v** and located at different points in space **x** can be described by an average number

$$dN = F(\mathbf{v}, \mathbf{x}) \, d^3\mathbf{v} \, d^3\mathbf{x} \tag{2.69}$$

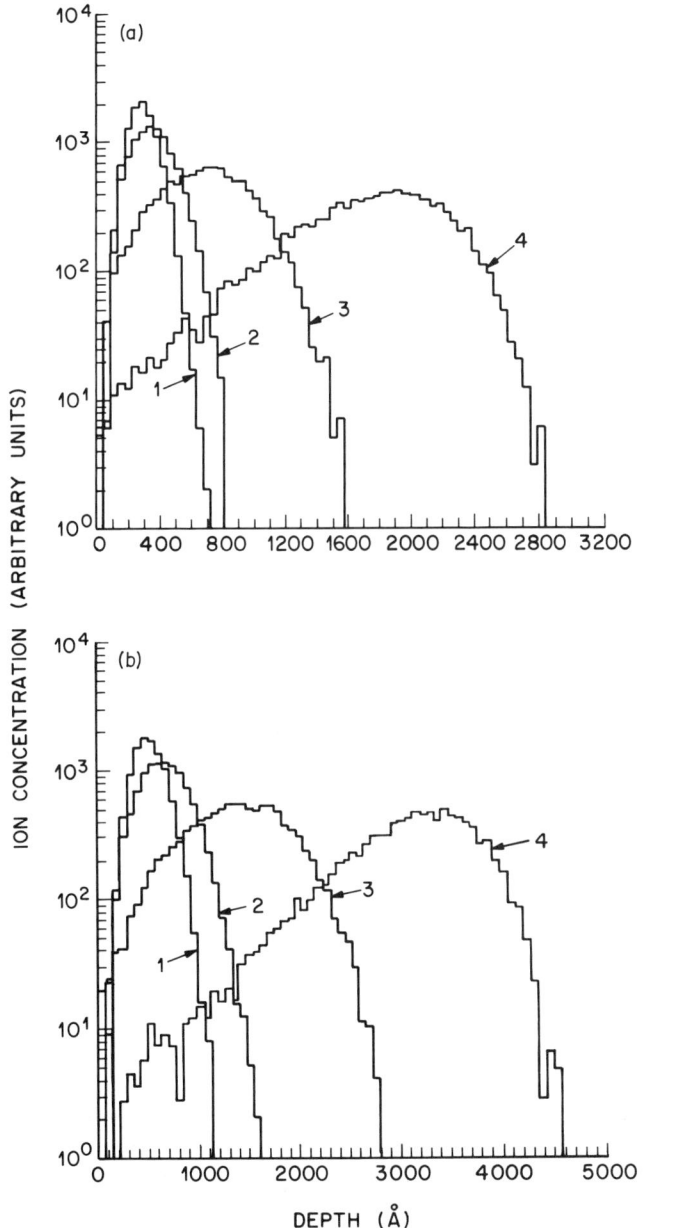

Fig. 20. Ion concentration as a function of depth for common elements implanted into silicon at (a) 50 and (b) 100 keV. 1, Sb; 2, As; 3, P; 4, B. All results are obtained simulating 10^4 ion trajectories.

for each differential element in phase space. There will be one such distribution for each projectile type.

The probability for a particle with velocity $\mathbf{v}$ to scatter into the interval $d^3\mathbf{v}'$ around $\mathbf{v}'$ during the time dt is given by

$$K(\mathbf{v} \to \mathbf{v}') d^3\mathbf{v}' dt = N_{scat} |\mathbf{v}| d\sigma(\mathbf{v} \to \mathbf{v}') dt, \quad (2.70)$$

where $K(\mathbf{v} \to \mathbf{v}')$ is the transition rate, N_{scat} is the density of scattering centers, and $d\sigma(\mathbf{v} \to \mathbf{v}')$ is the differential cross section.

Analogous to the kinetic theory of gases,[106] the average number of particles scattered into and out of a differential phase space element can be described by a Boltzmann transport equation for the distribution function $F(\mathbf{v}, \mathbf{x})$

$$\frac{\partial F(\mathbf{v}, \mathbf{x})}{\partial t} + \mathbf{v} \cdot \nabla F(\mathbf{v}, \mathbf{x}) = N_2 \int [d\sigma(\mathbf{v}' \to \mathbf{v}) |\mathbf{v}'| F(\mathbf{v}', \mathbf{x}')$$
$$- d\sigma(\mathbf{v} \to \mathbf{v}') |\mathbf{v}| F(\mathbf{v}, \mathbf{x})] + Q(\mathbf{v}, \mathbf{x}). \quad (2.71)$$

The quantity Q is a generation term that allows particles to be created from rest. If more than one projectile type is involved, subscripts should be added in Eq. (2.71).

Integration of this equation is carried out starting from $z = 0$ (the sample surface) and integrated for $z > 0$ with the initial condition

$$F(\mathbf{v}, 0) = N_D \delta(\mathbf{v} - \mathbf{v}_0), \quad (2.72)$$

where N_D is the total dose and $\mathbf{v}_0$ is the initial velocity of the incident beam.

The integration of Eq. (2.71) requires that the motion of each particle in the distribution be confined to a finite number of discrete states. Each state is defined by an energy E_i ($0 \le E_i \le E_0$) and an angle θ_j. In the work of Christel et al.,[81] θ_j has been limited to a range between 0 and $\pi/2$. Although this restriction removes backscattered particles from the final distribution, the results indicate no serious errors caused by this approximation, at least for the cases considered in Christel et al.[81] This approximation is expected to fail in the case of very light projectiles impinging on heavy targets ($M_1/M_2 \ll 1$). To keep computation times reasonable, 150 discrete elements (15 equally spaced energy states and 10 angular intervals) were used. The step size Δz was set to 1 Å. Giles and Gibbons[83] have developed a multipass algorithm where the region of interest is scanned iteratively until all particles have come to rest. At the end of the first pass, they obtain a concentration profile together with a set of matrices of backscattered ions. The second pass, however, is made from the target interior toward the surface, thus accounting for the motion of all ions backscattered in the original pass. Adding the stored backscatter distribution to the original profile, one proceeds until all ions have stopped.

In Fig. 21, BTE calculations using both the Wilson and Kalbitzer cross sections are compared with a Pearson IV distribution [see Eq. (2.95)] generated from LSS moments. The Kalbitzer cross section has been used in the LSS calculation. A dose of $10^{16}/\text{cm}^2$ phosphorus has been implanted at $E = 160$ keV. The experimental points are taken from Hirao et al.[107] For this condition, $\epsilon > \epsilon_2$, and nuclear and electronic mechanisms contribute to the total stopping. It is seen that the Wilson cross section gives better results than the Kalbitzer cross section and that the LSS result is too skewed.

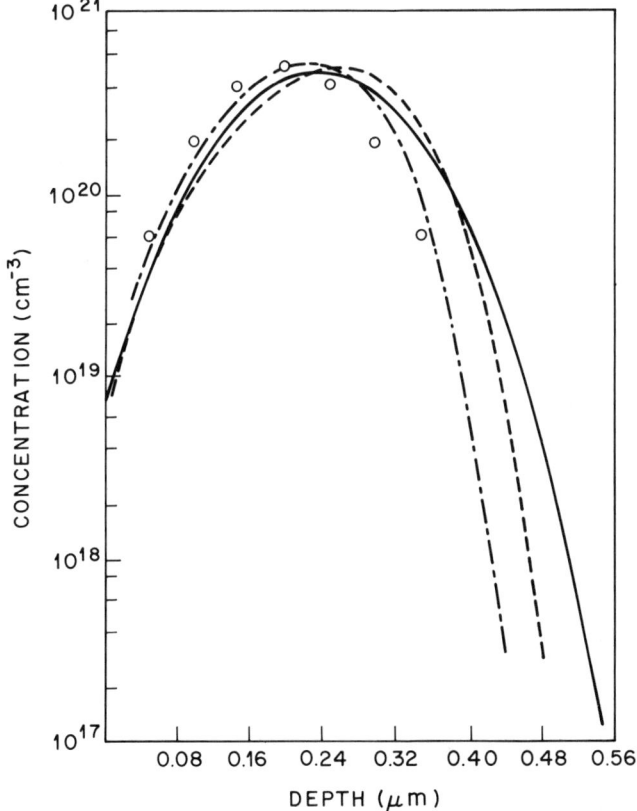

Fig. 21. Comparison of LSS and transport equation calculations and experimental results for the range profile of 160-keV phosphorus implanted into silicon to a dose of $10^{16}/\text{cm}^2$. The experimental data (○) are from Hirao et al.[107] —, LSS with Kalbitzer cross section; ---, BTE with Kalbitzer cross section; and –--–, BTE with Wilson cross section. (From Christel et al.[81])

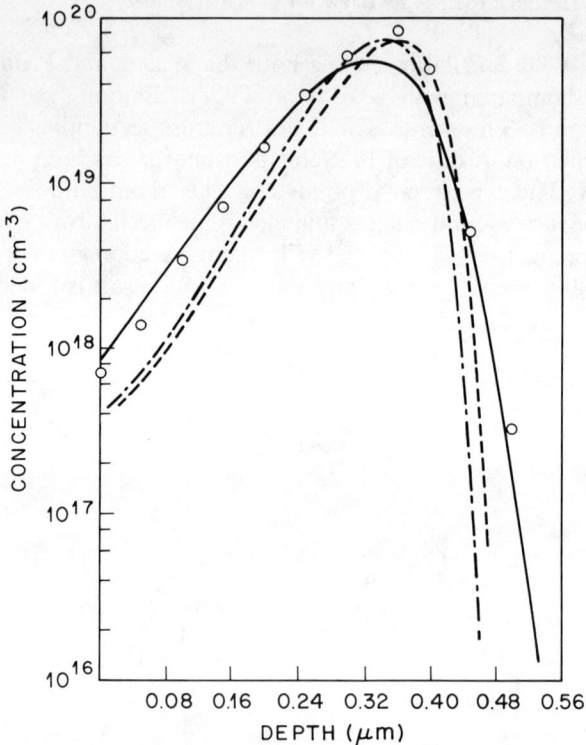

Fig. 22. Comparison of LSS and transport equation calculations and experimental results for the range profile of 100-keV boron implanted into silicon at a dose of $10^{15}/cm^2$. (—) LSS with Kalbitzer cross section, (--) transport equation with Kalbitzer cross section, (-·-·) same with Wilson cross section. The experimental points are from Hofker et al.[108] (From Christel et al.[81])

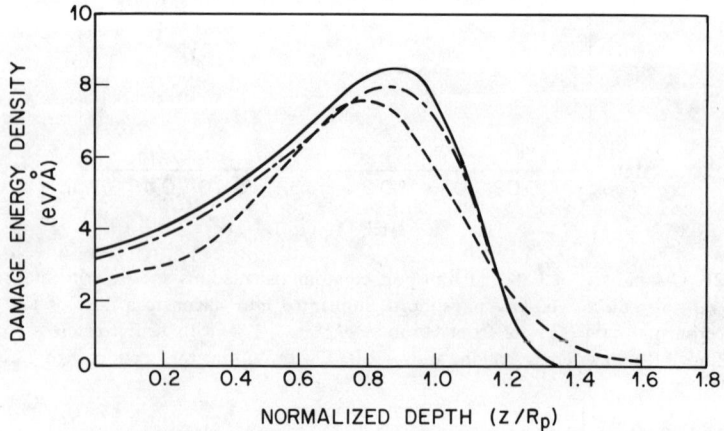

Fig. 23. As-deposited energy deposition profiles for 100-keV boron into silicon comparing the Brice and transport equation calculations. The abscissa is normalized to the projected range of the boron and the ordinate is the energy density per incident particle. —, Transport equation; —·—, transport equation with 30-eV threshold; ---, Brice. (From Christel et al.[81])

A similar result for boron is shown in Fig. 22. A dose of $10^{15}/cm^2$ is implanted at 100 keV into silicon. In this case, electronic stopping is dominant, which is reflected in the small differences in the results for the different cross sections. The experimental results are taken from Hofker et al.[108]

BTE calculations have been especially successful in calculating damage density and recoil range distributions. Only energy "lost" into nuclear processes is assumed to contribute to lattice disorder.[91] Figure 23 compares BTE results with transport equation results obtained by Brice.[109] The figure also shows the damage distribution by neglecting these events that transfer less than 30 keV of energy, accounting for the fact that a minimum amount of energy is required to remove an atom from its lattice position.

Kinchin and Pease[110] have published an expression for the number of atoms $N_d(E)$ that become displaced when the energy E is deposited into

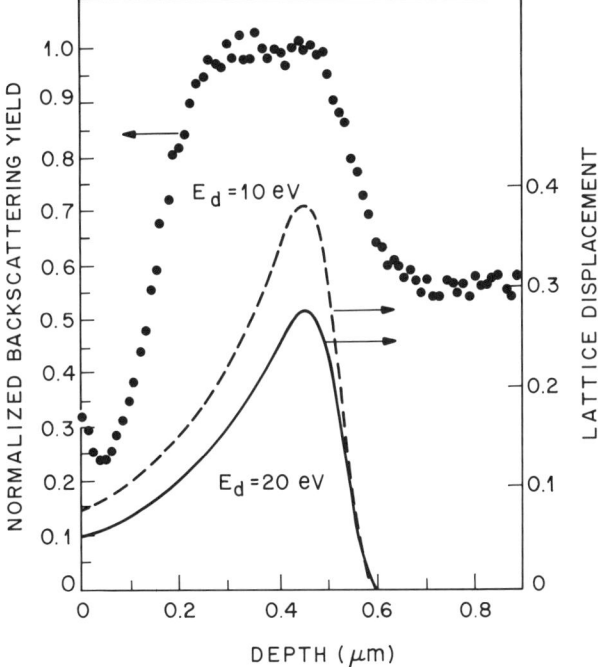

Fig. 24. Comparison of the experimental RBS results of North and Gibson[112] and a TE calculation[81] that shows the correlation between the experimentally observed edges of the buried amorphous layer and the calculated fractional displacement of the silicon lattice for an implantation of $3.6 \times 10^{15}/cm^2$ 150-keV B into silicon held at liquid nitrogen temperature. (From Christel et al.[81])

atomic processes

$$N_d = E/2E_d. \qquad (2.73)$$

A more detailed treatment by Sigmund[111] modifies Eq. (2.73) by an additional factor of 0.8. BTE calculations allow an accurate determination of the fraction of the lattice that is displaced at a certain depth during ion implantation. Figure 24 presents experimental disorder results of North and Gibson[112] obtained by channeling and backscattering of 2-MeV helium particles, resulting from a 150-keV $3 \times 10^{15}/\text{cm}^2$ boron implant into silicon at liquid nitrogen temperature. The theoretical results in Fig. 24 are obtained from a BTE calculation assuming displacement energies E_d of 10 and 20 keV. The fractional displacement at both edges of the amorphous layer is about 8.5% for $E_d = 20$ eV and about 12.5% for $E_d = 10$ eV, which suggests that a 10% lattice displacement makes the silicon substrate amorphous. For multi-

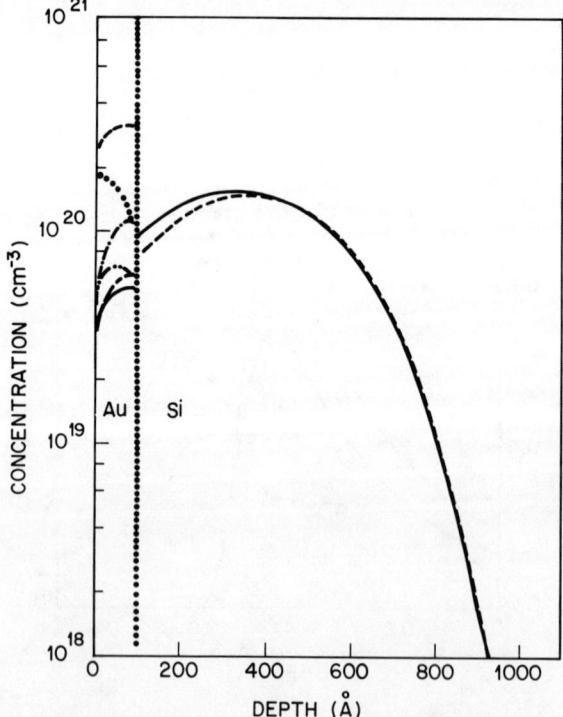

Fig. 25. Implanted profile for a 10-keV ^{11}B profile into 100 Å of Au on Si at a dose of $10^{15}/\text{cm}^2$ with the iteration pass as parameter. ----, One pass, 26% stopped; ······, two passes, 56% stopped; —·—·—, three passes, 71% stopped; —··—, four passes, 77% stopped; —·—, five passes, 79% stopped; —, six passes, 80% stopped. (From Giles and Gibbons.[83])

layered targets, BTE calculations have been very successful in the simulation of recoil effects. Giles and Gibbons[83] have modeled a low-energy ^{11}B implant ($E = 15$ keV) through a 100-Å layer of Au into Si. Figure 25 presents ^{11}B profiles in the Au–Si target as obtained by the multipass algorithm. While about 20% of the original dose are lost due to ^{11}B backscattering in the Au layer, the dose in the Si is slightly increased from 71 to 75% due to ^{11}B ions that were initially backscattered in the silicon and then returned because of backscattering at the Au–Si interface.

g. Range Distributions and Profile Construction

In the classical LSS theory of range distributions of low-energy ions in solids, a Gaussian form is assumed, determined by the parameters R_p, the projected range, and ΔR_p the standard deviation in the projected range

$$C(z) = \frac{C_\square}{\sqrt{2\pi}\,\Delta R_p} \exp\left[-\frac{(z - R_p)^2}{2\,\Delta R_p^2}\right], \qquad (2.74)$$

where $C_\square$ is the fluence, or ion dose per square centimeter. For this approximation, the peak impurity density will be

$$C_{\max} = \frac{C_\square}{\sqrt{2\pi}\,\Delta R_p} \simeq \frac{0.4 C_\square}{\Delta R_p}. \qquad (2.75)$$

Table VII is frequently useful to construct Gaussian profiles and to estimate the junction depth.[56]

For most cases of practical interest in silicon device fabrication (e.g., boron or arsenic in Si or SiO_2), experimentally determined profiles show considerable asymmetry.

We next summarize the most important approaches describing ion-implanted profiles.

(i) *Joined Half-Gaussian Distribution.* For cases of not too high asymmetry in the profile, the addition of a third moment can provide sufficient information to construct an accurate distribution.

TABLE VII

ORDINATES FOR THE GAUSSIAN DISTRIBUTION

$\left|\dfrac{z - R_p}{\Delta R_p}\right|$	0	1	2	2.15	3.03	3.72	4.29	4.80	5.26	5.68
$\dfrac{C(z)}{C_{\max}}$	1.0	0.606	0.135	10^{-1}	10^{-2}	10^{-3}	10^{-4}	10^{-5}	10^{-6}	10^{-7}

The third moment is a measure of the asymmetry and is defined by the relation

$$\gamma_1 = \left[\int_{-\infty}^{+\infty} (z - R_p)^3 f(z)\,dz\right] \bigg/ \Delta R_p^3 \quad (2.76)$$

with the normalized distribution function $f(z)$.

For negative values of γ_1 the profile has a larger slope deeper in the target. This is typical for conditions $M_1/M_2 < 1$, such as boron implantations in silicon. A positive value of γ_1 results in profiles that are skewed toward higher depths.

Gibbons et al.[56] have published a computational procedure to construct joined half-Gaussian profiles from the first three moments.

If the two profiles are joined at the model range R_M, the distribution is given by

$$f(z) = \frac{2}{(\Delta R_{p_1} + \Delta R_{p_2})\sqrt{2\pi}} \exp\left[-\frac{(z - R_M)^2}{2\Delta R_{p_1}^2}\right], \quad x \geq R_M, \quad (2.77a)$$

$$f(z) = \frac{2}{(\Delta R_{p_1} + \Delta R_{p_2})\sqrt{2\pi}} \exp\left[-\frac{(z - R_M)^2}{2\Delta R_{p_2}^2}\right], \quad x \leq R_M. \quad (2.77b)$$

As long as the third moment is less than the standard deviation, R_M, ΔR_{p_1} and ΔR_{p_2} are calculated from

$$R_M = R_p - 0.08(\Delta R_{p_2} - \Delta R_{p_1}) \quad (2.78)$$

$$\Delta R_p^2 = -0.64(\Delta R_{p_2} - \Delta R_{p_1})^2 - (\Delta R_{p_1}^2 - \Delta R_{p_1}\Delta R_{p_2} - \Delta R_{p_2}^2) \quad (2.79a)$$

$$CM_{3p} = (\Delta R_{p_2} - \Delta R_{p_1})(0.224\,\Delta R_{p_1}^2 - 0.352\,\Delta R_{p_1}\Delta R_{p_2} - 0.224\,\Delta R_{p_2}^2) \quad (2.79b)$$

Defining the variables

$$\Delta R_M = (\Delta R_{p_1} - \Delta R_{p_2})/2, \quad (2.80)$$

$$\Delta = (\Delta R_{p_1} - \Delta R_{p_2})/2, \quad (2.81)$$

we can solve Eqs. (2.78) and (2.79)

$$1 = \left(\frac{\Delta R_M}{\Delta R_p}\right)^2 + 0.44\left(\frac{\Delta}{\Delta R_p}\right)^2 \quad (2.82)$$

$$\frac{CM_{3p}}{\Delta R_p^3} = \frac{\Delta}{\Delta R_p}\left[0.8 - 0.256\left(\frac{\Delta}{\Delta R_p}\right)^2\right] \quad (2.83)$$

with the third central moment CM_{3p}.

The right-hand side of Eq. (2.83) has been tabulated.[56] Using this equation,

TABLE VIII
STANDARD DEVIATION FOR THE JOINED HALF-GAUSSIAN DISTRIBUTION

Third moment ratio	0.01	0.1	0.2	0.3	0.4	0.5	0.6	0.7	0.8	0.9
$\dfrac{\Delta R_{p_1}}{\Delta R_p}$	1.0	1.062	1.182	1.241	1.301	1.360	1.422	1.486	1.554	1.633
$\dfrac{\Delta R_{p_2}}{\Delta R_p}$	1.0	0.936	0.871	0.802	0.729	0.654	0.570	0.478	0.374	0.248

we compute $\Delta/\Delta R_p$ and solve Eq. (2.82). ΔR_{p_1} and ΔR_{p_2} are calculated from ΔR_M and Δ through Eqs. (2.80) and (2.81), and R_M is determined by Eq. (2.78).

The computation of $\Delta R_{p_1}/\Delta R_p$ and $\Delta R_{p_2}/\Delta R_p$ can be performed using linear interpolation and the values in Table VIII.

(ii) *Gram–Charlie Distribution.* A further method to construct distribution functions with higher moments uses the Gram–Charlie series type A, which essentially consists of a series expansion of the Gaussian function with Chebycheff–Hermite polynomials

$$f(z) = \sum_{j=0}^{\infty} C_j H_j(z) \exp\left(-\frac{z^2}{2}\right) \tag{2.84}$$

with

$$C_j = \int_{-\infty}^{+\infty} f(z) H_j(z)\, dz. \tag{2.85}$$

The polynomials H_j are defined by the derivatives of the Gaussian function

$$\left(-\frac{d}{dz}\right)^j \exp\left(-\frac{z^2}{2}\right) = H_j \exp\left(-\frac{z^2}{2}\right). \tag{2.86}$$

Substituting Eq. (2.86) in (2.85) results in an expansion for C_j. If one includes only terms up to the fourth order in this expansion, one obtains the Edgeworth distribution. The total concentration profile is then given by the product

$$C(z) = G(z)P(z),$$

where $G(z)$ is the Gaussian function and $P(z)$ is the Edgeworth distribution

$$P(z) = 1 + \frac{\gamma_1}{6}(z^3 - 3z) + \frac{\beta_2 - 3}{24}(z^4 - 6z^2 + 3)$$
$$+ \frac{\gamma_1^2}{72}(z^6 - 15z^4 + 45z^2 - 15). \tag{2.87}$$

Calculations and experimental values for the skewness γ_1 and the kurtosis β_2 have been obtained for a variety of cases over a limited energy range by Winterbon[79] and Hofker et al.[108]

Values for the skewness γ_1 have been tabulated in Gibbons et al.[56] for a large variety of ion–target combinations. Since β_2 is not known for most cases, the relation $\beta_2 = 3 + 5\gamma_1^2/3$ has been chosen that causes $P(z)$ to have the behavior $P(0) = 1$.

For this choice the total profile is given by

$$C(z) = \frac{C_\square}{\sqrt{2\pi}\,\Delta R_p} \exp\left(-\frac{z^2}{2}\right)\left[1 + \frac{\gamma_1}{6}(z^3 - 3z) + \frac{5\gamma_1^2}{72}(z^4 - 6z^2 - 3)\right.$$
$$\left. + \frac{\gamma_1^2}{72}(z^6 - 15z^4 - 45z^2 - 15)\right]. \tag{2.88}$$

(iii) *Pearson Distribution*. The Pearson system of univariate distributions[113] can be classified as solutions to the equation

$$(b_0 + b_1 x + b_2 x^2)(df/dx) = (x - a)f, \tag{2.89}$$

where $f(x)$ is the frequency function and $x = z - \bar{z}$ is the distance from the mean $\bar{z}$. The coefficients b_i, $i = 0, 1, 2$, can be expressed in terms of the moments $\mu_k = \langle x_k \rangle = \int x^k f\, dx$, $k = 1, \ldots, 4$. The four constants a, b_i can be represented by the first four moments

$$a = -\gamma_1 \Delta R_p(\beta_2 + 3)/A, \tag{2.90}$$

$$b_0 = -(\Delta R_p)^2(4\beta_2 - 3\gamma_1^2)/A, \tag{2.91}$$

$$b_1 = a, \tag{}$$

$$b_2 = -(2\beta_2 - 3\gamma_1^2 - \Delta R_p)/A, \tag{2.92}$$

$$A = 10\beta_2 - 12\gamma_1^2 - 18. \tag{2.93}$$

Different distributions are classified according to the behavior of the roots of

$$b_0 + b_1 x + b_2 x^2 = 0. \tag{2.94}$$

Table IX summarizes the different Pearson distributions relevant to a phenomenological description of ion implantation profiles. In this table, $f(x)$ is the solution to Eq. (2.89) and k is a normalization factor imposed by the requirement that

$$\int_{a_0}^{a_1} f(x)\, dx = 1, \tag{2.95}$$

TABLE IX

Pearson Distributions for Ion Ranges[a]

Type I (incomplete β-function of first kind)

$$f(x) = k(x - x_-)^{m_1}(x_+ - x)^{m_2}, \quad x_\pm = \frac{-b \pm \sqrt{d}}{2b_2},$$

$$m_{1,2} = \frac{1}{2b_2} = \frac{b_1(1 + 2b_2)}{2b_2\sqrt{d}} \quad (x_- \leq x \leq x_+, \ m_1 > 0, \ m_2 > 0)$$

Type III (incomplete Γ-function)

$$f(x) = k \exp\left(\frac{-x}{b_1}\right)(x_+ - x)^m, \quad x_+ = -b_0/b_1,$$

$$m = 1 - b_0/b_1^2 \quad (b_1 < 0, \ -\infty \leq x \leq x_+, \ x_+ > 0, \ m > 0)$$

Type IV

$$f(x) = k|b_0 + b_1 x - b_2 x^2|^{1/2b_2} \exp\left\{-\frac{(b_1/b_2) + 2b_1}{\sqrt{-d}} \tan^{-1}\left[\frac{2b_2 x + b_1}{\sqrt{-d}}\right]\right\}$$

$$\left(d < 0, \ \frac{1}{2b_2} < -\frac{5}{2}, \ -\infty \leq x \leq +\infty\right)$$

Type VI (incomplete β-function of second kind)

$$f(x) = k(x_- - x)^{m_1}(x_+ - x)^{m_2}, \quad x_\pm = \frac{-b_1 \pm \sqrt{d}}{2b_2},$$

$$m_{1,2} = \frac{1}{2b_2} \pm \frac{b_1(1 - 2b_2)}{2b_2\sqrt{d}} \quad (-\infty \leq x \leq x_+, \ m_1 > 0, \ m_1 + m_2 < -1)$$

[a] k = normalization constant, $d = b_1^2 - 4b_0 b_2$.

where

$$a_0 = \max(0, x_-), \quad a_1 = x_+ \text{ (Type I)},$$
$$a_0 = 0, \quad a_1 = x_+ \text{ (Type III)},$$
$$a_0 = 0, \quad a_1 = \infty \text{ (Type IV)},$$
$$a_0 = 0, \quad a_1 = x_- \text{ (Type VI)}.$$

This table also includes important conditions for various parameters [expressions in parentheses (.,.,.)] that have to be fulfilled to produce profiles with physical meaning. Using the results of a Monte Carlo simulation,[114] Petersen et al.[115] minimized the least-squares function

$$\phi = \sum_{i=1}^{N} \left(\Delta z f(z_i) - \frac{d_i}{N} \right)^2. \quad (2.96)$$

In Eq. (2.96), d_i is the number of implanted ions in histogram bin i at location z_i, $N = \sum_i d_i$ is the total number of nonreflected particles, Δz is the bin width, and N is the total number of bins. By using the basic solutions in Table IX to fit the Monte Carlo results, the most appropriate distribution type for a particular ion target energy combination can be obtained via the physical constraints.

Figure 26 is a plot of $\kappa = b_1^2/4b_0 b_2$ as a function of energy for $^{11}\text{B} \rightarrow \text{Si}$. The parameter b_2 has a zero around $E = 100$ keV, forcing a singularity in κ. For energy values below this singularity, the boron profile apparently is better fitted by a Type I Pearson distribution, while for large energy values it is better fitted by a Pearson VI distribution. These high-energy data are equivalent to Winterbon's result[116] and strongly support his findings.

Corresponding results for ^{31}P, ^{75}As, and ^{121}Sb are given in Fig. 27. For each element the parameter κ is outside $(0, 1)$ and the roots of Eq. (2.94) are

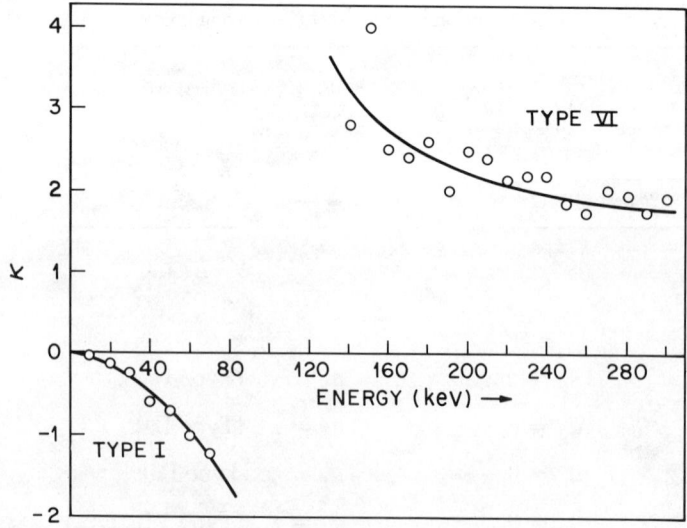

Fig. 26. $\kappa = b_1^2/4b_0 b_2$ as a function of energy for $^{11}\text{B} \rightarrow \text{Si}$. (From Petersen et al.[115])

real. For all cases shown, a Type I distribution would be adequate to fit these ion ranges.

Figure 28 supports these findings with typical results for two elements implanted into silicon. Curve a is the simulated Monte Carlo profile and the corresponding Pearson Type I fit for ^{121}Sb implanted into silicon at 100 keV. Curves b and c are similar data for boron, with (b) fitted Type I at 50 keV and (c) Type III at 100 keV.

(iv) *Two-Dimensional Profile Construction.* Up to now we have only considered one-dimensional distributions of ions. In device processing, however, it is quite common to implant into finite areas, which results in lateral doping distributions determined by the shape of the mask edge and the scattering of the ions during the slowing-down process.

The first calculations of lateral implantation profiles were presented by Furukawa et al.[76] They have assumed an infinitely steep and infinitely high mask edge and found that the lateral spread of the implanted ions can lead to a considerable extension of the doping profile under the mask edge.

The original work has been extended by Runge[117] allowing arbitrarily shaped mask edges to account for real processing conditions.

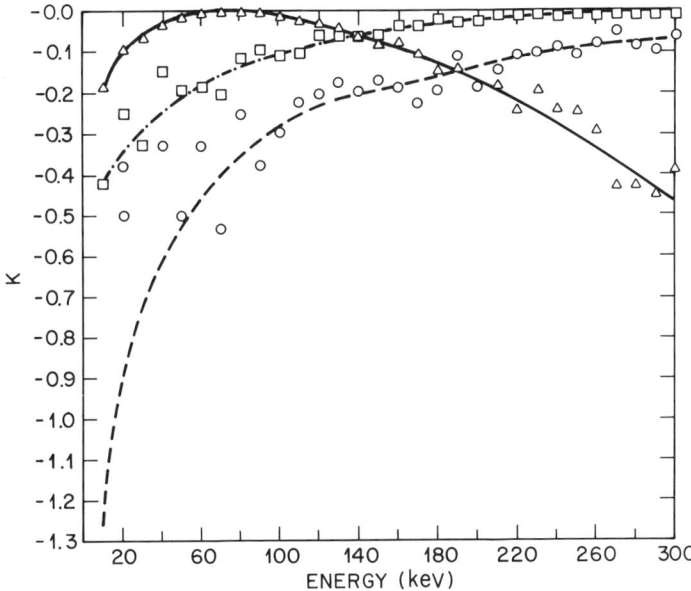

Fig. 27. $\kappa = b_1^2/4b_0b_2$ as a function of energy for ^{31}P ($\triangle$), ^{75}As ($\square$), and ^{121}Sb ($\bigcirc$) onto Si. $\triangle$, ^{31}P → Si; $\square$, ^{75}As → Si; $\bigcirc$, ^{121}Sb → Si. (From Petersen et al.[115])

The following assumptions are assumed to be valid (see Fig. 29).

(1) Ions entering a target at the points $(0, 0, 0)$ will come to rest at (x, y, z) with a Gaussian probability function

$$f(x, y, z) = \frac{1}{(2\pi)^{2/3} \Delta R_p \Delta X \Delta Y} \exp\left(-\frac{x^2}{2\Delta X^2} - \frac{y^2}{2\Delta Y^2} - \frac{(z - R_p)^2}{\Delta R_p^2}\right),$$
(2.97)

where ΔX and ΔY are the lateral standard deviations. In amorphous semiconductors, $\Delta X = \Delta Y$.

(2) The stopping power of the mask material is equal to the stopping power of the semiconductor. This assumption holds very well for the Si–SiO$_2$ system.

(3) Diffusion and channeling effects are ignored.

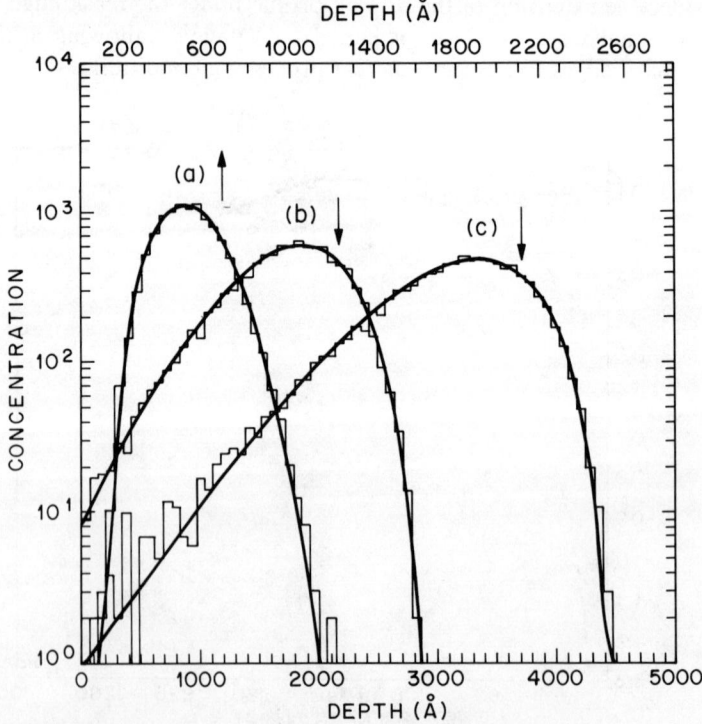

Fig. 28. Curve (a) is the simulated Monte Carlo profile and the corresponding Pearson Type I fit for ^{121}Sb implanted into silicon at 100 keV. Curves (b) and (c) are similar data for boron, with fitted Type I at 50 keV (b) and Type III at 100 keV (c). (From Petersen et al.[115])

PHYSICS OF VLSI PROCESSING AND PROCESS SIMULATION

The number of ions passing through the surface at (ξ, η, ζ) can be written as

$$C(\xi, \eta, \zeta) = C_{\square} \delta[\zeta - d_{ox}(\xi)], \tag{2.98}$$

where $C_{\square}$ is the dose and $d_{ox}(\xi)$ describes the shape of the mask edge.

The spatial distribution of all implanted ions, $C(x, y, z)$, is obtained by integration over all points of entry (ξ, η, ζ)

$$C(x, y, z) = \int_{-\infty}^{+\infty} \int_{-\infty}^{+\infty} \int_{-\infty}^{+\infty} C(\xi, \eta, \zeta) f(x - \xi, y - \eta, z - \zeta) \, d\xi \, d\eta \, d\zeta. \tag{2.99}$$

Substituting Eqs. (2.97) and (2.98) and integrating gives

$$C(x, y, z) = \frac{C_{max}}{\sqrt{2\pi}\,\Delta X} \exp\left(-\frac{(z - R_p)^2}{2 \Delta R_p^2}\right) \frac{1}{\sqrt{\pi}} \int_{\frac{x+a}{\sqrt{2}\Delta x}}^{\frac{x+a}{\sqrt{2}\Delta x}} \exp(-t^2) \, dt. \tag{2.100}$$

Three typical examples of the shape of an SiO_2 mask are given in Fig. 30: (a) the infinitely steep mask, (b) a 45° mask, and (c) a 60° mask edge. It is quite obvious that the lateral spread must not be neglected even in the case of a gentle slope of the mask edge. As a rule of thumb, one can assume that the implantation depth is roughly equal to the lateral extension of the implanted ions under the mask edge. Therefore, for the same doping depth the lateral

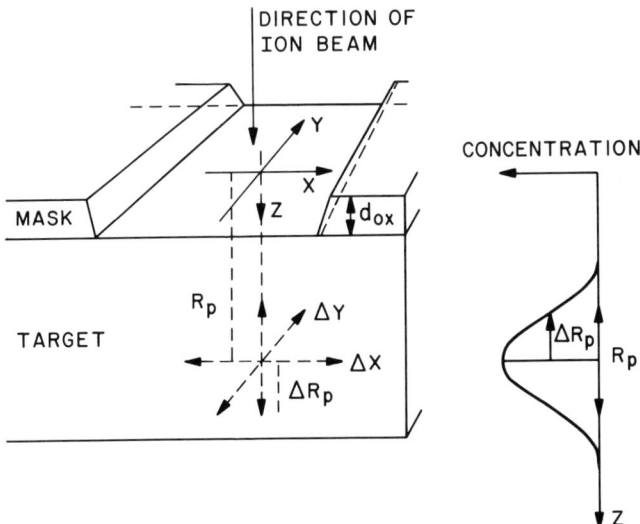

Fig. 29. Coordinate system and schematic geometry. (From Runge.[117])

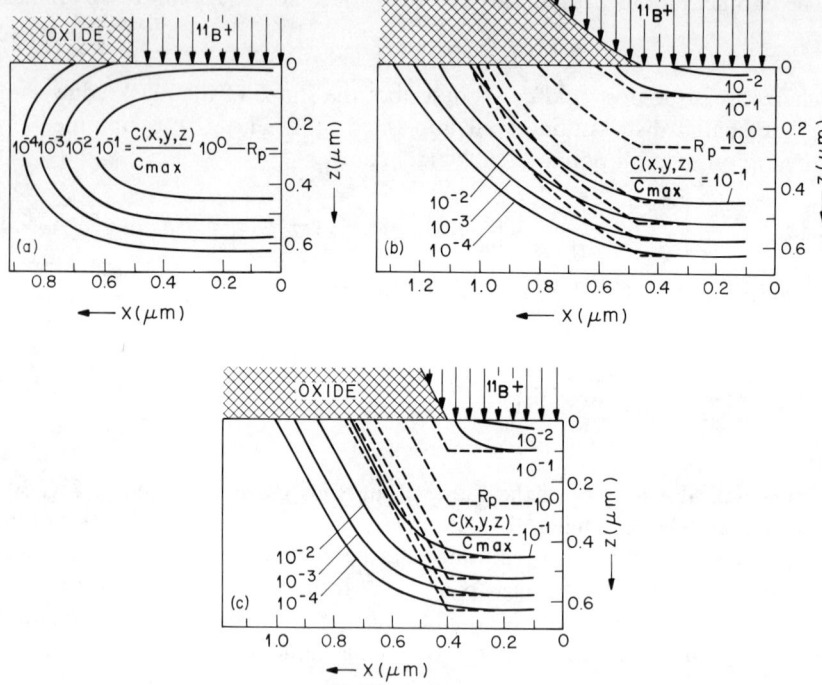

Fig. 30. Equidensities calculated from Eq. (2.100). 70keV $^{11}B \to Si$, $R_p = 0.271$ μm, $\Delta X = 0.1006$ μm, $\Delta R_p = 0.0824$ μm; (----) without lateral spread, (—) with lateral spread. (a) infinitely steep mask, (b) 45°, (c) 60°. (From Runge.[117])

extensions of ions implanted and diffused doping concentrations are of the same order of magnitude.

3. Diffusion

Solid-state diffusion is the physical mechanism responsible for impurity migration within the silicon crystal during high temperature ($T > 600°C$) fabrication steps. Together with epitaxy and ion implantation, it controls the conductivity type, the concentration level, and the distribution of impurities within localized regions.

In the following discussion, we summarize the basic equations and phenomena of impurity diffusion in silicon. A more thorough treatment can be found in another volume of this series[118] or in other review articles.[119-121]

The simulation of diffusion steps by either numerical or analytical means has become very important over the last few years to analyze and to predict impurity profiles in one or more dimensions. Several analytical solutions have been included in this section since they might prove useful in some situations when quick answers are needed or computer facilities are not available.

a. *Basic Equations*

In full generality, impurity diffusion in silicon is described by the following set of equations, in which the subscript i indicates the ith species.

(1) Flux equations for all charged particles:

$$\mathbf{J}_i = -D_i \nabla S_i + Z_i \mu_i S_i \mathbf{E}, \tag{2.101}$$

where S_i stands for the concentration of diffusing species (electrons, holes, donors, acceptors, interstitials, vacancies, etc.), D_i and μ_i are the corresponding diffusivities and mobilities, respectively, Z_i is the charge state ($+1$ for donors and -1 for acceptors), and $\mathbf{E}$ is the electric field.

(2) Continuity equations for all charged particles:

$$\partial S_i/\partial t + (1/q)\nabla \cdot \mathbf{J}_i = G_i, \tag{2.102}$$

where $\mathbf{J}_i$ is the particle flux of Eq. (2.101) and G_i stands for the generation–recombination rate.

(3) Poisson's equation:

$$\nabla \cdot (\epsilon \mathbf{E}) = q(p - n + N_D^+ - N_A^- + \text{other charged species}) \tag{2.103}$$

where n and p are the electron and hole density, respectively, and N_D^+ and N_A^- are the donor and acceptor concentrations.

At this point it is worthwhile to introduce several approximations. For the range of temperature of interest in diffusion (600–1200°C), electrons and holes can be assumed to be much "faster" than ions, that is,

$$\partial p/\partial t = \partial n/\partial t = 0,$$

which means that electrons and holes are much more mobile than impurity atoms. They diffuse ahead of the ions, thereby creating an electric dipole field that acts on the diffusing ions. We shall address this electric field enhancement below.

A second approximation is the assumption of the validity of nondegenerate statistics up to very high impurity concentrations. When dealing with degenerate silicon with doping levels in the range of 10^{19}–10^{21}/cm^3, the applicability of the one-band approximation to the band structure is questionable, especially in calculating the position of the Fermi level E_F.[122,123] For example, the one-band approximation and the assumption of complete

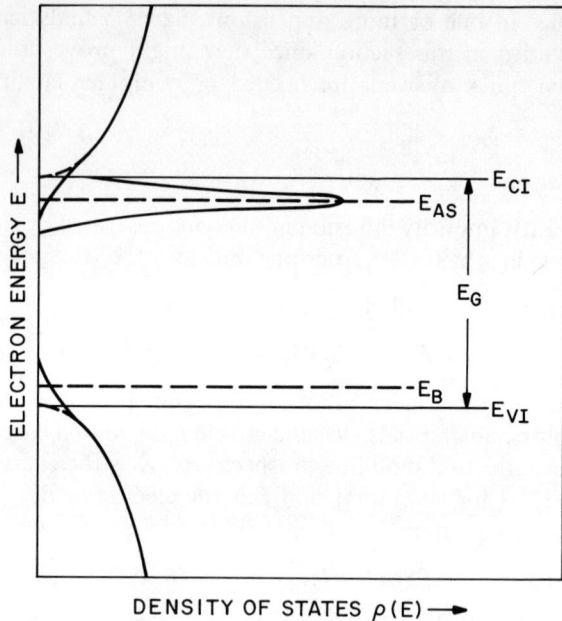

Fig. 31. Plots of the density of states functions versus electron energy for a heavily arsenic doped silicon. E_{CI} and E_{VI} are the conduction and valence band for intrinsic silicon.

ionization may give a Fermi energy E_F as high as 0.35 eV above the conduction band edge E_C. However, the one-band approximation is no longer valid for silicon at energies $E > E_C + 0.13$ eV, making this estimate of E_F unreliable.

Heavy doping effects[124-127] are characterized by the formation of band tails and an impurity band as illustrated in Fig. 31, which shows the density of states versus electron energy for a heavily doped arsenic and a lightly boron-doped silicon. In the following discussion the basis for this picture is described.

In the classical theory two parabolic density of states functions exist with well-defined band edges, the conduction band E_C and the valence band E_V, E_G being the intrinsic band gap. For the lightly doped case a discrete arsenic (donor) energy level E_{As} and a discrete boron (acceptor) energy level E_B exist. At high doping levels, however, this classical picture is no longer valid.

Kane[128] has shown that the statistical distribution of the impurity atoms results in a spatially fluctuating electric field. As a result the band edge varies with position, leading to

$$\rho_{\text{cond}}(E) = \frac{m_e^{*(3/2)} 2^{3/4} \sigma}{\pi^2 \hbar^3} f\left(\frac{E}{\sqrt{2}\,\sigma}\right), \tag{2.104}$$

where

$$f(x) = \frac{1}{\sqrt{\pi}} \int_{-\infty}^{x} \sqrt{x-y} \exp(-y^2)\, dy. \quad (2.105)$$

In Eq. (2.104) m_e^* is the effective electron mass and $\hbar$ is the normalized Planck constant. A similar expression can be written for the valence band tail. Impurity–impurity interaction causes the electron wave functions to overlap,[129] so that the discrete impurity level splits into a continuous Gaussian impurity band centered around the nondegenerate impurity level

$$\rho_{As}(E) = \frac{2C_{As}}{\sqrt{2\pi}\,\sigma_{eff}} \exp\left[-\frac{(E-E_{As})^2}{2\sigma_{eff}^2}\right] \quad (2.106)$$

with the arsenic concentration C_{As}.

Jain and van Overstraeten[124,125] and Jain[130] have incorporated Eqs. (2.104) and (2.106) in the calculation of the total electric field acting on the charged particles. Starting from the drift–diffusion expression for the current density (e.g., for electrons)

$$\mathbf{J}_n = \mathbf{J}_{diff} + \mathbf{J}_{drift},$$

the total electric field can be calculated from the thermal equilibrium condition, $\mathbf{J}_n = 0$,

$$\mathbf{E} = \frac{D_n}{\mu_n}\frac{1}{n}\nabla n. \quad (2.107)$$

For intrinsic conditions, the Einstein relation holds and we obtain

$$\mathbf{E} = \frac{kT}{q}\nabla(\ln n). \quad (2.108)$$

Under high doping conditions, however, the Einstein relation does not hold and the ratio D_n/μ_n is given by

$$\frac{D_n}{\mu_n} = \frac{(1/q)\int_{-\infty}^{+\infty}\rho(E)f\,dE}{\int_{-\infty}^{+\infty}\rho(E)(df/dE)\,dE}, \quad (2.109)$$

where

$$f = \left[1 + \exp\left(\frac{E-E_F}{kT}\right)\right]^{-1} \quad (2.110)$$

The function $\rho(E)$ is the total electron density of states, being the envelope of the conduction band density of states, Eq. (2.104), and the impurity band density of states, Eq. (2.106), E_F is the Fermi level, k Boltzmann's constant, T

the absolute temperature, and E the energy. Equation (2.109) can be rewritten as

$$\frac{D_n}{\mu_n} = \frac{1}{q} \frac{n}{\int_{-\infty}^{+\infty} \rho(E)(df/dE)\,dE}, \quad (2.111)$$

which, substituted into Eq. (2.107), gives

$$\mathbf{E} = -\frac{1}{q} \frac{\nabla n}{\int_{-\infty}^{+\infty} \rho(E)(df/dE)\,dE}. \quad (2.112)$$

Equation (2.112) describes the total electric field taking a generalized density of states into account.

In Fig. 32 the total electric field profiles are plotted for an arbitrary Gaussian donor (phosphorous) diffusion profile with a constant background

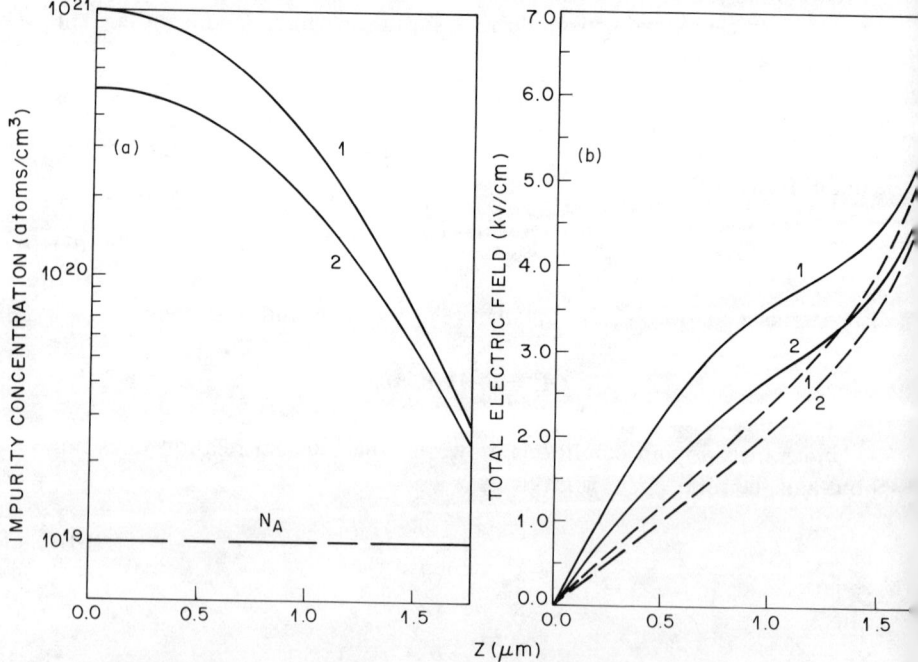

Fig. 32. Impurity concentrations and total electric fields as a function of depth. (a) The curves 1 and 2 are the two different donor impurity distributions into silicon with $N_A(=10^{19}$ atoms/cm^3) as the constant background doping level and (b) the curves 1 and 2 are the total electric field profiles at 900°C corresponding to curves 1 and 2 of (a). Curves denoted by solid lines include heavy doping effects while dashed curves are the classical results. (From Jain.[130])

(boron) doping $N_A = 10^{19}/cm^3$. For comparison, the intrinsic result for the electric field, Eq. (2.108) has been included in Fig. 32b. From the figure, we see that a decrease of the donor density also decreases the electric field. Taking heavy doping effects into account, the calculated field acting on the electrons is always greater than the intrinsic result. This means that the second extra quasi-electric field term due to the position dependent band structure is always positive and does not contribute much at lower doping levels. Thus, at lower doping levels the calculated electric field approaches the intrinsic result.

b. *The Quasi-Neutrality Condition and the Electric Field Effect*

A third approximation commonly used in the solution of diffusion problems is the assumption of quasi-neutrality during the diffusion process.[131] Mathematically, the quasi-neutrality condition can be expressed as (in the one-dimensional case)

$$dE/dz = 0 \qquad (2.113)$$

or

$$d^2\psi/dz^2 = 0 \qquad (2.114)$$

with the electrostatic potential defined by

$$E = -d\psi/dz. \qquad (2.115)$$

Hu[132] has investigated the diffusion problem in a thermodynamic treatment under the assumption of a vacancy mechanism. He analyzed the vacancy–impurity–semiconductor system based on an energy band model, in which the activity coefficients of the vacancy and of the impurity and the concentrations of the vacancy–impurity pairs are obtained. Assuming that the vacancy concentration is given by its equilibrium value, and the moderate impurity concentration, this theory predicts the impurity flux to be written as

$$J = -D_i \gamma_V^{-1}\left(1 + \frac{\partial \ln \gamma_A}{\partial \ln C_A}\right)\frac{\partial C}{\partial z}, \qquad (2.116)$$

where C is the impurity concentration, D_i is the corresponding diffusion coefficient, and γ_V and γ_A are the activity coefficients of the vacancy and impurity, respectively. The subscript i denotes a value for intrinsic material.

If an unperturbed band structure is assumed, the activity coefficients can be expressed as

$$\gamma_A \simeq (1 + \zeta_i)/(1 + \zeta) \qquad (2.117)$$

and

$$\gamma_V \cong (1 + \xi_i)/(1 + \xi), \qquad (2.118)$$

where

$$\zeta = g_A^{-1} \exp\left(\frac{E_A - E_F}{kT}\right), \qquad \xi = g_V \exp\left(\frac{E_F - E_i}{kT}\right). \qquad (2.119)$$

g_A and g_V are the degeneracy factors for the impurity and vacancy, respectively, and E_i is the first-vacancy level.

The second term in Eq. (2.116) is normally replaced by a field drift term [see Eq. (4.3) in Fair[118]]. According to Hu, this approximation having a field term is unnecessarily restrictive. It does not include effects such as strain interaction, ion pairing, etc., which would otherwise all be included in γ_A.

Equation (2.116) can be simplified further in the limiting case where the impurity concentration is dilute enough that clustering and complex formation are negligible, and the system is nondegenerate. For a shallow acceptor vacancy it follows that

$$\gamma_V^{-1} = \frac{1 + \xi}{1 + \xi_i} \simeq \frac{\xi}{\xi_i} = \exp\left(\frac{E_F - E_i}{kT}\right) = \frac{n}{n_i} \qquad (2.120)$$

$$= \frac{C_A}{2n_i} + \left(1 + \frac{C_A^2}{4n_i^2}\right)^{1/2} \qquad (2.121)$$

$$\frac{\partial \ln \gamma_A}{\partial \ln C_A} \simeq \frac{C_A}{\sqrt{C_A^2 + 4n_i^2}}. \qquad (2.122)$$

The step from Eq. (2.116) to Eq. (2.122) is valid only if C_A is a well-defined function of the space coordinates and if γ_A is a unique function of C_A.

Hu[123] and Lowney[133] have investigated the validity of the quasi-neutrality condition by solving the Poisson equation, Eq. (2.103). In normalized form, Eq. (2.103) becomes

$$\frac{d^2 u}{d\xi^2} = \left(\frac{\lambda}{L_D}\right)^2 \left(\sinh u + \frac{C_A}{2n_i}\right), \qquad (2.123)$$

where u is the normalized potential $(E_C - E_F)/kT$; $\xi = z/\lambda$, where λ is the impurity diffusion length, and L_D is the intrinsic Debye length.

Hu[123] considered the case $C_A(z) = C_A(0)\,\text{erfc}(z/\lambda)$ with $C_A(0)/2n_i = 5$ as an example. The results of the numerical solution of Eq. (2.123) subject to the boundary conditions

$$du/d\xi = 0 \quad \text{at} \quad z = 0 \quad \text{and} \quad z = \infty \qquad (2.124)$$

are given in Figs. 33 and 34. It is evident that the local charge neutrality condition is reasonably good for $\lambda/L_D \sim 6$, except for the region very close to the surface. At a temperature $T = 1000°C$, the intrinsic density $n_i \simeq$

$10^{19}/\mathrm{cm}^3$ and $L_D = 1.87 \times 10^{-7}$ cm, which means that the neutrality condition is reasonable for $\lambda \geq 1.1 \times 10^{-6}$ cm.

Lowney[133] has obtained a similar result as shown in Fig. 35a and b. Assuming a Gaussian distribution of the form $C(z) = C_0 \exp(-z^2/z_0^2)$ to simulate an ion-implanted profile, he calculated the electric field enhancement factor in Eq. (2.116) by numerically solving Eq. (2.123). For Fig. 35a, $z_0 = 10^{-6}$ cm and for Fig. 35b, $z_0 = 5 \times 10^{-5}$ cm. The narrow distribution represents, for example, a low-energy implant into silicon. The other distribution corresponds to a typical high-energy implant situation. The approximate solution, given by Eq. (2.122), is also shown in the figures.

The agreement between the numerical solution and the approximate result is good near $z = 0$ and for $z > 0.04$ μm in Fig. 35a. However, strong deviations occur in the portion of the profile between 0.01 and 0.04 μm. Agreement is very good for all z in Fig. 35b. Therefore, as the distribution

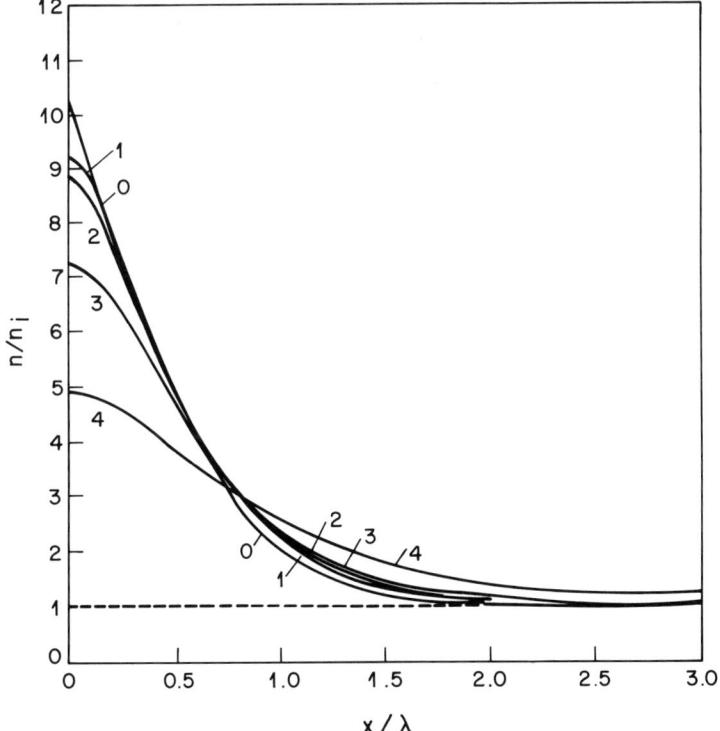

Fig. 33. Local normalized electron concentration $n(x)/n_i$ for a sample erfc profile for various ratios of diffusion length to Debye length. $C_A(0)/2n_i = 5$; 0, neutrality approximate; 1–4, exact calculation; (a) $\lambda/L_D = 6$; (2) $\lambda/L_D = 4$; (3) $\lambda/L_D = 2$; (4) $\lambda/L_D = 1$. (From Hu.[123])

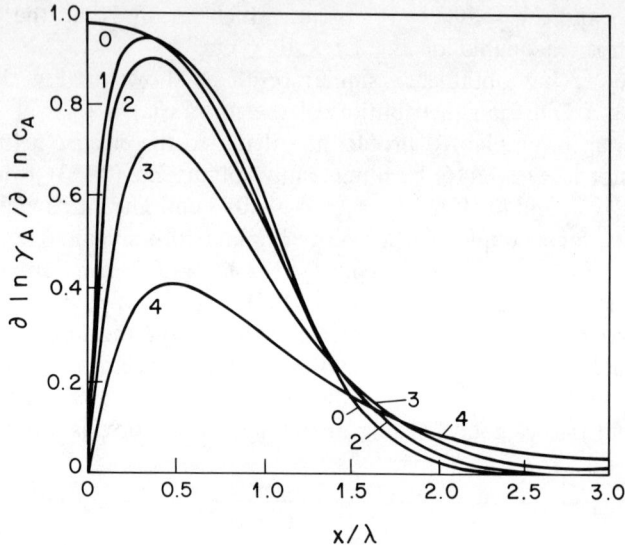

Fig. 34. The internal field enhancement factor $\partial \ln \gamma_A / \partial \ln C_A$ for a sample erfc impurity profile for various values of the ratio of diffusion length to Debye length. $C_A(0)/2n_i = 5$; 0, neutrality approximation; 1–4, exact calculation; (1) $\lambda/L_D = 6$; (2) $\lambda/L_D = 4$; (3) $\lambda/L_D = 2$; (4) $\lambda/L_D = 1$. (From Hu.[123])

spreads out during diffusion, the deviations from the approximate solution become negligible. However, it can be expected that for the shallow junctions necessary in VLSI devices, the deviations arising from steep impurity profiles can no longer be neglected.

Assuming nondegenerate conditions and quasi-neutrality, the drift-diffusion form of the current flow equation can be simplified by combining both terms into one by the following derivation.[134,135] In the case of arsenic diffusion, $D \approx D_i(n/n_i)$, and

$$\begin{aligned}
C_{As}\frac{dD}{dz} &= C_{As}D_i \frac{d}{dz}\left(\frac{n}{n_i}\right) \\
&= C_{As}D_i \frac{n}{n_i}\frac{d}{dz}\ln\left(\frac{n}{n_i}\right) \\
&= C_{As}D \frac{d}{dz}\ln\left(\frac{n}{n_i}\right) \\
&= C_{As}\left[D\frac{q}{kT}\left(\frac{kT}{q}\frac{d}{dz}\ln\frac{n}{n_i}\right)\right] \\
&= -C_{As}\mu E,
\end{aligned} \qquad (2.125)$$

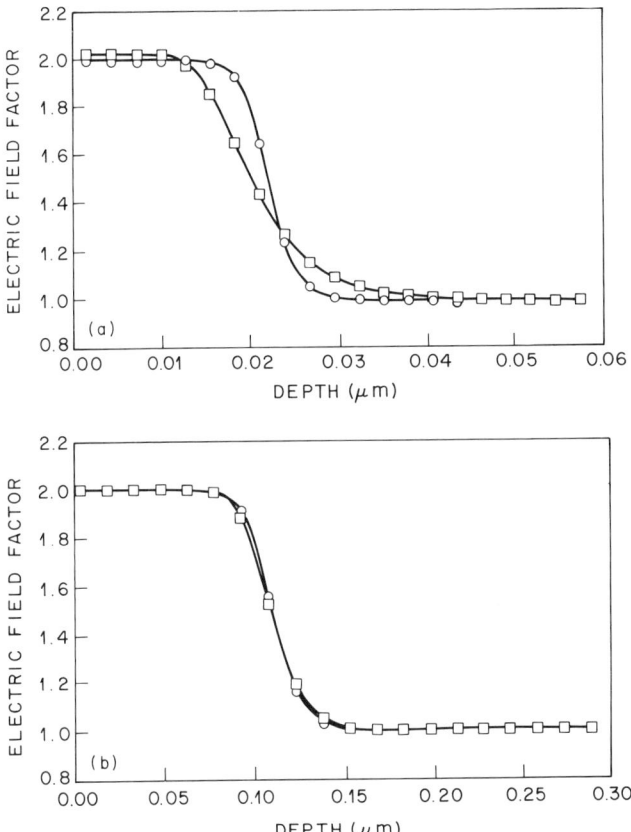

Fig. 35. Comparison of the exact and approximate solutions for Gaussian profiles with $N_0 = 10^{20}/cm^3$, $T = 973°K$; (a) $x_0 = 0.01$ μm, (b) $x_0 = 0.05$ μm. □, exact solution; ○, approximate solution. (From Lowney.[133] This figure was originally presented at the Spring 1974 Meeting of The Electrochemical Society, Inc., held in St. Louis, Missouri.)

where we have used the Einstein relation and Eq. (2.108). From Eq. (2.125) it follows that

$$J = -D\frac{dC_{As}}{dz} + \mu E C_{As}$$

$$= -D\frac{dC_{As}}{dz} - C_{As}\frac{dD}{dz}$$

$$= -\frac{d}{dz}(DC_{As}). \quad (2.126)$$

With some approximation, the diffusion flux has been simplified to Eq.

(2.126), which is much easier to treat numerically.[136] Equation (2.126) is the form used in most process simulation programs.

Apart from very shallow profiles, the effect of the built-in field is, in many cases, not very significant.[137] Under conditions of double diffusion of oppositely charged ions, however, the field effect can be rather dramatic and has a considerable influence on the final device structure. As an example, consider the case when n-type dopant (P, As) is diffused after a p-type dopant (B). Suppose that the boron has been diffused first from the surface (or implanted and then annealed) to give a profile as in Fig. 36a and an electric field due to the subsequent diffusion that is constant to a depth a and zero for $z > a$ (Fig. 36b). If the field has the direction and strength to overcome the

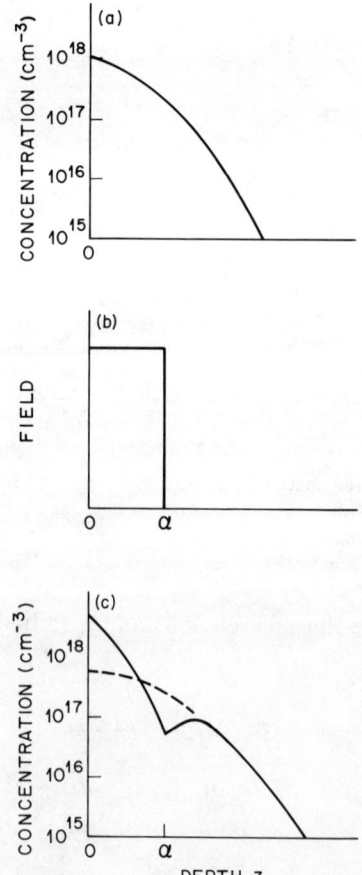

Fig. 36. Simple model of dip production by an electric field. (From Willoughby.[120] Copyright North-Holland Physics Publishing, Amsterdam, 1981.)

concentration gradient, forcing the impurity flux toward the surface for $0 < z < a$, then this region appears as a sink for impurities in the region $z > a$. The profile will become depleted close to a and piled up near the surface. The result is a dip in the profile centered at $z = a$ as shown in Fig. 36c. The field may also reduce the penetration of the first dopant and cause a shallower junction than expected.

Hu and Schmidt[131] were the first to study coupled diffusion phenomena. In the case of a system of k donors and l acceptors, Eq. (2.108) is generalized to

$$E(z, t) = \frac{kT}{q} \frac{d}{dz} \ln f, \qquad (2.127)$$

where

$$f = \frac{\sum C_k - \sum C_l}{2n_i} + \left[1 + \left(\frac{\sum C_k - \sum C_l}{2n_i} \right)^2 \right]^{1/2}, \qquad (2.128)$$

assuming $np = n_i^2$ and local charge neutrality. Noting that

$$\frac{\partial f}{\partial C_k} = -\frac{\partial f}{\partial C_l}, \qquad (2.129)$$

and applying the Einstein relation, one can write the flux, Eq. (2.101),

$$J_i = -D_i \frac{dC_i}{dz} - D_i Z_i C_i \frac{\partial \ln f}{\partial C_k} \left(\sum \frac{dC_k}{dz} - \sum \frac{dC_l}{dz} \right). \qquad (2.130)$$

For a two-component system with acceptors C_A and donors C_D, Eq. (2.131) reduces to

$$J_A(z, t) = -D_A^i f_A h_A \frac{\partial C_A}{\partial z} - D_A^i f_A (h_A - 1) \frac{\partial C_D}{\partial z} \qquad (2.131)$$

and

$$J_D(z, t) = -D_D^i f_D h_D \frac{\partial C_D}{\partial z} + D_D^i f_D (h_D - 1) \frac{\partial C_A}{\partial z}, \qquad (2.132)$$

where D_D^i and D_A^i are the intrinsic donor and acceptor diffusivities, respectively. The electric field effect is contained in the parameters h_A and h_D, which are, assuming full ionization and nondegenerate statistics,

$$h_A(h_D) = 1 + \frac{C_A(C_D)}{\sqrt{(C_D - C_A)^2 + 4n_i^2}}. \qquad (2.133)$$

The factors f_A and f_D are related to the diffusion by charged vacancies. Assuming that acceptors diffuse mainly via neutral vacancies and donors by

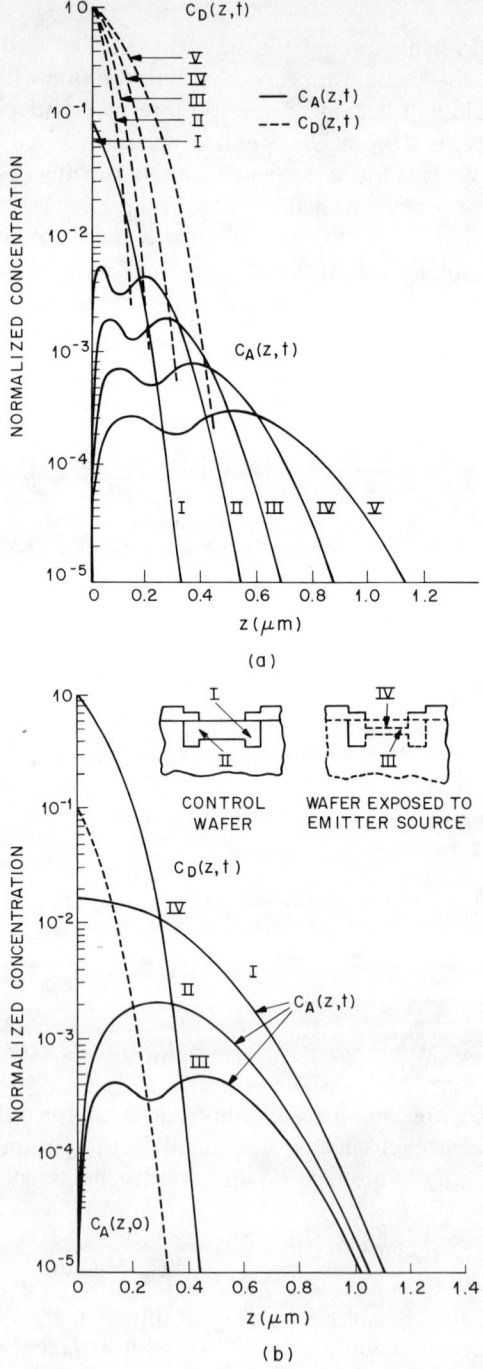

Fig. 37. Behavior of boron impurity profile under a zero surface concentration boundary condition during constant source emitter diffusion. $D_A^i = 0.12 \times 10^{-12}$ cm^2/sec, $D_D^i = 0.12 \times 10^{-13}$ cm^2/sec. (a) $t = 0$(I), 500 sec (II), 1000 sec (III), 2000 sec (IV), 4000 sec (V). (b) Blocking

singly negatively charged vacancies, we obtain for full ionization and nondegenerate statistics

$$f_A = 1 \qquad (2.134)$$

and

$$f_D = \frac{C_D - C_A}{2n_i} + \left[\frac{(C_D - C_A)^2}{4n_i^2} + 1\right]^{1/2}. \qquad (2.135)$$

The factor f_D is the ratio of the total electron concentration to the intrinsic concentration. Equations (2.131) to (2.135) have been solved numerically for a variety of different conditions. Let us consider the case of the emitter–base formation in bipolar processing.[131] The boundary conditions for the solution of Eqs. (2.130) and (2.131) depend on the source of the second diffusion or on the properties of the gaseous ambient. The boundary conditions for the second (donor) diffusion are given by

$$C_D(z, t) = 0 \quad \text{as } z \to \infty \quad \text{for } t = 0 \qquad (2.136)$$

$$C_D(z, t) = C_D(0, 0). \qquad (2.137)$$

If the emitter diffusion is accompanied by the formation of a glass layer, a blocking boundary is appropriate for the acceptor concentration given by

$$\frac{\partial C_A}{\partial z} = \left(1 - \frac{1}{h_A}\right)\frac{\partial C_D}{\partial z} \quad \text{at } z = 0 \quad \text{for all } t. \qquad (2.138)$$

If the semiconductor is exposed to the vapor phase of the emitter impurity, some fraction of C_A will evaporate. This evaporation flux is either controlled by the diffusion in the solid or by the transport through the gaseous boundary layer above the surface of the solid.

Another type of condition would arise if the initial donor profile were obtained by ion implantation.

Figure 37 shows the influence of different boundary conditions on the final profiles in a double diffusion calculation. The curves represent the results for the blocking boundary case with no interaction between the impurities (curve I), a zero surface concentration case without interaction (curve II), and a zero surface concentration case showing full interaction between donor and acceptors (curve III). Curve II represents a hypothetical control experiment in which a base layer is prediffused, a masking layer is then deposited, and a window is opened. Heat treatment simulating the emitter diffusion causes out-diffusion and results in an apparent junction "retardation."

---

boundary, noninteracting (I); $C_A(0, t) = 0$, noninteracting (II); $C_A(0, t) = 0$ interacting with C_D (III), $t = 3000$ sec. (From Hu and Schmidt.[131])

Equations (2.130) and (2.131) have been simplified using the same procedure that led to Eq. (2.125). We obtain

$$J_A(z, t) = - D_A \frac{\partial C_A}{\partial z}, \qquad D_A = D_A^i \frac{1 + \beta(n_i/n)}{1 + \beta} \qquad (2.139)$$

with $\beta \sim 10$ and, in the case of an arsenic donor,

$$J_D(z, t) = 2 D_{As} \, \partial C_D / \partial z. \qquad (2.140)$$

The interactive diffusion of boron from a flat initial profile in the presence of an arsenic gradient can also be reproduced without formal consideration of an internal electric field. An example is shown in Fig. 38, which compares experimentally determined profiles[138] with those calculated from Eqs. (2.140) and (2.125) with the diffusion coefficients given by the right-hand side of Eq. (2.139) and $D_{As} = D_i C_{As}/n_i$. Arsenic was diffused into an initially flat profile of boron (5×10^{18} cm^3) in silicon for 2 hr at 1000°C from a constant As surface concentration. The simpler calculation from Eq. (2.125) gives as good an agreement with the data as the more complicated one from Eq. (2.140). This result contradicts speculations published by Lowney and Larrabee,[139] who questioned the applicability of Eq. (2.125) in coupled diffusion problems.

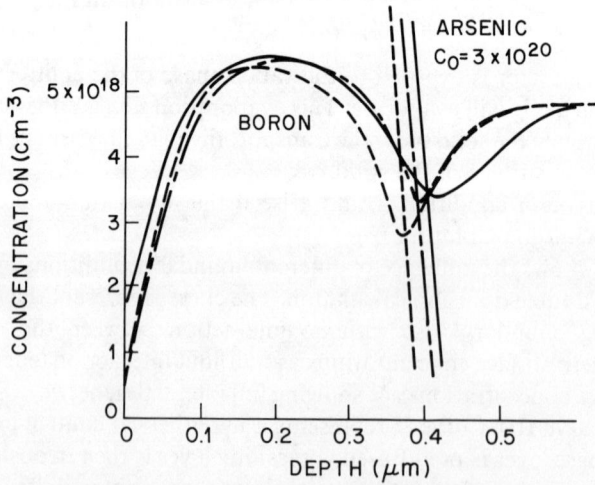

Fig. 38. Comparison of experimentally determined profiles (—) with calculations using Eq. (2.140) (- - -) and Eq. (2.125) (—·—·—) with D given by Eq. (2.139) and $D_{As} = D_i (C_{Ar}/n_i)$ for a time of 2 hr at 1000°C. (From Morehead.[134] This figure was originally presented at the Spring PRO Meeting of The Electrochemical Society, Inc. held in St. Louis, Missouri.)

c. *Point Defects and Diffusion Models*

For temperatures $T > 0$, minimization of the free energy of a crystal in thermal equilibrium requires the presence of point defects in the lattice.

The concentration of neutral point defects, C_x^{eq}, is given by

$$C_x^{eq} = n_H \exp[-(G_{fx}/kT)], \qquad (2.141)$$

where the subscript x stands for either vacancies (V) or interstitials (I), n_H is the concentration of Si atoms in the lattice, and G_{fx} is the free energy of defect formation. The diffusivity D_x is given by

$$D_x = a^2 v/8 \exp[-(G_{mx}/kT)], \qquad (2.142)$$

where a is the lattice constant, v is the lattice vibration frequency, and G_{mx} denotes the free energy of defect migration.

Self-diffusion of silicon atoms can occur via both self-interstitials and vacancies and the self-diffusion coefficient can be written as

$$D^{self} = \frac{f}{n_H}(C_V^{eq} D_V^{self} - C_I^{self} D_I^{self}), \qquad (2.143)$$

where f is the correlation factor of the diamond lattice. From Eq. (2.143), it is evident that the diffusion process depends not on the concentration of the point defects alone but rather the products $C_x^{eq} D_x$.

Figure 39 shows values for the components $D_I C_I^{eq}$ and $D_V C_V^{eq}$ calculated from experimental results.[140-144] Self-interstitials contribute more to self-diffusion for temperatures above 1000°C and vacancies contribute more for lower temperatures.

For a purely vacancy-dominated process, one assumes $C_V^{eq} \gg C_I^{eq}$ and $D_V^{eq} C_V^{eq} \gg D_I^{eq} C_V^{eq}$, while for a purely self-interstitial model the reverse relations hold.

The diffusion of a substitutional impurity is described by a generalization of Eq. (2.143)[145]

$$D = D_I^i \frac{C_I}{C_I^{eq}} - D_V^i \frac{C_V}{C_V^{eq}}, \qquad (2.144)$$

where D_I^i and D_V^i are the intrinsic diffusivities of the impurity atom. This equation only holds in the case that neither C_I^{eq} and C_V^{eq} vanish.

Defining the fractional interstitialcy component

$$f_I = D_I^i/D^i, \qquad (2.145)$$

the diffusivity can be rewritten as

$$\frac{D}{D^i} = f_I \frac{C_I}{C_I^{eq}} + (1-f_I)\frac{C_V}{C_V^{eq}}. \qquad (2.146)$$

The question of whether interstitials or vacancies dominate the diffusion of substitutional impurities in silicon is currently a very active area of research.[146–154] An analysis of the diffusion behavior of group III and group V dopants shows that in thermal equilibrium as well as under oxidizing conditions both vacancies and interstitials are present.

The purely vacancy-dominated model of self- and impurity diffusion in silicon, as pioneered by Fair,[118] forms the basis for many of the published

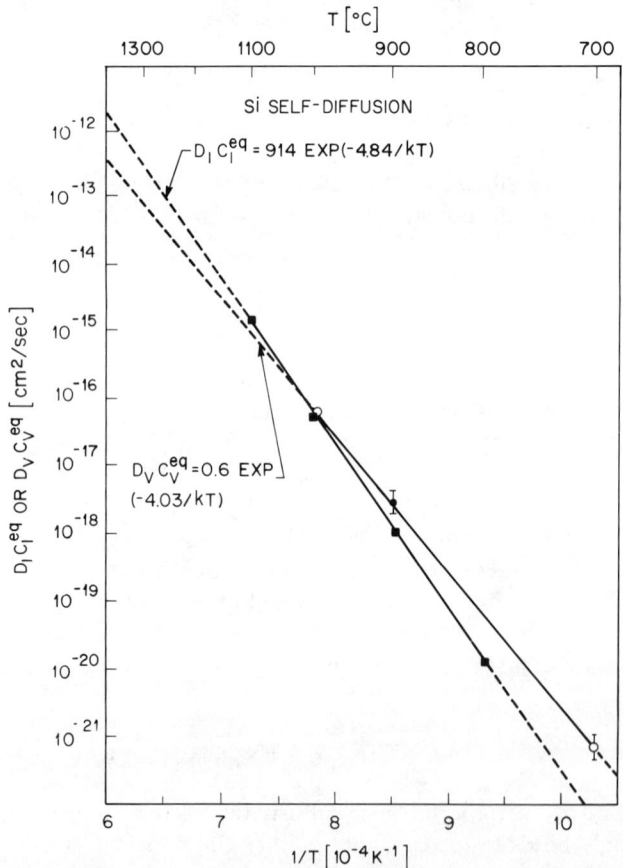

Fig. 39. Components $D_I C_I^{eq}$ and $D_V C_V^{eq}$ of Si self-diffusion versus $1/T$ calculated from the diffusion of Au into dislocation-free Si and from the diffusion and precipitation of Ni in dislocated Si. ■, Stolwijk et al.;[141] ○, Morehead et al.;[142] ⌽, Wilcox et al.;[143] ●, Kitagawa et al.[144] (From Gösele and Tan.[151] Reprinted by permission of the publisher from The nature of point defects and their influence on diffusion processes in silicon at high temperatures by U. Gösele and T. Y. Tan, "Defects in Semiconductors II," S. Mahajan and J. W. Corbett, eds., p. 45. Copyright 1983 by Elsevier Science Publishing Co., Inc.)

process simulation programs.[19,136,155,156] In another volume of this series, Fair has presented the vacancy model in detail.[118]

d. *Anomalous Diffusion Phenomena: OED and ORD*

The diffusion of dopants in silicon is often carried out under oxidizing conditions. During the oxidation process, a thin film of amorphous silicon forms at the silicon surface. Oxygen from the gas phase diffuses in the form of O_2 molecules through holes in the SiO_2 network toward the SiO_2–Si interface to form new SiO_2 material. This formation is associated with a large (100%) volume increase. At sufficiently high temperatures, the reaction is made possible by viscoelastic flow of the oxide toward the surface of the SiO_2 film.[151,157]

Defining self-interstitial and vacancy supersaturation ratios[146]

$$S_I = \frac{C_I - C_I^{eq}}{C_I^{eq}}, \qquad (2.147)$$

$$S_V = \frac{C_V - C_V^{eq}}{C_V^{eq}}, \qquad (2.148)$$

Eq. (2.146) can be rewritten by introducing the normalized diffusion

$$\Delta_{ox} = (D_{ox} - D)/D = f_I S_I + (1 - f_I) S_V, \qquad (2.149)$$

where D_{ox} is the impurity diffusivity under oxidizing conditions.

A small percentage of the volume increase during oxide formation is due to the injection of excess self-interstitials from the SiO_2–Si interface into the

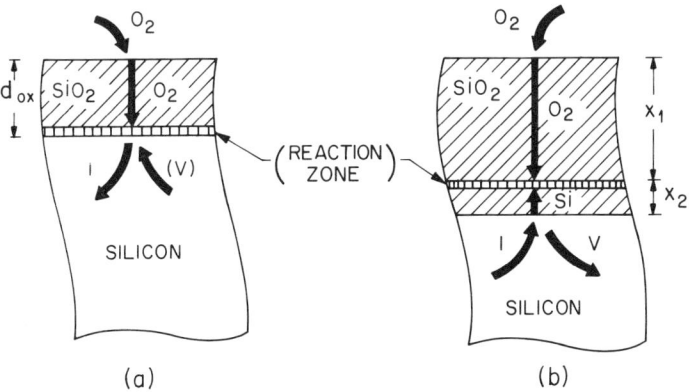

Fig. 40. Point defect generation (or absorption) mechanism during surface oxidation of silicon; (a) for thin oxides and/or moderate temperatures; (b) for thick oxides and/or high temperatures. (From Gösele and Tan.[151])

Si bulk.[158,159] Figure 40a shows the various reactions. After equilibrium has been reached, no measurable physical difference exists between I injection and V absorption. The magnitude of the induced S_I decreases with increasing temperature because the viscous flow of the oxide is enhanced.

For thicker oxides and/or higher temperatures, the reaction zone shifts from the interface to the interior of the insulator film (Fig. 40b). Silicon is transported to this reaction zone either by self-diffusion[160] or the diffusion of SiO molecules.[161] Both mechanisms lead to injection of vacancies and/or self-interstitial absorption, resulting in vacancy supersaturation $S_V > 0$ and interstitial undersaturation $S_I < 0$. For a given temperature, S_I starts with a positive value, decreases with time, and may even become negative.

Prussin[162] and Sirtl[163] have suggested that vacancies and interstitials maintain local equilibrium via vacancy–interstitial pair creation and recombination in the bulk. External influences changing the equilibrium values C_V^{eq} and C_I^{eq} result in local point defect equilibrium

$$C_V C_I = C_V^{eq} C_I^{eq} \qquad (2.150)$$

describing the relation

$$I + V \to 0. \qquad (2.151)$$

For times that are short compared to the relaxation time τ_{eq} needed to establish equilibrium, S_I and S_V are assumed to be independent of each other, related by,[164]

$$S_V = - S_I/(1 + S_I) \qquad (2.152)$$

and equivalently,

$$S_I = -S_V/(1 + S_V), \qquad (2.153)$$

which leads to

$$\Delta_{ox} = (2f_I + f_I S_I - 1) S_I/(1 + S_I). \qquad (2.154)$$

Figure 41 shows Δ_{ox} as a function of the interstitial supersaturation ratio for three values of f_I. Figures 42–45 compare experiments and calculations obtained from Eq. (2.154) for As, P, B, and Sb. With the exception of Sb, the other elements exhibit oxidation-enhanced diffusion (OED). In Fig. 45 the theoretical curve is a fit to long-term oxidation-reduced diffusivity (ORD) data for Sb. While this result is satisfactory, considerable discrepancy exists for short oxidation times ($T \lesssim 60$ min). Antoniadis and Moskovitz[147,148] have demonstrated that for this time regime, Sb data do not follow a constant f_I relationship, which also means that Eq. (2.150) is not valid and C_V/C_V^{eq} has a strong influence on the diffusivity.

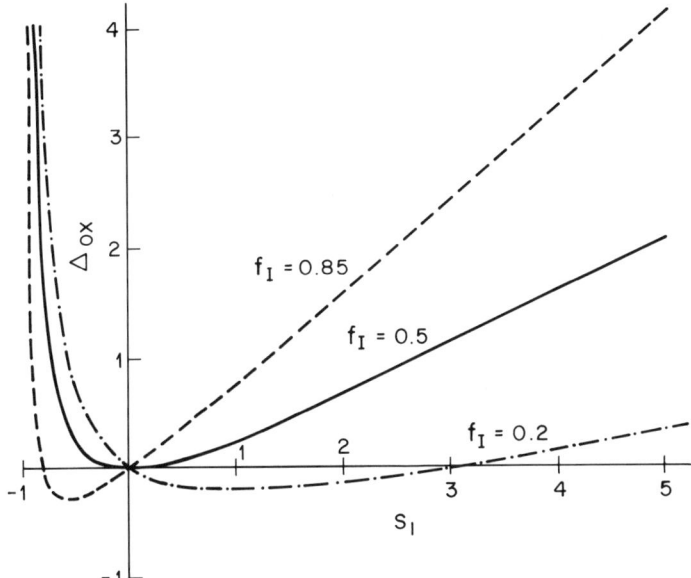

Fig. 41. Plot of Eq. (2.154) as a function of the self-interstitial supersaturation ratio S_I for three values of f_I. The prediction assumes that I and V coexist in thermal equilibrium and interact to fulfill Eq. (2.150). (From Tan et al.[146])

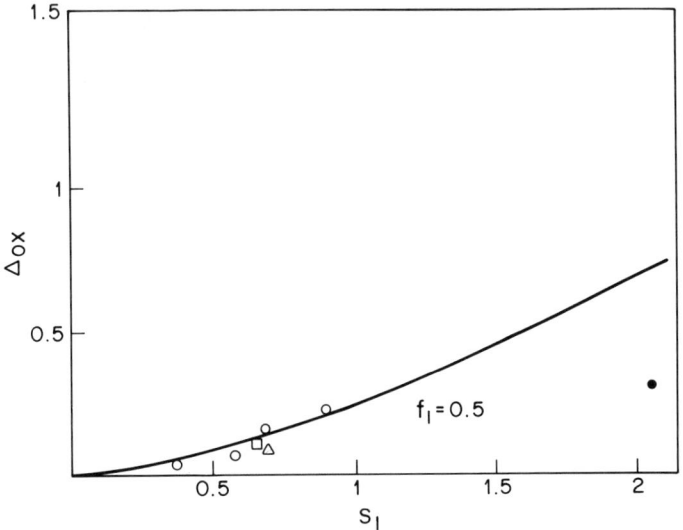

Fig. 42. Fit of available As OED data with Eq. (2.154). ○, Tan and Ginsberg;[167] □, Mizuo and Higuchi;[165] ●, △, Antoniadis et al.[166] (From Tan et al.[146])

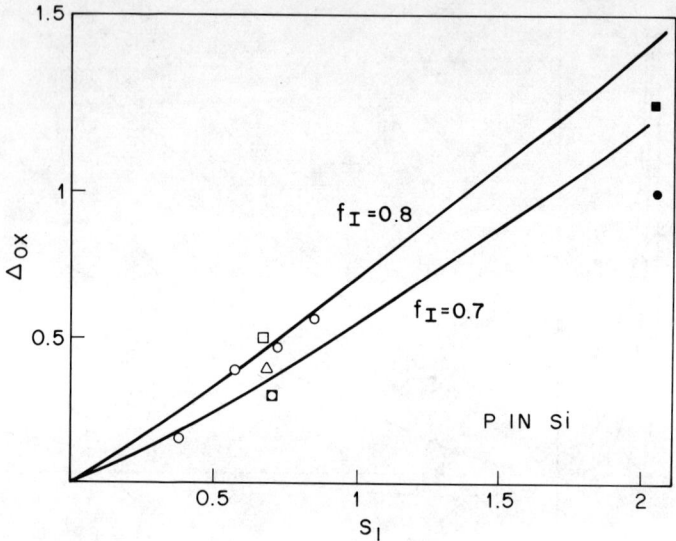

Fig. 43. Fit of available P OED data. ○, Tan and Ginsberg;[167] □, Mizuo and Higuchi;[165] ●, △, Antoniadis et al.;[166] ▫, ◻, Lin et al.[168] (From Tan et al.[146])

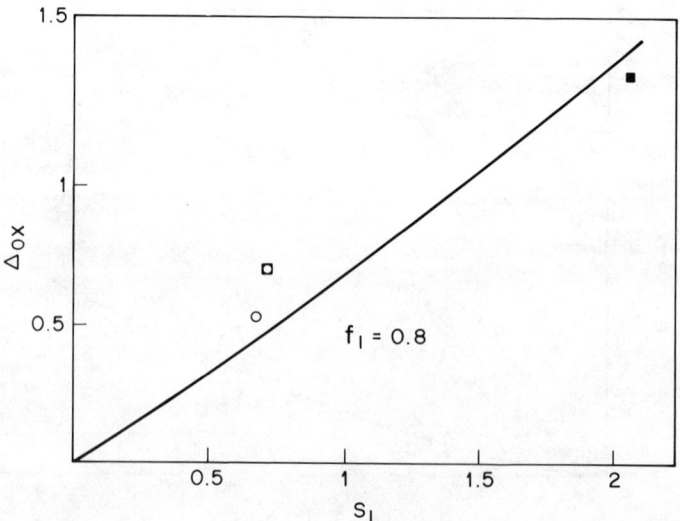

Fig. 44. Fit of available B OED data. ○, Mizuo and Higuchi;[165] ▫, ◻, Lin et al.[168] (From Tan et al.[146])

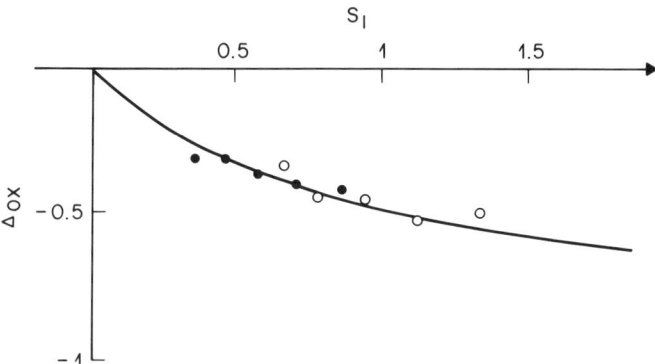

Fig. 45. Fit of available Sb ORD data. ○, Mizuo and Higuchi;[165] ●, Tan and Ginsberg.[167] (From Tan et al.[146])

Figure 46 indicates that for 1100°C and up to 60 min, C_V has a strong influence. The Antoniadis–Moskovitz model assumes that the diffusion of Sb is dominated by a vacancy mechanism, and that the I–V recombination in Eq. (2.151) is dominated by a rate constant K that is so small that the excess self-interstitials do not instantaneously decrease the vacancy concentration. For longer diffusion times, the reaction in Eq. (2.151) consumes vacancies and the diffusivity drops until the steady state is reached [Eq. (2.150)].

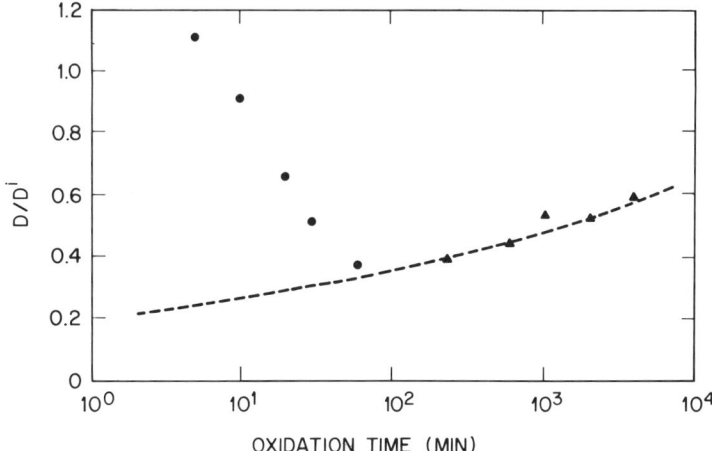

Fig. 46. Experimental results and calculations of normalized Sb diffusivity during oxidation for $f_I = 0.01$ and $C_V/C_V^{eq} = C_I/C_I^{eq}$. The dashed line gives the equilibrium result for $f_I = 0.01$. ●, Antoniadis and Moskovitz;[147] ▲, Mizuo and Higuchi.[165] (From Antoniadis and Moskovitz.[148])

During the transient period, Eq. (2.150) has to be replaced by

$$dC_V/dt = KC_I^{eq}C_V^{eq} - KC_IC_V, \tag{2.155}$$

where K is the rate constant. The concentration of self-interstitials can be written as

$$C_I - C_I^{eq} = K_I(dd_{ox}/dt)^n \tag{2.156}$$

with the oxide thickness d_{ox} and the proportionality constant K_I. For oxidation times less than or equal to 60 min,

$$\frac{C_I}{C_I^{eq}} = 1 + \frac{K_I}{C_I^{eq}t}\int_0^t \left(\frac{dd_{ox}}{d\tau}\right)^{0.5} d\tau \tag{2.157}$$

and for longer times,

$$\frac{C_I}{C_I^{eq}} = 1 + \frac{K_I}{C_I^{eq}} B^{0.5}[A^2 + 4B(t - \tau)]^{-0.25}, \tag{2.158}$$

where A and B are the linear and parabolic rate constants in the Deal–Grove equation (see oxidation Subsection 3.f) and τ is a parameter. Substituting

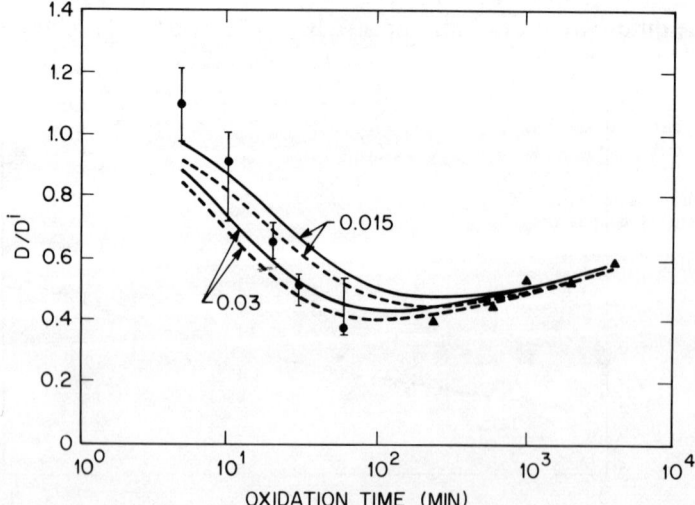

Fig. 47. Theoretical calculations and experimental results of the normalized Sb diffusivity during oxidation for two different values of KC_I^{eq} per minute. ●, Antoniadis and Moskovitz,[147] ▲, Mizuo and Higuchi;[165] ——, $f_I = 0.015$; ----, $f_I = 0$. (From Antoniadis and Moskovitz.[148])

Eqs. (2.157) and (2.158) into (2.156) leads to linear time-dependent equations that can be solved numerically. Substituting the result into Eq. (2.146) yields

$$\frac{D}{D^i} = \frac{f_I}{t}\int_0^t \frac{C_I}{C_I^{eq}}dt' + \frac{1-f_I}{t}\int_0^t \frac{C_V}{C_V^{eq}}dt'. \tag{2.159}$$

Figure 47 shows solutions of Eq. (2.159) for Sb with different values for KC_I^{eq} and experimental results. Very good agreement exists between the theoretical result and the measurements, especially for the case $f_I = 0.015$.

e. *Diffusion Anomalies: Arsenic and Phosphorus*

(i) *Arsenic.* In high-dose As implants, a substantial reduction of the electrical carrier concentration occurs that can be explained by the kinetics of As clustering. Considerable discrepancy between the theoretical models and the actual diffusivity develops at arsenic concentrations above $3 \times 10^{20}/$ cm^3, and there is even a decrease of diffusivity with arsenic concentration at higher concentrations.[169] The original model assumed two arsenic atoms per cluster.[170] Later on, a four-arsenic-atom cluster model was proposed,[122] which was largely based on the vapor pressure data of Sandhu and Reuter.[171] The cluster is envisaged to consist of four arsenic atoms forming a tetrahedron either with a normal interstitial site or with a silicon atom at its center. The formation of two As–As covalent bonds makes the complex electrically inactive.

Guerrero *et al.*[172] have presented a model for As clustering in single-crystal silicon that allows for the participation of arbitrary numbers of As ions, electrons, and arbitrarily charged vacancies.

Assuming that m arsenic ions and k electrons (or vacancies) form a cluster with the electric charge $r = m - k$, the reaction for clustering/declustering can be written as

$$m\text{As}^+ + k e^- \rightleftarrows \text{As}_m^r. \tag{2.160}$$

The law of mass action for this reaction is given by

$$C_C = k_{eq} C_{As}^m n^k, \tag{2.161}$$

where k_{eq} is the equilibrium constant depending on the temperature and C_{As}, C_C, and n are the high-temperature equilibrium concentrations of As$^+$, the clusters, and the electrons, respectively. The total As concentration C_T consists of the positive As ions and the As in the clusters

$$C_T = C_{As} + mC_C = C_{As} + mk_{eq}C_{As}^m n^k. \tag{2.162}$$

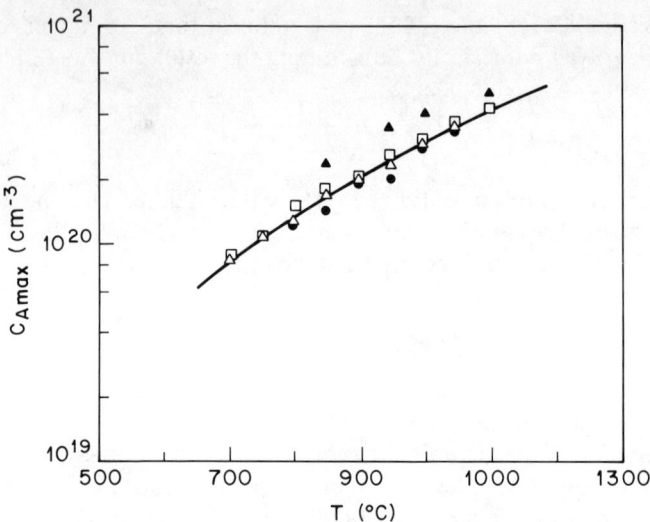

Fig. 48. Electrical solubility C_{Amax} as a function of temperature. Symbols denote experimental results. (From Guerrero et al.[172] Reprinted by permission of the publisher. The Electrochemical Society, Inc.)

Furthermore, electrical neutrality leads to

$$n = C_{As} + rC_C = rk_{eq}C_{As}^m n^k. \qquad (2.163)$$

For C_T given, C_{As}, C_C and n can be calculated from these equations.

To find the optimum cluster parameters, Guerrero et al.[172] have analyzed experimental results for the electrical solubility $C_{A,max}$ as a function of temperature. Their results are shown in Fig. 48. Assuming a number k' of charged particles and m As atoms in the cluster, the temperature dependence of the reaction constant k_{eq} can be derived from the results in Fig. 48 according to

$$C_{As,max} = (1/rk_{eq})^{1/m}. \qquad (2.164)$$

From the existence of $C_{As,max}$, it can be concluded that electron and or vacancy charges take part in the cluster formation. Furthermore, as no decrease in C_{As} with increasing C_T has been observed, exactly one charge (e^- or V^-) participates.[172]

Figures 49a and b present experimental results of $\alpha = C_T/C_{As} - 1$ for two temperatures. The curves correspond to $k' = 1$ and $m = 2, 3, 4$.

Fig. 49. Ratio $\alpha = C_T/C_A - 1$ as a function of the electrically active concentration at (a) $T = 1000°C$ and (b) $950°C$ and $k' = 1$. In both figures, symbols denote experimental results. (From Guerrero et al.[172] Reprinted by permission of the publisher, The Electrochemical Society, Inc.

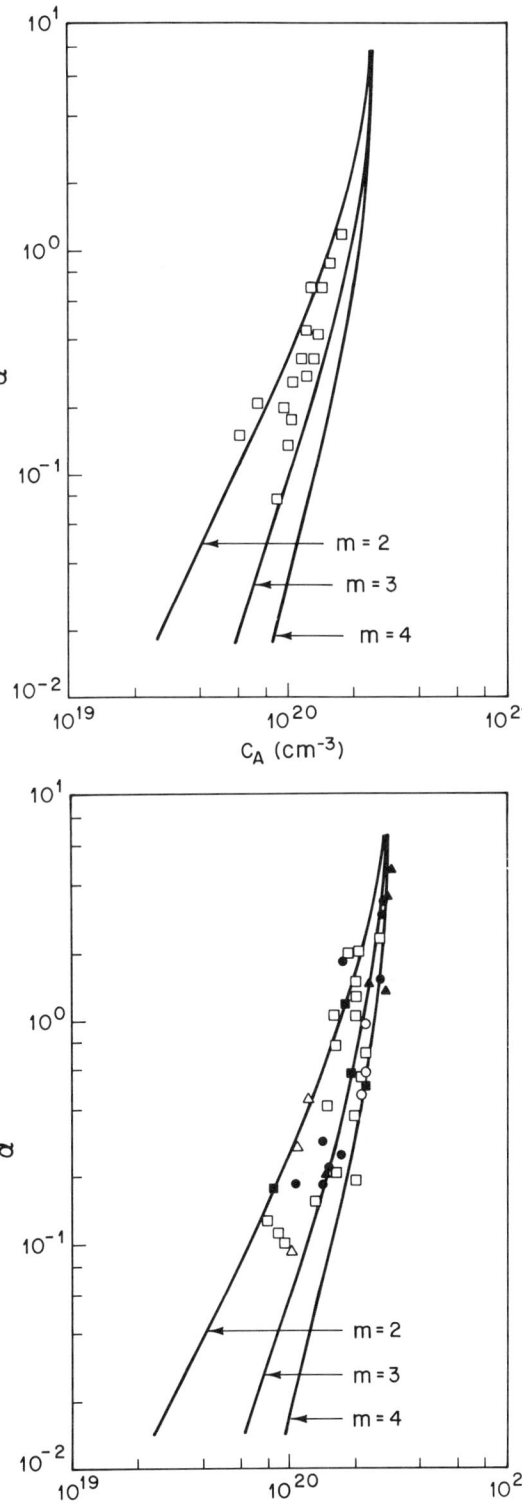

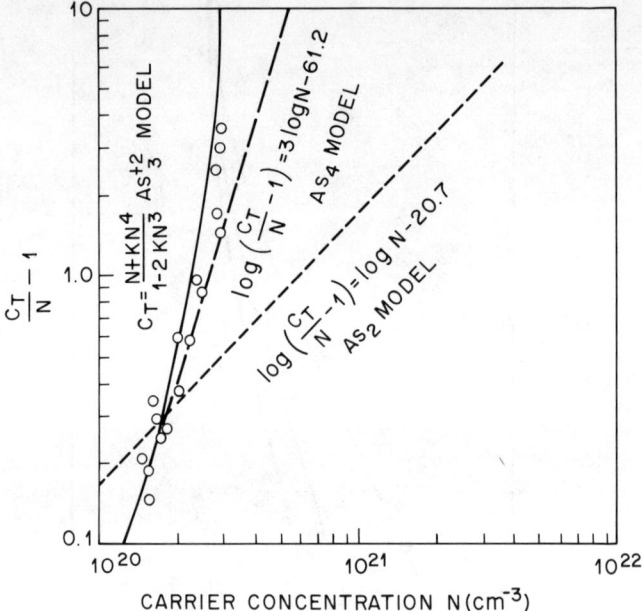

Fig. 50. Comparison of clustering models for As in Si after $T = 1000°C$ annealing. ○, experiment; C_T, total As; N, electrical carrier concentration. (From Tsai et al.[135])

Tsai et al.[135] have proposed a clustering model that strongly supports the results of Guerrero et al.[172] They have compared arsenic concentrations as a function of carrier concentration for three clustering models. Figure 50 compares arsenic concentrations as a function of carrier concentration for three different models for arsenic clustering in silicon. It is evident from these results that both the two- and the four-arsenic cluster models fail to fit experimental results. The atomic concentration C_T has been determined by neutron activation and the electrical carrier concentration N at room temperature by the Hall effect technique.

The new model assumes that three arsenic atoms and one electron cluster together at annealing temperatures, which can be expressed as

$$3As^+ + e^- \underset{\text{Annealing}}{\overset{\text{Temperatures}}{\rightleftharpoons}} As_3^{+2} \overset{25°C}{\rightleftharpoons} As_3. \quad (2.165)$$

The clustered atoms are electrically active at annealing temperatures and they become neutral at room temperature. The arsenic clustering coefficient k_{eq} is defined as

$$k_{eq} = [As_3^{+2}]/[As^+]^3 n, \quad (2.166)$$

where n is the carrier concentration at annealing temperature

$$n \cong [\text{As}^+] + 2[\text{As}_3^{+2}], \tag{2.167}$$

and $[\text{As}_3^{+2}]$ is the clustered As concentration, which, using Eq. (2.166), can be written as

$$[\text{As}_3^{+2}] = k_{eq}N^4/(1 - 2N^3 k_{eq}). \tag{2.168}$$

The total arsenic concentration C_T is then

$$C_T = [\text{As}^+] + 3[\text{As}_3^{+2}] = \frac{N + k_{eq}N^4}{1 - 2k_{eq}N^3}. \tag{2.169}$$

As shown in Fig. 50, Eq. (2.169) fits the experimental data very well at $T = 1000°C$.

The electrical solubility limit can be defined when the denominator in Eq. (2.169) goes to zero, or

$$C_{\text{As,max}} = (2k_{eq})^{-1/3}. \tag{2.170}$$

The model accounts for the fourth-order behavior found experimentally, because the concentration of As clusters is proportional to N^4.

One assumption underlying the preceding analysis is that equilibrium (or quasi-equilibrium) exists in a small region surrounding the complex. The rate of formation of arsenic clusters in this small region is diffusion controlled, but is probably much faster than the macroscopic diffusion. This condition prevails when the dimension of the small region is much smaller than the diffusion length. If this is not the case, it becomes necessary to solve the diffusion problem by including terms in the continuity equations that represent the kinetics of clustering.[122]

In real cases the kinetics of As clustering is very important, as shown in the work of Tsai et al.[135] At low annealing temperature ($T \sim 800°C$), the time of the heat treatment is usually much shorter than the time needed to arrive at equilibrium, which means that the carrier concentration must be determined by the kinetics of As. The kinetic equation of As clustering is

$$(d/dt)[\text{As}_3^{+2}] = k_C[\text{As}^+]^3 n - k_D[\text{As}_3^{+2}] \tag{2.171}$$

with

$$k_C/k_D = k_{eq}. \tag{2.172}$$

k_C and k_D are the clustering and declustering coefficients, respectively. It was found that

$$k_{eq} = 1.258 \times 10^{-70} \exp(2.062/kT) \tag{2.173}$$

and

$$k_D = 2.6226 \times 10^{10} \exp(-3.35/kT) \qquad (2.174)$$

fit the data in Fig. 50 extremely well.

Many physical phenomena can be explained with this model in terms of As clustering. As an example, Fig. 51 shows that a sample doped with As to a concentration of $2 \times 10^{21}/\text{cm}^3$ reaches equilibrium at $T = 1000°C$ with a carrier concentration $N = 2.81 \times 10^{20}/\text{cm}^3$. From the clustering model, Eq. (2.171), one can predict that subsequent anneals at lower temperatures will decrease the carrier concentration, as shown in the figure. Furthermore, the time required to reach equilibrium rapidly increases with decreasing temperature. With a constant annealing time of 30 min, the sample annealed at $T = 800°C$ shows the lowest carrier concentration.

(ii) *Phosphorus.* The diffusion process of P in Si is still not fully understood. For the case of very high concentrations, the impurity profiles display plateau, kink, and tail regions. Furthermore, the anomaly of the emitter-push effect[122] has been observed for phosphorus only.

Figure 52 shows two P profiles obtained with $POCl_3$ sources corresponding to two different surface chemical concentrations.[173]

Several models have been proposed to account for the diffusion anomalies; Schwettmann and Kendall[174,175] published a model based on three assumptions:

(1) Phosphorus diffuses via vacancy–phosphorus pairs (E centers), which exist in two charge states;

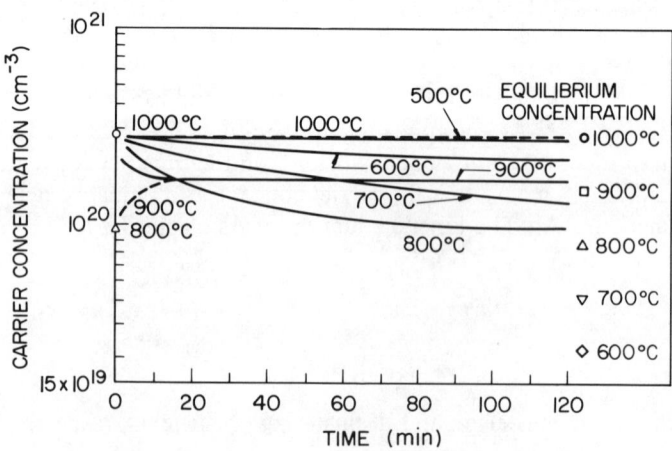

Fig. 51. Carrier concentration as a function of time. The equilibrium carrier concentrations are plotted at the right-hand side. The curvature of each curve represents the time to reach equilibrium. $C_T = 2 \times 10^{21}$ cm^{-3}. (From Tsai *et al.*[135])

(2) the negatively charged E center discharges at the kink concentration, but does not dissociate; and

(3) donors are compensated electrically by E centers.

Yoshida[176-178] also assumed P diffusion via E centers, but his model allowed E-center dissociation in the region beyond the kink.

The model of Fair and Tsai[179] is the most widely used today, partly because it has been incorporated into SUPREM.[19] A thorough presentation of this model can be found in another volume of this series.[118]

Hu et al.[180] have recently reviewed the salient features of P diffusion in Si and phenomena associated with it. They propose that P diffuses via a dual mechanism; i.e., a mixture of vacancy and interstitialcy mechanism with the interstitialcy mechanism being the dominant part. Under the assumption that the rate of change from an interstitialcy mechanism to the vacancy

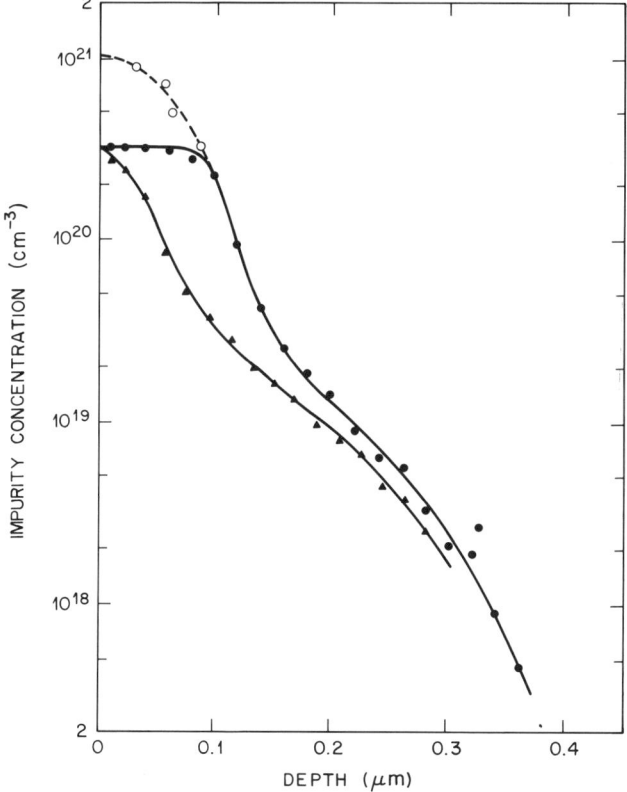

Fig. 52. Two examples of phosphorus diffusion profiles with different surface concentrations. (---) chemical concentration, (—) electrically active concentration. ▲, $POCl_3$, 0.05%; ○, ●, $POCl_3$, 0.27%; predeposition 920°C, 7 min. (From Solmi et al.[173])

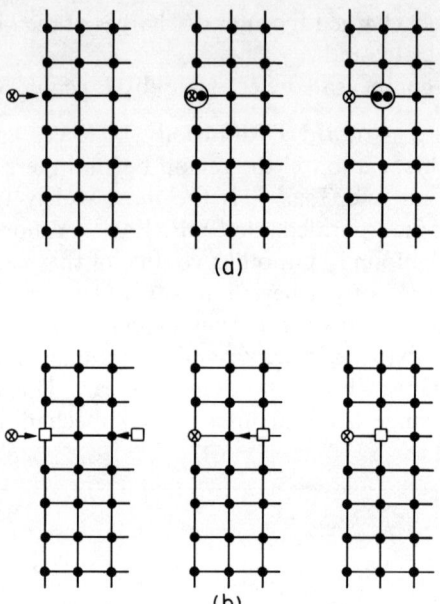

Fig. 53. Conceptual picture of (a) the generation of a silicon self-interstitial and (b) the consumption of two vacancies as a phosphorus atom enters the silicon lattice. ⊗, Phosphorus atom; ●, silicon atom; ⊗●, phosphorus interstitialcy; ●●, self-interstitialcy; □, vacancy; ⊗□, E-center. (From Hu et al.[180] Reprinted by permission of the publisher, The Electrochemical Society, Inc.)

mechanism is small, the typical kink, two-zone concentration profile can be obtained. The conversion also acts as a source of excess interstitials during the phosphorus diffusion.

Figure 53 describes schematically the processes involved. For a P atom to enter the lattice as an interstitialcy, it has to attach itself to a lattice site by sharing it with a Si atom already there. This interstitialcy becomes immediately mobile, until it converts into a substitutional atom by ejecting the interstitial. This conversion can either happen as

$$\text{Si} + \text{P}_\text{I} \underset{k_1'}{\overset{k_1}{\rightleftarrows}} \text{P}_\text{S} + \text{I} \qquad (2.175\text{a})$$

or

$$\text{V} + \text{P}_\text{I} \underset{k_2'}{\overset{k_2}{\rightleftarrows}} \text{P}_\text{S} + \text{Si}, \qquad (2.175\text{b})$$

where Si, V, P_S, and P_I denote a lattice atom, a vacancy, the self-interstitial phosphorus, and the substitutional and interstitialcy phosphorus, respectively. Tan and Gösele[181] have published further data to augment the paper of Hu et al.[180]

f. *Oxidation and Segregation*

For one-dimensional problems, silicon oxidation has been successfully modeled by the linear–parabolic formulation of Deal and Grove.[182] In a first approximation, the growth rate of SiO_2 on top of Si can be obtained by integrating the oxygen flux that diffuses through an SiO_2 layer of thickness d_{ox} to react at the surface

$$d_{ox}(t)/B + Ad_{ox}(t)/B = t + \tau, \quad (2.176)$$

where B and B/A are the parabolic and linear reaction constants, respectively. The factor τ can be considered as a correction factor describing the initial phase of oxidation (e.g., a natural oxide).

For very short oxidation, $t \ll A/4B$, Eq. (2.176) can be simplified to

$$d_{ox}(t) = (B/A)(t + \tau) \quad (2.177)$$

with the linear rate constant

$$B/A = (k_s h C^*)/(k_s + h)N_1. \quad (2.178)$$

C^* is the oxidant concentration at the SiO_2 surface, k_s is the chemical surface reaction rate constant for silicon, h the transport coefficient in the gas phase, and N_1 the number of oxidant molecules incorporated into a unit volume of growing oxide.

For long times, $t \gg A^2/4B$, Eq. (2.176) becomes

$$d_{ox}^2(t) = Bt. \quad (2.179)$$

Heavily doped samples can exhibit oxidation rates that are considerably higher than those for lightly doped silicon. It is well known that the linear growth rate B/A can increase by more than an order of magnitude in the case of highly phosphorus-doped samples, whereas the parabolic rate constant B is affected only slightly.

Ho and Plummer[183] have suggested a model that accounts for the increase in B/A by assuming that the total vacancy concentration increases with increasing doping level. Under intrinsic doping conditions,

$$\left(\frac{B}{A}\right)_I = R_1 + K_0 C_{VT}^I = C_1 \exp\left(-\frac{2.0 \text{ eV}}{kT}\right), \quad (2.180)$$

where C_{VT} is the total vacancy concentration and K_0 is a proportionality constant. R_1 accounts for nonvacancy-dominated interface reactions such as the incorporation of an oxygen atom onto a silicon site.

Under extrinsic conditions, $n \neq n_i^2$, and the total vacancy concentration increases with the doping level since the Fermi level shifts. We obtain

$$\frac{B}{A} = \left(\frac{B}{A}\right)_i \left[1 + \frac{K}{C_1} C_{VT}^i \exp\left(\frac{2.0 \text{ eV}}{kT}\right)\left(\frac{C_{VT}}{C_{VT}^i} - 1\right)\right]. \quad (2.181)$$

For very high doping levels, $C_{VT} > C_{VT}^i$ and the term in brackets becomes larger than unity. C_{VT} is given by

$$C_{VT} = \frac{1 + [V^+]\left(\frac{n_i}{n}\right) + [V^-]\left(\frac{n}{n_i}\right) + [V^{2-}]\left(\frac{n}{n_i}\right)^2}{1 + [V^+] + [V^-] + [V^{2-}]}. \quad (2.182)$$

with the vacancy concentrations

$$[V^+] = \exp\left(\frac{E^+ - E_i}{kT}\right), \quad (2.183a)$$

$$[V^-] = \exp\left(\frac{E_i - E^-}{kJ}\right), \quad (2.183b)$$

$$[V^{2-}] = \exp\left(\frac{2E_i - E^- - E^{2-}}{kT}\right). \quad (2.183c)$$

The agreement between theory and experiment is good. The model has been used with great success in a variety of process simulation programs.

A significant amount of effort has been directed toward finding efficient solutions to the general oxidation–redistribution problem.[184] Early attempts have used analytical techniques to solve oxidation cases under special conditions.[185-190] Such solutions do not generalize to arbitrary initial impurity profiles, concentration-dependent diffusivities, nonconstant segregation coefficients, etc.

Early numerical solutions used coordinate transformation methods[191] or relied on Green's function techniques,[192,193] but eliminated any reference to the oxide, thus ignoring effects related to the interaction between the oxide and the silicon.

Accurate numerical solutions require that both SiO_2 and Si be treated. Several authors[194-196] have published techniques and results of such numerical calculations. Kraft[194] has used the differential form of the diffusion equation; i.e.,

$$\frac{\partial C_1}{\partial t} = D_1 \frac{\partial^2 C_1}{\partial z^2} + (1 - \alpha) v_{ox} \frac{\partial C_1}{\partial z} \quad (2.184)$$

in SiO_2 and

$$\frac{\partial C_2}{\partial t} = D_2 \frac{\partial^2 C_2}{\partial z^2} \qquad (2.185)$$

in silicon. The factor α is the ratio of oxidized silicon to the final SiO_2 volume, and $v_{ox} = d_{ox}$ is the velocity of the moving interface relative to the oxide.

The drift term in Eq. (2.184) represents the convection occurring in the SiO_2 and it can be eliminated by a transformation to a moving coordinate system

$$\bar{z} = z + d_{ox}(t)(1 - \alpha) \qquad (2.186)$$

resulting in

$$\frac{\partial C_1}{\partial t} = D_1 \frac{\partial^2 C_1}{\partial \bar{z}^2} \quad \text{for} \quad 0 \leq \bar{z} \leq \bar{z}_0(t). \qquad (2.187)$$

Antoniadis et al.[195] have used the equation of impurity conservation, which, in the absence of generation–recombination terms, can be written as

$$\oint_{S(t)} \mathbf{F} \cdot \mathbf{n} \, dA = -\frac{d}{dt} \int_{V(t)} C \, dV, \qquad (2.188)$$

where $\mathbf{F}$ is the flux vector, $\mathbf{n}$ the outward unit vector, C the concentration, and $S(t)$ and $V(t)$ are the closed surface and volume, respectively. Equation (2.188) can be rewritten in the form

$$H(t) = (d/dt)G(t) \qquad (2.189a)$$

with

$$H(t) = \oint_{S(t)} \mathbf{F} \cdot \mathbf{n} \, dA \qquad (2.189b)$$

and

$$G(t) = -\int_{V(t)} C \, dV. \qquad (2.189c)$$

Two competing fluxes exist at the Si–SiO_2 interface. The motion of the interface during oxidation induces a flux F_b across the interface given by

$$F_b = -v_{ox}(C_{ox} - \alpha C_{Si}). \qquad (2.190)$$

The redistribution of dopant during oxidation is largely controlled by the segregation process at the moving interface. Under thermodynamic equilibrium conditions, this process is characterized by the equilibrium segregation coefficient m_{eq}, defined as

$$m_{eq} = (C_{Si}/C_{SiO_2})|_{eq}. \qquad (2.191)$$

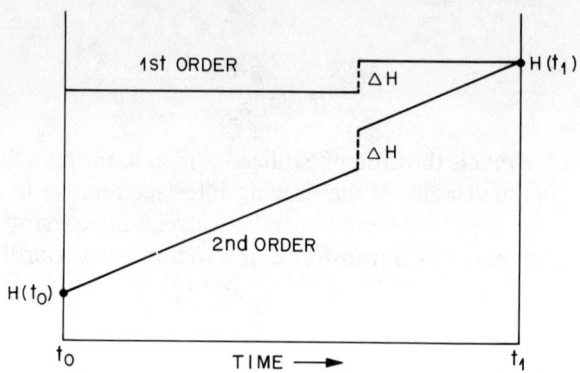

Fig. 54. Implicit first-order and second-order approximation of the integrand in Eq. (2.169). (From Dutton and Antoniadis.[184])

Equilibrium of the segregation process can be achieved by introducing a flux F_S across the interface given by

$$F_S = h_s[C_{ox} - (C_{Si}/m_{eq})], \qquad (2.192)$$

where h_s is a kinetic factor having the dimensions of a velocity.

Thermodynamic equilibrium is reached only when the condition $h_s \gg v_{ox}$ is fulfilled, which might not always be true since v_{ox} is inversely proportional to the square root of time.

From Eq. (2.190) or (2.192), it is evident that there exists a jump discontinuity of the impurity at the interface that moves through space during oxidation. This will introduce a discontinuity when the interface crosses the surface.

This problem can be avoided by numerical integration of Eq. (2.189):

$$\int_{t_0}^{t_1} H(t)\,dt = G(t_1) - G(t_0). \qquad (2.193)$$

Solving the integral by a first-order method, for continuous H, we obtain

$$H(t_1) = \frac{G(t_1) - G(t_0)}{t_1 - t_0}. \qquad (2.194)$$

When a discontinuity occurs at the time t in the interval $[t_0, t_1]$, as shown in Fig. 54, the approximation of the integral leads to

$$H(t_1) + \frac{t_1 - t}{t_1 - t_0} \Delta H = \frac{G(t_1) - G(t_0)}{t_1 - t_0}. \qquad (2.195)$$

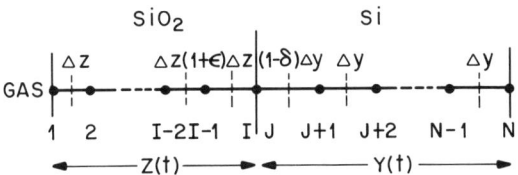

Fig. 55. Discretization method of the Si–SiO$_2$ region used in SUPREM. I and J are chosen so that $(I = \epsilon) \Delta z = z(t)$ and $-\frac{1}{2} \leq \epsilon < \frac{1}{2}$. J and δ are chosen so that $(J - \delta) \Delta y = \delta z(t)$ and $-\frac{1}{2} \leq \delta < \frac{1}{2}$. N remains fixed. The total number of nodes at any time is given by $N - J(t) - I(t) - 1$. (From Dutton and Antoniadis.[184])

Solving the integral by the trapezoidal method leads to an implicit second-order method. For continuous H, we obtain

$$\frac{1}{2}[H(t_1) + H(t_0)] = \frac{G(t_1) - G(t_0)}{t_1 - t_0} \qquad (2.196)$$

and for a jump discontinuity

$$\frac{1}{2}[H(t_1) - H(t_0)] + \frac{t_1 - t}{t_1 - t_0} \Delta H = \frac{G(t_1) - G(t_0)}{t_1 - t_0}. \qquad (2.197)$$

The spatial discretization is performed by subdividing the region and defining a set of grid points. The functions H and G are then approximated by node concentrations. In one dimension, Eq. (2.188) can be rewritten as

$$\int_{t_0}^{t_1} [F(z_2, t) - F(z_1, t)] \, dt = -\left[\int_{z_1(t_1)}^{z_2(t_1)} C(z_1, t_1) \, dz - \int_{z_1(t_0)}^{z_2(t_0)} C(z_1, t_0) \, dt \right]. \qquad (2.198)$$

where z_1 and z_2 correspond to $S(t)$ in Eq. (2.188).

Figure 55 illustrates the discretization technique implemented in SUPREM.[19] The moving SiO$_2$–Si interface is always on a grid point and the grid points on either side are a minimum specified distance away from the interface node. The subregion boundaries [z_1 and z_2 in Eq. (2.198)] are then placed at the midpoints between the nodes. The spatial discretization is now combined with the discretization in time, Eqs. (2.193) or (2.196).

Dutton and Antoniadis[184] have compared the numerical methods presented here with an analytical example.[186] The numerical solutions have been obtained with and without the interfacial induced flux.

The error in the impurity concentration at the interface, C_{Si}, and in the amount of impurities "lost" into SiO$_2$ during oxidation, Q_{ox}, is shown as a

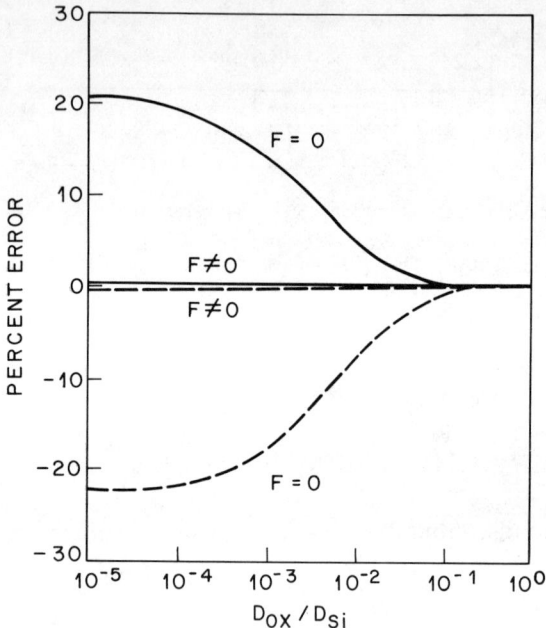

Fig. 56. Percentage error in C_{si} and Q_{ox} versus diffusivity ratio D_{ox}/D_{si}. Results shown are for $F = 0$ and $F \neq 0$. —, C_{si}; --, Q_{ox}. (From Dutton and Antoniadis.[184])

function of the impurity diffusivity ratio, D_{ox}/D_{Si}, in Fig. 56. The theoretical results yield $C_{Si} = 5.535$ and $Q_{ox} = 1.844$ in these cases. The results indicate the induced interfacial flux does not have much effect when the diffusivity ratio is near unity. For a decreasing ratio, however, errors of up to 20% occur if the interfacial flux is not included.

Whereas oxidation simulation in one dimension has been successful, no well-established theory is available at this point that would allow a first-principles simulation of two-dimensional oxidation phenomena. A typical example would be the lateral oxidation under a Si_3N_4 mask, giving rise to the bird's-beak phenomenon.

Several simplifying models have been proposed assuming one-dimensional approximations or using coordinate transformation methods.

Dutton et al.[196] and Chin et al.[197] obtain a quasi-two-dimensional profile by assuming that the oxide layer grows only in a vertical direction in a semi-recessed oxidation. They have generalized the original, one-dimensional Huang and Welliver[198] model, which assumes that the solution for the impurity distribution during thermal oxidation can be divided into two parts

$$C(z, t) = C_1(z, t) + C_2(z, t), \qquad (2.199)$$

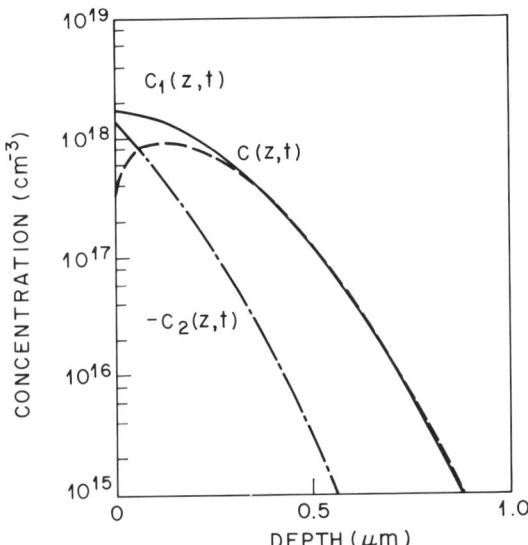

Fig. 57. Boron profile $C(z, t)$ calculated numerically and $C_1(z, t)$ calculated analytically. The difference is $C_2(z, t)$. (From Lee et al.[199] Reprinted by permission of the publisher, The Electrochemical Society, Inc.)

where C_1 is the impurity distribution resulting from thermal diffusion in an inert ambient and C_2 is the impurity diffusion resulting from the loss into the oxide due to the moving interface while growing an oxide of thickness d_{ox} in the presence of a segregation effect. Using Eq. (2.199), we have the interface boundary condition

$$D \frac{\partial C_2}{\partial z}\bigg|_{z=0} = \left(\frac{1}{m} - \alpha\right)[C_1(0, t) + C_2(0, t)]\frac{\partial d_{ox}}{\partial t} - D \frac{\partial C_1}{\partial z}\bigg|_{z=0}. \quad (2.200)$$

Substituting an exact solution for $C_1(z, t)$ into this equation, we can obtain the boundary condition for $C_2(z, t)$ at $z = 0$.

Although the exact analytical solution for $C_2(z, t)$ has not yet been obtained, C_2 can be calculated "backwards" if $C(z, t)$ is obtained by numerical means and C_1 analytically.

For a boron implant (80 keV, $10^{14}/cm^2$), Lee et al.[199] have calculated C_2 after the initial profile has been oxidized for 2 hr in wet ambient at $T = 1000°C$. The result is shown in Fig. 57, where $C(z, t)$ is the numerical and $C_1(z, t)$ is the analytical profile for the Gaussian initial condition. The difference between C and C_1 is $C_2(z, t)$, which has a negative value for all z and decreases rapidly in magnitude with increasing z. The total impurity profile, $C(z, t)$, is a well-behaved function with a single maximum in concentration and decreases monotonically on either side.

In the case of two dimensions it is assumed that the final impurity profile can be represented as

$$\bar{C}(x, y, z) = \bar{C}_1(x, y, z) + \bar{C}_2(x, y, z), \qquad (2.201)$$

where $\bar{C}_1$ and $\bar{C}_2$ have the same meaning as C_1 and C_2 have in Eq. (2.199). $\bar{C}_1$ is written as

$$\bar{C}_1(x, y, z) = C_x(x, t)C_1(z, t) \qquad (2.202)$$

with

$$C_x(x, t) = \sqrt{\pi Dt}\left[\text{erf}\frac{x + (W/2)}{2\sqrt{Dt}} - \text{erf}\frac{x - (W/2)}{2\sqrt{Dt}}\right], \qquad (2.203)$$

where W is the mask opening and C_1 is the same as in the one-dimensional case. The initial condition for the implanted profile is the same as the one given in Eq. (2.100). $\bar{C}_2(x, y, z)$ is obtained analogous to Eq. (2.200),

$$D\frac{\partial \bar{C}_2(x, z, t)}{\partial z}\bigg|_{z=0} = \left(\frac{1}{m} - \alpha\right)\bar{C}_2(x, 0, t)\frac{\partial d_{ox}}{\partial t} - D\frac{\partial \bar{C}_2(x, z, t)}{\partial z}\bigg|_{z=0}. \qquad (2.204)$$

The validity of this approach is discussed in Lee et al.[199] Since this equation is a boundary condition for the z direction only which has to hold for any x and t, $\bar{C}_2(x, y, z)$ will have the same x dependence as $\bar{C}_1(x, y, z)$, namely,

$$\bar{C}_2(x, z, t) = C_x(x, t)C_2(z, t) \qquad (2.205)$$

with C_2 given by the one-dimensional solution. The total profile is then

$$\bar{C}(x, z, t) = C_x(x, t)[C_1(z, t) + C_2(z, t)] - C_x(x, t)C(z, t). \qquad (2.206)$$

The preceding theory has been incorporated into the process simulation program SUPRA.[197] Figure 58 shows the result of the simulation of a MOS process through the definition of the polysilicon gate. After the silicon substrate has been etched, an 800-Å pad oxide, a 700-Å Si_3N_4 layer, and a 2-μm photoresist layer are deposited and selectively etched. The field region is implanted with boron ($5 \times 10^{12}/cm^2$, 100 keV) and subsequently oxidized for 3 hr at $T = 1000°C$ in wet oxygen. The nitride is then stripped and an unmasked enhancement implant is performed. Contours of constant boron concentration ranging from 10^{15} cm to $10^{17}/cm^3$ are plotted.

Penumalli[200] has used a coordinate transformation method to obtain approximate numerical solutions to the two-dimensional oxidation problem. The diffusion equation and the associated boundary conditions are trans-

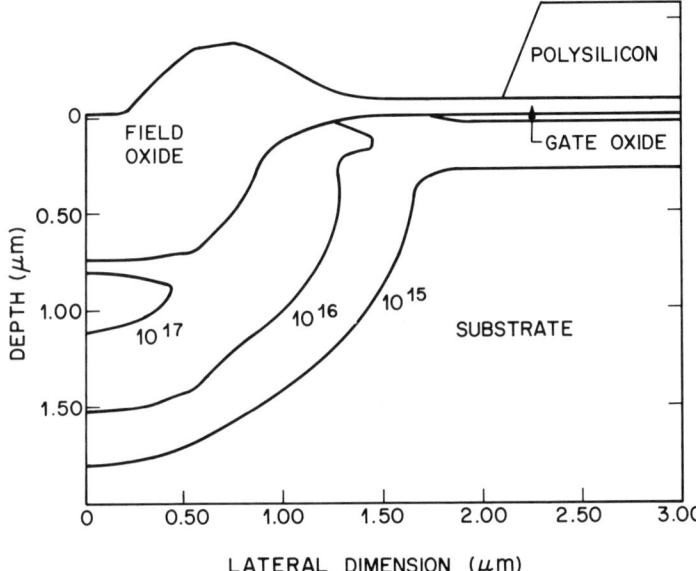

Fig. 58. SUPRA result for MOS process simulation up to the polysilicon gate definition. For explanation see text. (From Chin et al.[197])

formed from the physical domain to a coordinate system where the moving boundary remains stationary in time. Figure 59 illustrates this transformation schematically. With this approach the solution domain is simplified at the expense of complicating the underlying equations, which can be solved by straightforward numerical methods.

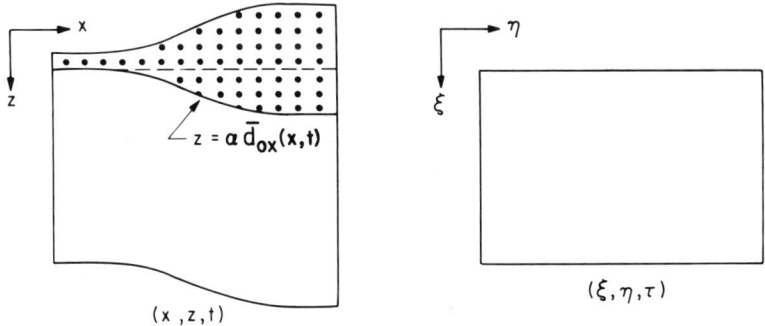

Fig. 59. Simulation regions (a) before and (b) after coordinate transformation. (From Penumalli.[200])

The coordinate transformation is defined by:

(1) The Si–SiO$_2$ interface boundary given by $z = \alpha \bar{d}_{ox}(x, t)$ be transformed to $\xi = 0$.

(2) The boundary deep in the substrate given by $z_l + \alpha \bar{d}_{ox}(x, t)$ be transformed into $\xi = x_l$.

(3) The planes of symmetry given by $\pm x_l/2$ be transformed to $\eta = \pm x_l/2$.

Mathematically, (1) to (3) can be represented as

$$\xi = z - \alpha \bar{d}_{ox}(x, t), \qquad \eta = x, \qquad \tau = t, \qquad (2.207)$$

where

$$\bar{d}_{ox}(x, t) = (d_{ox}(t)/2) \operatorname{erfc} \sqrt{2x/k d_{ox}(t)} \qquad (2.208)$$

and k_l is the ratio of lateral to vertical oxidation. Applying Eqs. (2.207) to the diffusion equation yields the following equations in the transformed plane

$$\frac{\partial C}{\partial \tau} = V(\dot{\bar{d}}_{ox} - D\bar{d}''_{ox})\frac{\partial C}{\partial \xi} + [1 + (\alpha \bar{d}'_{ox})^2]\frac{\partial}{\partial \xi}\left(D\frac{\partial C}{\partial \xi}\right)$$

$$+ \frac{\partial}{\partial \eta}\left(D\frac{\partial C}{\partial \eta}\right) - \alpha \bar{d}'_{ox}\left[\frac{\partial}{\partial \xi}\left(D\frac{\partial C}{\partial \eta}\right) + \frac{\partial}{\partial \eta}\left(D\frac{\partial C}{\partial \xi}\right)\right], \qquad (2.209)$$

where

$$\dot{\bar{d}}_{ox} = \frac{\partial \bar{d}_{ox}}{\partial \tau}, \qquad \bar{d}'_{ox} = \frac{\partial \bar{d}_{ox}}{\partial \eta}, \qquad \bar{d}''_{ox} = \frac{\partial^2 \bar{d}_{ox}}{\partial \eta^2}.$$

The spatial derivatives in Eq. (2.209) are discretized using finite differences. The diffusion terms are approximated by central differences and the convection terms by upwind differences. The backward Euler method is used for the time integration.

Figure 60 shows the region and the boron profile before (a) and after (b) local oxidation. In Fig. 60a, the as-implanted boron profile is shown. Oxidizing this profile for several hours in wet and dry atmospheres not only redistributes the boron considerably, but also results in the bird's-beak geometry in Fig. 60b.

The models discussed up to now are essentially two-dimensional extensions of the one-dimensional kinetic model of Deal and Grove. Wilson[201] has attacked the problem quite differently by reformulating oxidation as a Stefan problem[202] involving the solution of a diffusion equation that describes the diffusion of the oxidizing species through the oxide layer and of two partial

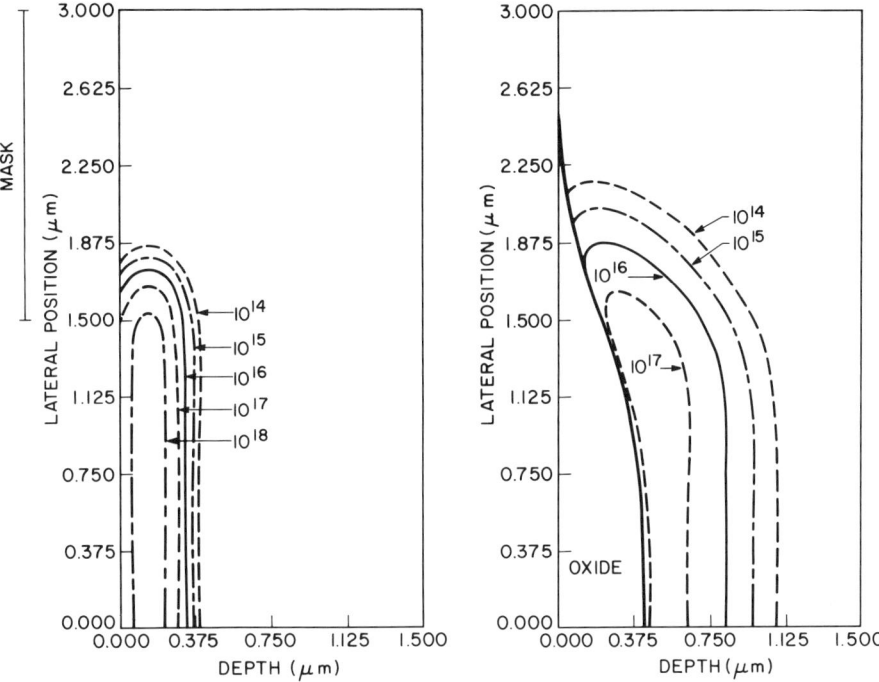

Fig. 60. (a) As-implanted and (b) oxidized boron profile under field oxide. (From Penumalli.[200])

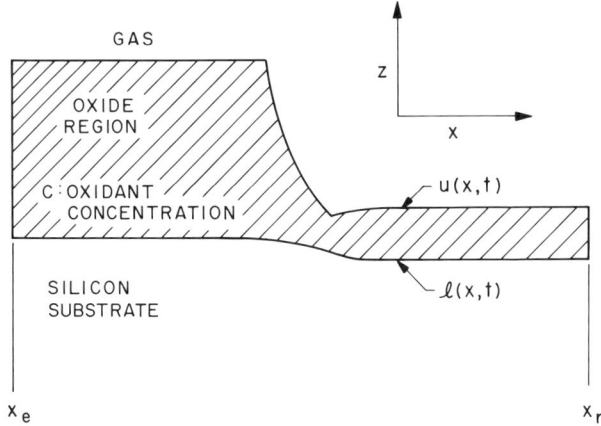

Fig. 61. Model configuration and definition of variables. (From Wilson.[201] Reprinted by permission of the publisher, The Electrochemical Society, Inc.)

differential equations that describe the motions of two interfaces: the gas–oxide interface and the Si–SiO$_2$ region are modeled. The gas–oxide interface is given by $z = u(x, t)$ and the Si–SiO$_2$ interface by $z = \ell(x, t)$ (see Fig. 61). The region boundaries x_ℓ and x_r are far away from the region of interest.

The oxidizing species diffuses through the oxide layer and reacts at the Si–SiO interface. The concentration C of the oxidizing species satisfies the diffusion equation

$$\partial C / \partial t = D \nabla^2 C \qquad (2.210)$$

with the diffusivity D.

At the gas–oxide interface, the normal component of the flux is

$$D(\partial C / \partial n) = h(C^* - C) \quad \text{for} \quad z = u(x, t) \qquad (2.211)$$

and at the Si–SiO$_2$ interface it is

$$-D(\partial C / \partial n) = k_s C \quad \text{for} \quad z = \ell(x, t). \qquad (2.212)$$

The factors h and k_s are the rate constants for oxygen transfer at the two interfaces and C^* is the equilibrium concentration of the oxidant in the oxide. At the vertical boundaries

$$\partial C / \partial n = 0 \quad \text{for} \quad x = x_\ell \quad \text{and} \quad x = x_r. \qquad (2.213)$$

The original positions of the two interfaces are given by

$$\begin{aligned} u_0(x) &= u(x, 0) \\ \ell_0(x) &= \ell(x, 0). \end{aligned} \qquad (2.214)$$

As silicon is oxidized, the lower boundary moves down and the upper boundary moves up. Wilson has assumed that each element of oxide moves straight up in the z direction, thus ignoring any "flow" of the oxide during the oxidation process. It follows that

$$\alpha(\partial u / \partial t) = -(1 - \alpha)(\partial \ell / \partial t) \qquad (2.215)$$

and

$$\alpha[u(x, t) - u_0(x)] = -(1 - \alpha)[\ell(x, t) - \ell_0(x)]. \qquad (2.216)$$

If C_1 is the total number of oxidant molecules incorporated into a unit volume of the oxide layer, then

$$C \int_{x_\ell}^{x_r} \left(\frac{\partial u}{\partial t} - \frac{\partial \ell}{\partial t} \right) dx = \int_{x_\ell}^{x_r} \left(-D \frac{\partial C}{\partial n} \right) \bigg|_{\ell(x,t)} ds = \int_{x_\ell}^{x_r} k_s C \frac{ds}{dx} dx, \qquad (2.217)$$

where ds represents an infinitesimal arc length element along the lower boundary. Thus,

$$-\frac{C_1}{\alpha}\frac{\partial \ell(x,t)}{\partial t} = k_s C[x, \ell(x,t), t]\left[1 + \left(\frac{\partial \ell}{\partial x}\right)^2\right]^{1/2}. \qquad (2.218)$$

Equations (2.210), (2.215), and (2.218) together with the boundary conditions constitute a two-dimensional Stefan problem. The diffusion equation, Eq. (2.210), can be simplified by using the quasi-static approximation

$$\nabla^2 C = 0, \qquad (2.219)$$

as did Deal and Grove in the one-dimensional version of the problem.

The solution procedure is as follows:

(1) Assume an initial oxide configuration at $t = 0$.
(2) Solve Laplace's equation, Eq. (2.219) for the concentration C in that region.
(3) Evaluate the concentration at the lower boundary ℓ and substitute into Eq. (2.218).
(4) Solve Eqs. (2.218) and (2.215) to find new positions of the lower and upper boundaries at $t = t_1$.
(5) Go back to step 2 until the cycle is completed.

The actual implementation of the numerical method is nontrivial and involves a number of delicate issues.

Figure 62 shows computational results for an oxide layer grown in a window with a field oxide slope of 60°. The dashed lines indicate the initial configuration. Qualitatively, very good agreement is observed between calculated results and experimental data.

A serious shortcoming of the Wilson model is the neglect of oxide flow during oxidation, which is caused by the assumption that the oxide is allowed to grow only vertically.

(i) *The Tan–Gösele Model.* As we have already discussed, the free volume required for SiO_2 growth is supplied by viscoelastic flow of the newly formed oxide near the surface. In this context the generation and injection of interstitials is not the result of the chemical reaction leading to the growth, but rather a consequence of the stress buildup.

The growth of a new SiO_2 layer across the interface amounts to the insertion of oxygen atoms in the bond-centered positions of Si atoms, which, if unrelaxed, would lead to a linear expansion of about 30% in x, y, and z directions.[157] In the z direction this linear expansion can happen directly by lifting the free SiO_2 surface. In the x–y plane the required expansion cannot be realized that easily. Initially, the SiO_2 material experiences elastic plane strain with an associated compressive plane stress τ, which transports the excess material upward. This transport phenomenon can be described as viscoelastic flow. As soon as enough strain has developed, the flow will start until all the strain is fully released, i.e., all extra material has moved to its upper layers. The flow should mainly be confined to a thin oxide region of width w at the interface, because newly formed SiO_2 has a much lower viscocity than "annealed" material in upper layers.

The flow (or strain) rate $d\gamma/dt$ induced by the stress τ is given by

$$d\gamma/dt \sim \tau^n. \tag{2.220}$$

Under steady-state conditions, the required volume expansion rate is equal to the relaxation rate due to viscoelastic flow. Since this required volume expansion rate is directly proportional to the flux of indiffusing oxygen

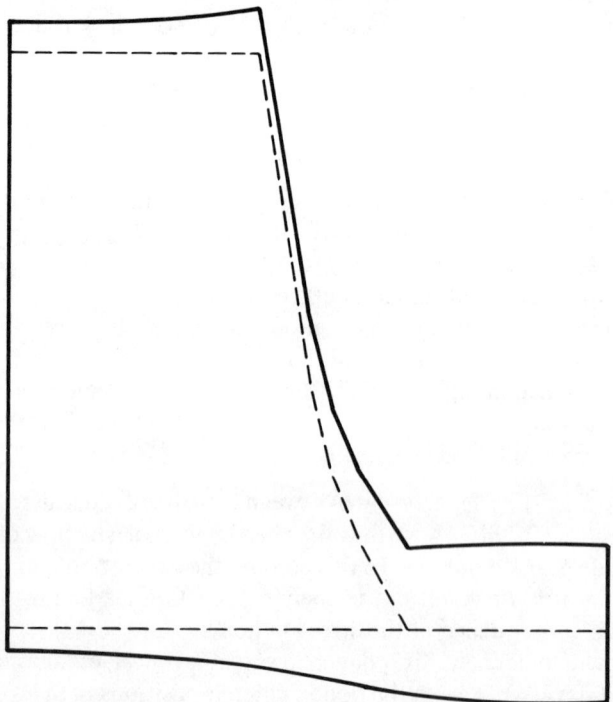

Fig. 62. Computational results for an oxide layer with an initial wall angle of 60°. The dashed lines indicate the initial configuration. (From Wilson.[201] Reprinted by permission of the publisher, The Electrochemical Society, Inc.)

(i.e., the oxidation rate)

$$dy/dt \sim dz/dt \qquad (2.221)$$

or

$$\tau \sim (dz/dt)^{1/n}. \qquad (2.222)$$

It is the presence of this stress τ that gives rise to a higher concentration of interstitials, namely, $S_I \sim \tau$ under dynamical equilibrium conditions at the interface. The term τ^n can be written as $\tau/\bar{\eta}(\tau)$, where $\bar{\eta}$ is an averaged viscocity that depends on the state of the material and/or the stress level. While an amorphous solid has a time-independent viscosity η_∞ and a simple linear relationship exists between $d\tau/dt$ and τ ($n = 1$), freshly formed and unannealed SiO_2 has a smaller initial viscosity η_i, which increases in time toward η_∞.

Charitat and Martinez[203] have recently critiqued the Tan–Gösele model. They have studied the stress evolution in the oxide taking into account the relaxation of viscous flow in the SiO_2.

Eernisse[204] was the first to demonstrate viscous flow of thermally grown SiO_2 at temperatures as low as 960°C. He examined dry and steam oxides and he found viscocities similar to synthetic fused silica.[205]

Chin et al.[206-208] published the first general two-dimensional oxidation model capable of simulating bird's-beak geometries that includes the viscous flow of the oxide. In a manner analogous to the Tan–Gösele model, they split the oxidation process into two distinct mechanisms: the flow of oxygen through SiO_2, and (2) the flow of the oxide due to the volume expansion and associated stress buildup. They assume that the Si_3N_4 mask is an ideal elastic body and ignore the elasticity of the oxide. As in Wilson,[201] oxygen diffusion is simplified using the quasi-static approximation [Eq. (2.219)] with the additional assumption that the diffusivity is independent of concentration and stress.

For temperatures above 960°C, the hydrodynamic equation

$$\delta(\partial \mathbf{v}/\partial t) = -\nabla P + \mu \nabla^2 \mathbf{v} + \mathbf{F} \qquad (2.223)$$

describes the flow, where δ, μ, and P are the density, viscosity, and pressure, respectively, $\mathbf{v}$ is the velocity, and $\mathbf{F}$ is the gravitational force. For SiO_2, Eq. (2.223) reduces to a creeping-flow equation

$$\mu \nabla^2 \mathbf{v} = \nabla P, \qquad (2.224)$$

where the viscous force is balanced by the pressure gradient. Furthermore, the force has to satisfy the incompressibility condition

$$\nabla \cdot \mathbf{v} = 0. \qquad (2.225)$$

The two-dimensional model has been applied to the problem of semirecessed oxidation. The solution of Eqs. (2.223)–(2.225) is discussed in Chin et al.[207]

Simulation results for the growth of two field oxides (0.7 and 0.53 μm) at 1025°C in a wet ambient are shown in Fig. 63a and b. The initial pad oxide has a thickness of 480 Å, and the nitride layer thicknesses are 250 and 1700 Å, respectively.

Figures 64a and b present the flow pattern for both cases, revealing the complicated nature of two-dimensional oxide growth. The oxide begins to move in a direction normal to the interface. In the case of the thin nitride layer, this motion more or less continues during the later stages of oxide growth. The growth of the oxide with the thick Si_3N_4 layer is strongly affected by the nitride. The compressive stress due to the nitride reduces growth by forcing SiO_2 to flow toward the open surface. This becomes more evident by analyzing the stress results in Fig. 65. Under the thin nitride the stress is tensile and it becomes compressive on the SiO_2 surface. This stress is transferred toward the silicon surface, thereby changing the flow pattern (Fig. 64b).

Matsumoto and Fukuma[209] have extended the model by Chin et al.[206-208] to lower temperatures. They assume that both SiO_2 and Si_3N_4 are linear

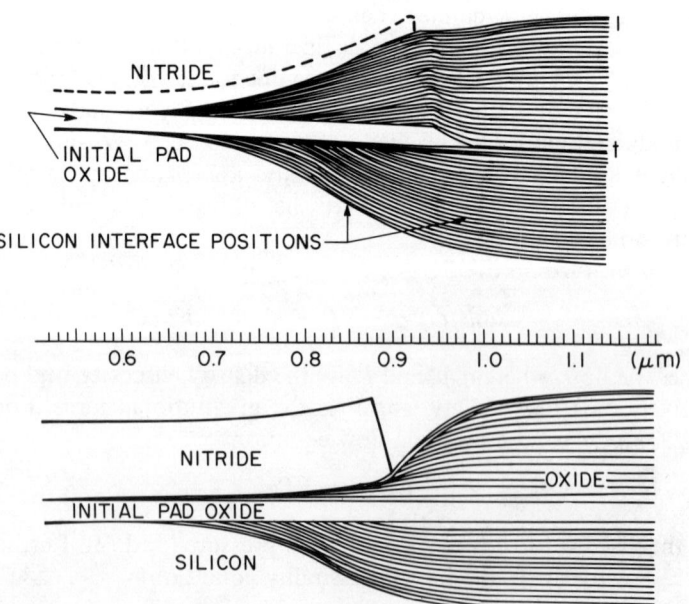

Fig. 63. Simulated results of isolation oxide growth for (a) a thin and (b) a thick nitride mask. (From Chin et al.[206] © 1982 IEEE.)

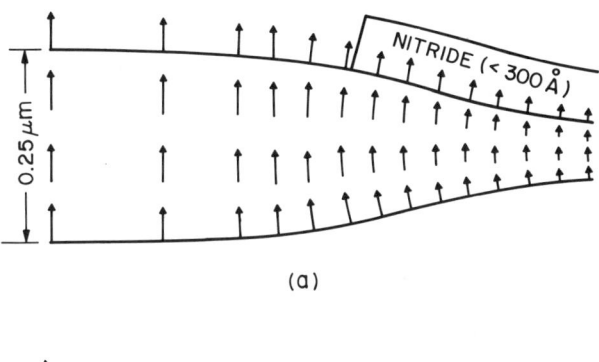
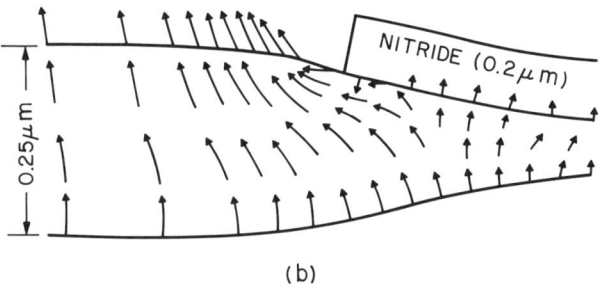

Fig. 64. Velocity field distribution illustrating the oxide flow for (a) a thin and (b) a thick nitride mask. The arrows represent the direction as well as the magnitude of the velocity vector. (From Chin et al.[206] © 1982 IEEE.)

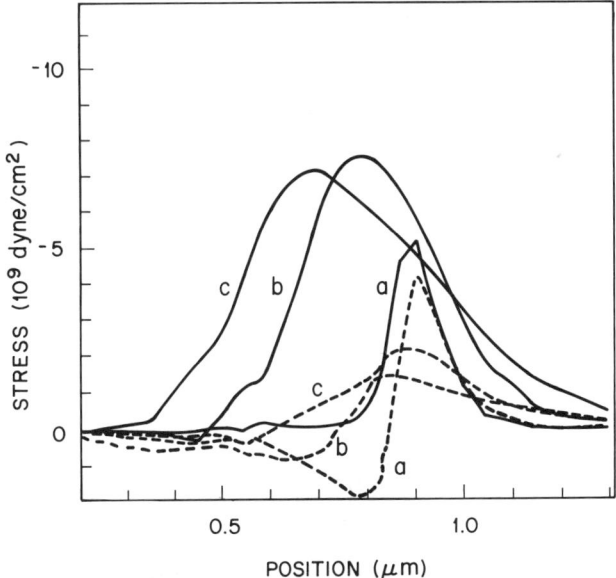

Fig. 65. Stress distribution along the silicon interface for different oxidation times (wet, at 1000°C) and two nitride thicknesses. Nitride thickness: - - -, 0.025 μm; —, 0.15 μm. (a) 6.5 min, (b) 29 min, (c) 60 min. (From Chin et al.[206] © 1982 IEEE.)

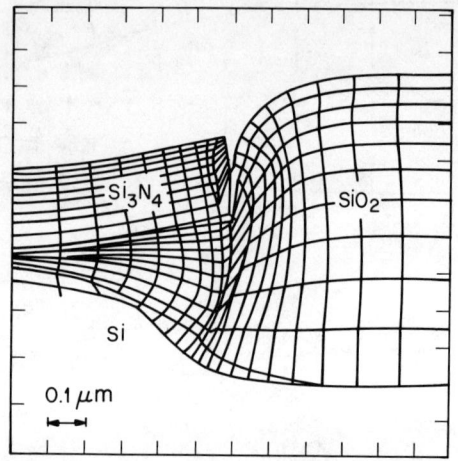

Fig. 66. Final shape for a local oxide with an initial pad oxide thickness of 500 Å. (From Matsumoto and Fukuma.[209] © 1983 IEEE.)

viscoelastic bodies. A typical result for a long oxidation (4 hr, 1000°C, wet) is shown in Fig. 66. The authors claim good agreement with experimental results.

g. Analytical Solutions in One and More Dimensions

Although numerical solutions to the diffusion equation can be obtained for a variety of different problems by using publicly available software, analytical solutions are still very helpful to provide quick estimates (or to check computer programs). In the following we summarize the solutions for a variety of cases that are of interest in silicon processing. For all cases we assume that the diffusion coefficient is independent of concentration. More complicated solutions can be found in standard textbooks on diffusion and heat conduction.[31,33]

(i) *Constant Surface Concentration.* We require a solution of

$$\partial C/\partial t = D(\partial^2 C/\partial z^2) \qquad (2.226)$$

subject to the boundary conditions

$$C = 0 \quad \text{for} \quad z > 0 \quad \text{at} \quad t = 0, \qquad (2.227a)$$

$$C = C_0 \quad \text{for} \quad z = 0 \quad \text{at} \quad t > 0. \qquad (2.227b)$$

The solution is

$$C(z, t) = C_0 \, \text{erfc}(2/\sqrt{Dt}). \qquad (2.228)$$

PHYSICS OF VLSI PROCESSING AND PROCESS SIMULATION

The amount of material that has entered the solid in time t is given by

$$M = 2C_0(Dt/\pi)^{1/2}. \tag{2.229}$$

The problem of the semi-infinite medium can be generalized to the case where the concentration is C_0 throughout the region, initially, and the surface is maintained at a constant concentration C_1. The solution is

$$\frac{C - C_1}{C_0 - C_1} = \text{erf}\left(\frac{z}{2\sqrt{Dt}}\right). \tag{2.230}$$

The special case, Eq. (2.228), is obvious.

(ii) *Instantaneous Source.* A thin, uniformly doped region at the silicon surface is used as a doping source with initial impurity concentration C_0. Under the conditions that the total amount of diffusing substance ϕ remains constant and the boundary condition

$$\partial C/\partial z = 0 \quad \text{for} \quad z = 0. \tag{2.231}$$

For these conditions, Eq. (2.226) has the solution

$$C(z, t) = \frac{\phi}{\sqrt{\pi Dt}} \exp\left(-\frac{z^2}{4Dt}\right). \tag{2.232}$$

(iii) *Surface Evaporation Condition.* During annealing steps, loss of impurities occurs by surface evaporation. To treat this problem mathematically, it is assumed that the rate of exchange of dopant between the solid and the gas phase is directly proportional to the difference between the actual concentration C_s at the surface at any time and the concentration C_0 that would be in equilibrium with the vapor pressure in the atmosphere far from the surface. This gives the boundary condition

$$-D(\partial C/\partial z) = h(C_0 - C_s) \quad \text{at} \quad z = 0, \tag{2.233}$$

where the factor h is the evaporation coefficient. Using Eq. (2.233), and assuming that the concentration in the semi-infinite medium is initially C_2 everywhere, the solution is

$$\frac{C - C_2}{C_0 - C_2} = \text{erfc}\frac{z}{2\sqrt{Dt}} - \exp(\alpha z + \alpha^2 Dt)\,\text{erfc}\left[\frac{z}{2\sqrt{Dt}} + \alpha\sqrt{Dt}\right] \tag{2.234}$$

with $\alpha = h/D$.

The rate at which the total amount of diffusing substrate Q changes in time is given by

$$\frac{dQ}{dt} = -\left(D\frac{\partial C}{\partial z}\right)\bigg|_{z=0} = h(C_0 - C_s) \tag{2.235}$$

and by inserting (2.234) and integration with respect to t

$$Q = \left(\frac{C_0 - C_2}{\alpha}\right)\left[\exp(\alpha^2 Dt)\operatorname{erfc}\frac{\alpha}{\sqrt{Dt}} - 1 + \frac{2}{\sqrt{\pi}}\alpha\sqrt{Dt}\right]. \quad (2.236)$$

(iv) *Diffusion from a Thick Film.* This case considers out-diffusion from a film deposited on the surface, when the thickness of the film cannot be ignored, i.e., the thickness is not small compared to $\sqrt{Dt}$.

If the film is initially confined to the region $-s < z < +s$, this problem has the solution

$$C = \frac{C'}{2}\left[\operatorname{erf}\frac{s+z}{2\sqrt{dt}} + \operatorname{erf}\frac{s-z}{2\sqrt{Dt}}\right], \quad (2.237)$$

where C' is the initial concentration in the film. The solution is symmetrical around $z = 0$.

(v) *The Infinite Composite Medium.* Here we consider systems in which two media are present. The region $z > 0$ is of one substance in which the diffusion coefficient is D_1, and the region $z < 0$ has the diffusion coefficient D_2.

Initially, the region $z > 0$ is at a uniform concentration C_0, and the concentration in $z < 0$ is zero. If C_1 denotes the concentration in $z > 0$ and C_2 in $z < 0$, the interface boundary conditions may be written as

$$C_2/C_1 = m, \quad (2.238)$$

$$D_1(\partial C_1/\partial z) = D_2(\partial C_2/\partial z), \quad z = 0, \quad (2.239)$$

where m is the segregation coefficient when final equilibrium is reached. Equation (2.239) expresses the fact that there is no accumulation of dopant at the interface. The solutions are

$$C_1 = \frac{C_0}{1 + m\sqrt{D_2/D_1}}\left[1 + m\sqrt{D_2/D_1}\operatorname{erf}\frac{z}{2\sqrt{D_1 t}}\right], \quad (2.240)$$

$$C_2 = \frac{mC_0}{1 + m\sqrt{D_2/D_1}}\operatorname{erfc}\frac{|z|}{2\sqrt{D_2 t}} \quad (2.241)$$

As the diffusion proceeds, the concentrations at the interface remain constant at the values

$$C_1 = \frac{C_0}{1 + m\sqrt{D_2/D_1}} \quad \text{and} \quad C_2 = \frac{mC_0}{1 + m\sqrt{D_2/D_1}}. \quad (2.242)$$

The problem becomes more complicated if we assume that there is a "contact resistance" at $z = 0$, which replaces Eq. (2.238) by

$$D_1(\partial C_1/\partial z) + h(C_2 - C_1) = 0 \quad (2.243)$$

with Eq. (2.239) still valid. The solutions are

$$C_1 = \frac{C_0}{1 + \sqrt{D_2/D_1}} \left\{ 1 + \sqrt{D_2/D_1} \left[\operatorname{erf} \frac{z}{2\sqrt{D_1 t}} + \exp(\alpha_1 z + \alpha_1^2 Dt) \right. \right.$$
$$\left. \left. \times \operatorname{erfc}\left(\frac{z}{2\sqrt{D_1 t}} + \alpha_1 \sqrt{D_1 t} \right) \right] \right\} \quad (2.244)$$

$$C_2 = \frac{C_0}{1 + \sqrt{D_2/D_1}} \left[\operatorname{erfc} \frac{|z|}{2\sqrt{D_2 t}} - \exp(\alpha_2 z + \alpha_2^2 D_2 t) \right.$$
$$\left. \times \operatorname{erfc}\left(\frac{|z|}{2\sqrt{D_2 t}} + \alpha_2 \sqrt{D_2 t} \right) \right] \quad (2.245)$$

with

$$\alpha_1 = (h/D_1)(1 + \sqrt{D_1/D_2}) \quad \text{and} \quad h_2 = (h/D_2)(1 + \sqrt{D_2/D_1}). \quad (2.246)$$

(vi) *Impurity Redistribution During Oxidation.* Grove et al.[185] have obtained a solution to Eqs. (2.184) and (2.185). They have used two different coordinate systems whose origins are fixed with respect to the moving Si–SiO$_2$ interface. This way, the diffusion equation can be expressed in a simple form, avoiding the velocity term in Eq. (2.184). Figure 67 shows the coordinate system. The coordinate x describes the distance within the oxide,

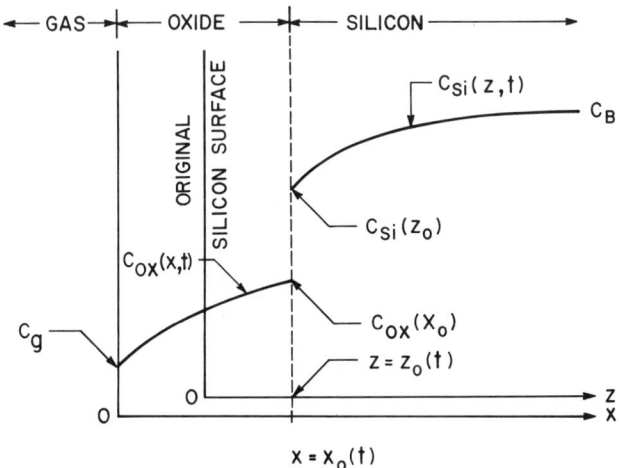

Fig. 67. Coordinate systems used to describe impurity redistribution. (From "MOS (Metal Oxide Semiconductor) Physics and Technology," by E. Nicollian and J. R. Brews.[210] Copyright © 1982 John Wiley & Sons, Inc. Reprinted by permission of John Wiley & Sons, Inc.)

measured from the oxide surface; and the coordinate z describes the distance within the silicon, measured from the initial silicon surface.

The impurity concentration within the oxide C_{ox} is determined by a solution to

$$\frac{\partial C_{ox}(x,t)}{\partial t} = D_{ox} \frac{\partial^2 C_{ox}(x,t)}{\partial x^2}, \qquad 0 < x < d_{ox}, \qquad (2.247)$$

where D_{ox} is the diffusion coefficient in the oxide and $d_{ox} = d_{ox}(t)$ is the oxide thickness. Similarly, the impurity concentration within the silicon C_{Si} is given by the solution of

$$\frac{\partial C_{Si}(z,t)}{\partial t} = D_{Si} \frac{\partial^2 C_{Si}(z,t)}{\partial z^2}, \qquad z > z_0(t), \qquad (2.248)$$

where D_{Si} is the diffusion coefficient in the silicon and $z_0(t)$ is the position of the interface at time t relative to the initial silicon surface. Equations (2.247) and (2.248) have been solved under the following conditions:

$$C_{Si}(z, 0) = C_B, \qquad (2.249)$$

$$C_{ox}(0, t) = C_g, \qquad (2.250)$$

$$C_{Si}(z, t) \to C_B \quad \text{as} \quad z \to \infty, \qquad (2.251)$$

$$C_{Si}(z_0, t)/C_{ox}(d_{ox}, t) = m, \qquad (2.252)$$

$$C_{Si}(z_0, t) \frac{\partial d_{ox}}{\partial t}\left(\alpha - \frac{1}{m}\right) = D_{ox} \frac{\partial C_{ox}}{\partial x}\bigg|_{x=d_{ox}} - D_{Si} \frac{\partial C_{Si}}{\partial z}\bigg|_{t=z_0}, \qquad (2.253)$$

to yield in the oxide

$$\frac{C_{ox}(x,t) - C_g}{C_{ox}(d_{ox}, t) - C_g} = \frac{\operatorname{erf}(x)/(2\sqrt{D_{ox} t})}{\operatorname{erf}(\sqrt{B/4D_{ox}})}, \qquad 0 < x < d_{ox}, \qquad (2.254)$$

and in the silicon

$$\frac{C_{Si}(z, t) - C_B}{C_{Si}(z_0, t) - C_B} = \frac{\operatorname{erfc}\left(\dfrac{z}{2\sqrt{D_{Si} t}}\right)}{\operatorname{erfc}(\alpha \sqrt{B/4D_{Si}})}, \qquad z > z_0, \qquad (2.255)$$

where C_g is the impurity concentration at the SiO_2 surface and it has been assumed that $d_{ox}(t) = \sqrt{Bt}$.

Av-Ron et al.[186] have considered more general growth rates, $d_{ox}(t) \sim t^n$ for $n \neq 1/2$. They have given results for dopant redistribution during postoxidation anneals, using Eq. (2.254) and Eq. (2.255) as initial conditions.

(vii) *Two-Dimensional Diffusion of Impurities Near a Mask Edge.* Kennedy and O'Brien[211] have given an analytical solution to the two-dimensional diffusion equation in the vicinity of a mask edge. The geometrical configuration of the diffusion problem is given in Fig. 68.

A semi-infinite block of semiconductor material is covered with an impenetrable "ideal" mask on half of its surface ($\phi = \pi$); on the other half a constant impurity concentration C_0 is predeposited at the time $t = -0$.

The impurity concentration $C = C(r, \phi, t)$ satisfies the diffusion equation in polar coordinates

$$D \nabla^2 C = D \left(\frac{\partial^2 C}{\partial r^2} + \frac{1}{r} \frac{\partial C}{\partial r} + \frac{1}{r^2} \frac{\partial^2 C}{\partial \phi^2} \right) = \frac{\partial C}{\partial t} \quad (2.256)$$

with the following boundary conditions

$$C = C_0 \quad \text{at} \quad \phi = 0, \quad \partial C/\partial y = 0 \quad \text{at} \quad \phi = \pi, \quad C = 0 \quad \text{at} \quad t = 0. \quad (2.257)$$

The diffusion coefficient D is assumed to be concentration independent.

It is a tedious task to solve Eq. (2.256) under the conditions of Eq. (2.257). Kennedy and O'Brien have attacked the problem by Laplace transformation. Cherednichenko et al.[212] used the method of separation of variables that is somewhat simpler and gives the same result. They expressed their final result in terms of infinite sums of hypergeometric series

$$C(r, \phi, t) = C_0 \left\{ 1 - \frac{2}{\pi} \sum_{n=0}^{\infty} \sin[(n + \tfrac{1}{2})\phi] \frac{\Gamma(n + \tfrac{1}{2})}{2\Gamma(n + \tfrac{3}{2})} z^{(n + 1/2)/2} \right.$$
$$\left. \times M \left(\frac{n + \tfrac{1}{2}}{2}, n + \tfrac{3}{2}, -z \right) \right\} \quad (2.258)$$

with $z = r^2/4Dt$.

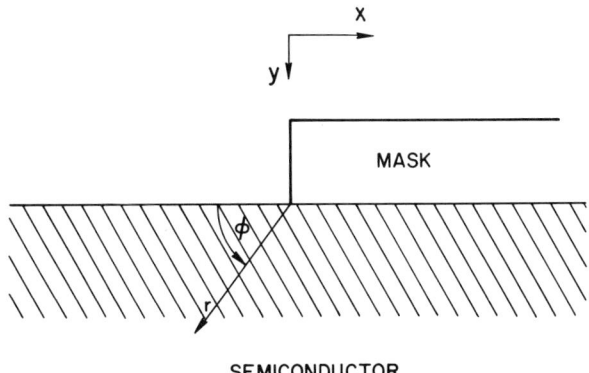

Fig. 68. Problem geometry for constant surface concentration diffusion in two dimensions.

(viii) *Instantaneous Source diffusion in Three Dimensions.* The solution to the instantaneous source diffusion, Eq. (2.232), has been generalized to three dimensions by Kennedy and O'Brien.[211] It can be written as

$$C(x, y, z, t) = \frac{Q}{4\sqrt{\pi Dt}} \exp\left(\frac{-z^2}{4Dt}\right)\left(1 + \text{erf}\frac{x}{2\sqrt{Dt}}\right)\left(1 + \text{erf}\frac{y}{2\sqrt{Dt}}\right) \quad (2.259)$$

Figure 69 illustrates calculated contours of constant impurity atom concentration for both a constant surface concentration process, Eq. (2.258), and

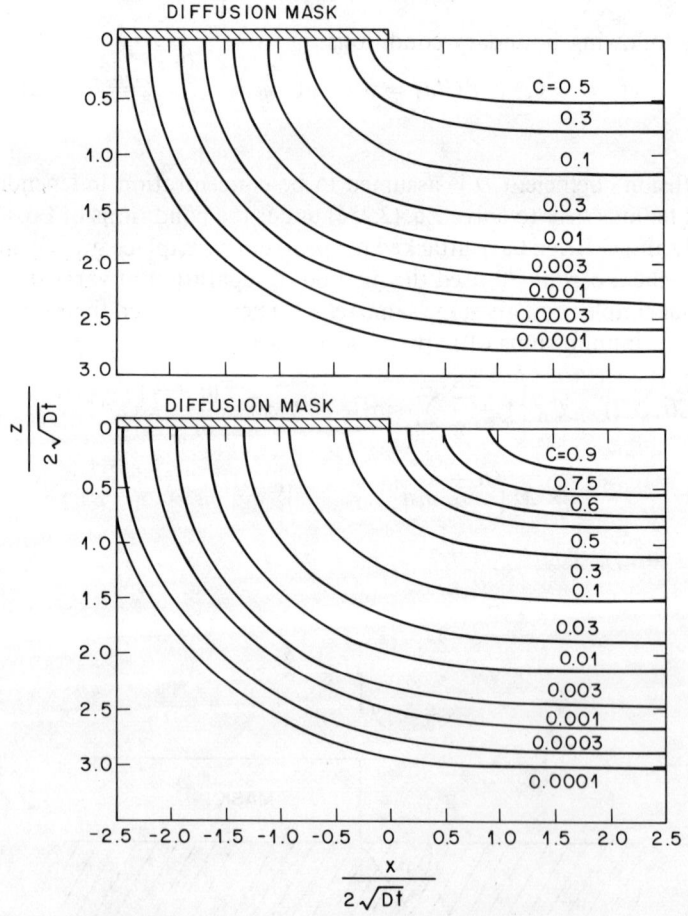

Fig. 69. Calculated contours of constant impurity concentration around a mask edge in two dimension: (a) constant surface concentration case, (b) instantaneous-source diffusion process. (From Kennedy and O'Brien.[211] Copyright 1965 by International Business Machines Corporation; reprinted with permission.)

an instantaneous source diffusion process, Eq. (2.259). Assuming a constant doping level in the bulk the results indicate the families of *pn* junctions that would result for various levels of bulk doping. To obtain the results in Fig. 69b, Eq. (2.259) has been reduced to two space variables by setting $y = \infty$.

h. *A Two-Dimensional Numerical Example*

In this section we consider an example of numerical process simulation in two dimensions. Although analytical solutions can be quite helpful, the analysis of complete process sequences can only be performed by numerical means. With the increased trend toward reduced dimensions and higher packing density, numerical simulations in one dimension are no longer accurate enough.

The example we describe is the Twin-Tub CMOS technology recently introduced.[213] CMOS technology is the candidate for VLSI circuits providing inherently low static power capabilities, and, at the same time, achieving lower power delay products than NMOS technology with comparable design rules. In the Twin-Tub process, separate tubs for the *p*- and the *n*-channel transistor allow improved device behavior. Figure 70 shows a cross section of a CMOS inverter fabricated using the Twin-Tub technology. A critical step of this process is the formation of the source–drain areas of both the *n*- and the *p*-channel transistor. In an effort to save masking steps, all the sources and drains are implanted nonselectively with boron and the n^+ sources and drains are selectively implanted and overdoped with a higher dose of phosphorus. Both the boron and the phosphorus implant are performed through the gate oxide. After implant, several temperature treatments (glass reflow, getter, etc.) are performed that redistribute the original profiles considerably.

Figure 71 shows the boron and phosphorus profiles in the n^+ source–drain region in vertical direction after implant (curves 1), after reflow (curves 2), and at the end of the process (curves 3). After implantation, the tail of the boron

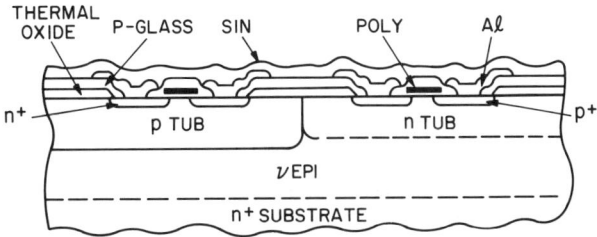

Fig. 70. Cross section of Twin-Tub CMOS inverter at the end of the process. (From Parillo *et al.*[213] © 1980 IEEE.)

profile overcompensates the phosphorus concentration for depths exceeding 0.25 μm. The diffusion of phosphorus and boron is an ideal example illustrating coupled diffusion phenomena (see Fig. 36). During the diffusion process, the electric field caused by the concentration gradient in the phosphorus profile does not allow the boron to outdiffuse from under the n^+ profile. Figure 72 shows a surface plot (a) and the contour plot (b) of the final phosphorus profile in the vicinity of the poly-edge of the gate. The maximum concentration is close to 10^{20} cm^3. The vertical junction depth is 1.12 μm, and the ratio of lateral to vertical junction depth is 0.75.

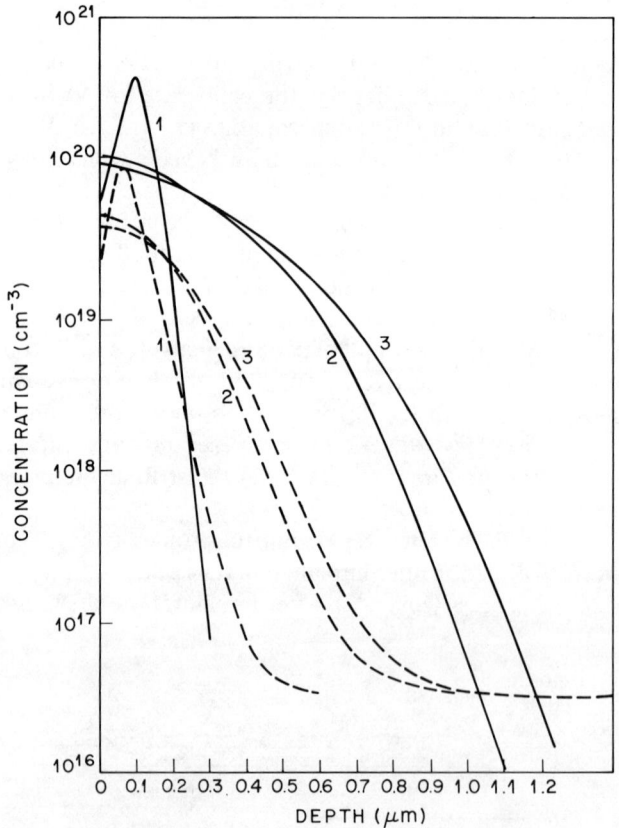

Fig. 71. Phosphorus (—) and boron (· · ·) profiles in n^+p-junction region in vertical direction at various stages during processing: (1) after implant; (2) after flow (3) final profiles. —, phosphorus; - - - -, boron.

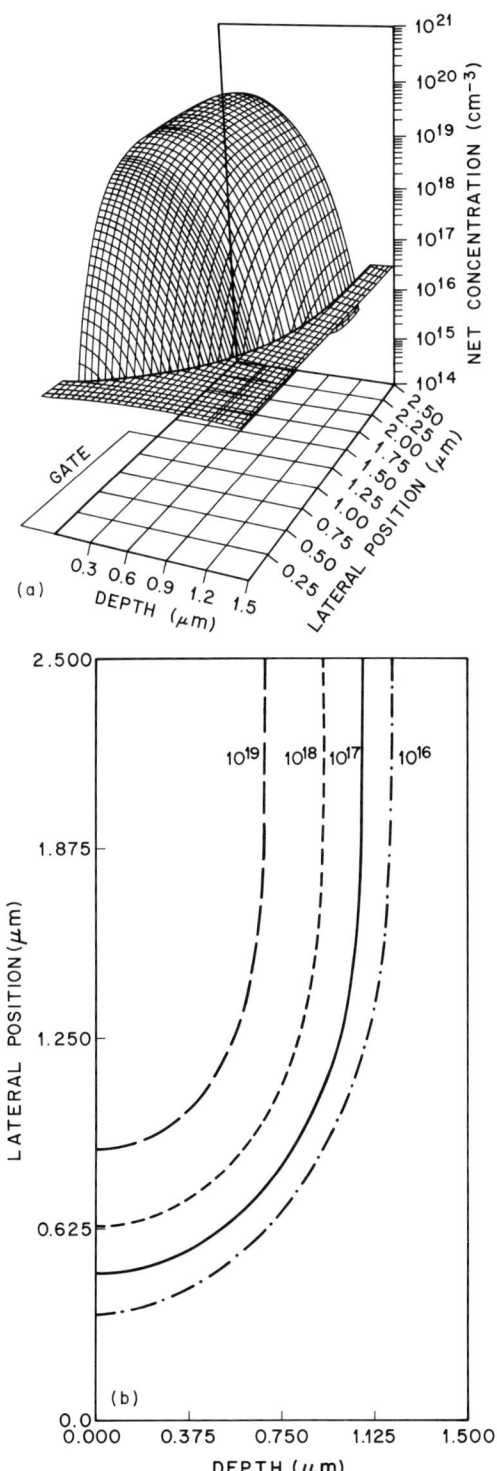

Fig. 72. (a) Surface plot and (b) contour plot of final phosphorus profile in the vicinity of the polysilicon gate edge.

4. Lithography

Microelectronic fabrication technology is based on the use of lithography to create very small features and patterns that make up an integrated circuit.

The processes used for pattern definition have evolved over the years from contact printing of negative resists with wet chemical etching to a transition phase with projection printing, negative and positive resists, and wet chemical and dry plasma etching. Originally, the trend to reduce linewidths was (and is) completely dominated by optical lithography. This trend will certainly continue to 0.5-μm linewidths, but there will be considerable problems with diffraction, reflectance, depth of focus, mask-to-wafer spacing, dust particles, and linewidth uniformity.

Direct electron-beam (e beam) writing is the optimum lithography technique to define very small structures in the submicrometer and even subsubmicrometer range. The ultimate resolution achievable is limited by electron scattering in the resist and the substrate to approximately 0.01 μm.

In terms of resolution, x-ray and ion-beam lithography are very promising. X-ray lithography does not have the problems of dust particles, diffraction, or interference; its ultimate resolution is limited by secondary electron generation to about 500 Å. Ion-beam lithography is basically ion implantation using finely focused ion beams.

This section deals with the basic physical principles of and the numerical models for optical, e-beam, x-ray, and ion-beam lithography.

a. *Optical Lithography*

(i) *Contact and Proximity Printing.* In contact and proximity printing, the wafer is brought in intimate contact with or close proximity to the mask. When the feature sizes and the distance between the mask and the wafer become comparable to the wavelength of the exposing light, the diffraction from the mask constitutes an electromagnetic diffraction problem. In the following we consider different approaches to the solution of this diffraction problem. Basically, the field has to be evaluated in close vicinity to the mask, either by rigorous solutions of Maxwell equations or by special approximations relevant to the near-field diffraction problem.[214]

The diffraction from the mask pattern is dependent on the polarization of the exposing light. For each polarization, three diffracted field components are produced. In the case of a transverse electric (TE) polarization, a TE field in the same direction, an orthogonal tangential magnetic (TM) field, and an axial magnetic field are produced in the diffracted beam. From the two

PHYSICS OF VLSI PROCESSING AND PROCESS SIMULATION

Maxwell equations

$$-\dot{\mathbf{B}} = \nabla \times \mathbf{E} \tag{2.260}$$

$$\dot{\mathbf{D}} = \nabla \times \mathbf{H} \tag{2.261}$$

with

$$\mathbf{B} = \mu \mathbf{H} \tag{2.262a}$$

$$\mathbf{D} = \epsilon \mathbf{E} \tag{2.262b}$$

we can obtain the components, assuming a time dependence of the form $\exp(i\omega t)$,

$$H_x = \frac{i}{\omega \mu} \frac{\partial E_y}{\partial z}, \tag{2.263}$$

$$H_z = -\frac{i}{\omega \mu} \frac{\partial E_y}{\partial x}, \tag{2.264}$$

where $\mathbf{E}$ is the electric field strength, $\mathbf{B}$ the magnetic flux density, $\mathbf{H}$ the magnetic field strength, ϵ the permittivity, and μ the permeability.

For photoresist exposure, only the square of the amplitude of the electric field strength, $|\mathbf{E}|^2$, has to be determined, since photoresists react only to the intensity of the electric field. Under the assumption of an ideal mask (infinitely thin, perfectly conducting), any one-dimensional pattern can be constructed from a combination of TE waves incident on slits and half-planes. The TE field goes to zero on the opaque part of the mask, which results in vanishing interaction between slits. In the case of TM exposure, interaction between slits occurs that does not allow superposition. However, the TM case can be derived from a TE result of a complementary mask pattern.

No rigorous solution of Eqs. (2.260) and (2.261) has been published. Heitman and van der Berg[215] have studied the diffraction of a plane electromagnetic wave by a semi-infinite screen in one of the plane interfaces of a layered medium. The screen is assumed to be infinitely thin and perfectly conducting. The authors have solved two separate two-dimensional scalar problems, the case of E polarization and H polarization. The configuration considered (Fig. 73) consists of $N - 1$ parallel layers that are separated by N plane interfaces. An ideal screen is located in one of the interfaces. The incident plane wave varies sinusoidally in time and propagates through the x, z plane, which makes both the configuration and the incident wave independent of y. Similar to Eqs. (2.263) and (2.264), the electromagnetic field consists of two uncoupled parts; in one of them E_y, H_x, and H_z are unequal to zero (E polarization, TE case); in the other H_y, E_x, and E_y differ from zero (H polarization, TM case.) From Eqs. (2.260)–(2.262) we obtain, in the TE case

$$E_y = u, \tag{2.265}$$

$$H_x = \frac{i}{\omega\mu}\frac{\partial u}{\partial z}, \tag{2.266}$$

$$H_z = -\frac{i}{\omega\mu}\frac{\partial u}{\partial x}, \tag{2.267}$$

and in the TM case,

$$H_y = u, \tag{2.268}$$

$$E_x = -\frac{i}{\omega\epsilon}\frac{\partial u}{\partial z}, \tag{2.269}$$

$$E_z = \frac{i}{\omega\epsilon}\frac{\partial u}{\partial x}, \tag{2.270}$$

where u is the scalar wave function to be determined. The function u satisfies

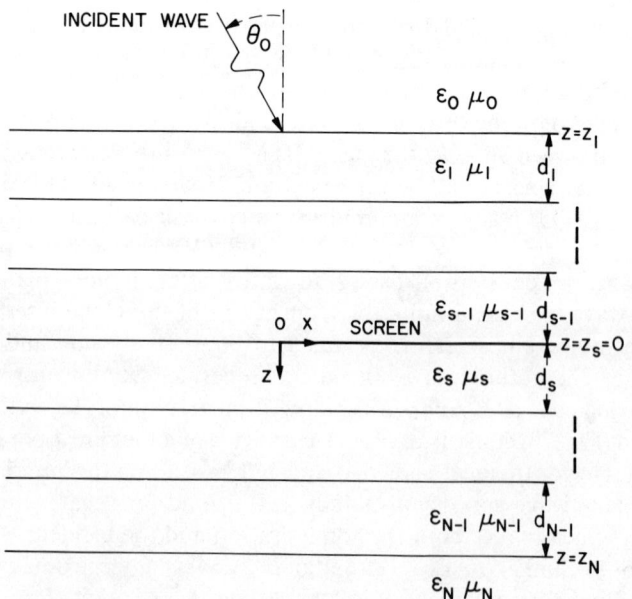

Fig. 73. Configuration with $N - 1$ layers and a semi-infinite screen at $z = 0$. (From Heitman and van der Berg.[215])

the two-dimensional Helmholtz equation

$$\frac{\partial^2 u}{\partial x^2} + \frac{\partial^2 u}{\partial z^2} + k^2 u = 0 \qquad (2.271)$$

with $k = \omega\sqrt{\epsilon\mu}$. The incident wave is generated by a source at infinity in the region $-\infty < z < z_1$ and has unit amplitude and zero phase at $x = 0$, $z = z_1$. In each layer the incident field consists of a superposition of plane waves all having the same x dependence $\exp[ik_0 \sin(\theta_0 x)]$, where $k_0 = 2\pi/\lambda_0$ is the wave number and λ_0 the wavelength in a medium with ϵ_0, μ_0; and θ_0 is the angle of incidence, $-\pi/2 \leq \theta_0 \leq \pi/2$. The diffracted field is also written as a continuous superposition of plane waves, represented by expressions of the form

$$u^d(x, z) = \frac{1}{2\pi} \int \exp(i\alpha x)\{P(\alpha) \exp[i\gamma(\alpha)z]$$
$$+ Q(\alpha) \exp[-i\gamma(\alpha)z]\}\, d\alpha, \qquad (2.272)$$

where $\gamma(\alpha)$ is given by

$$\gamma(\alpha) = \sqrt{k^2 - \alpha^2} \qquad (2.273)$$

with $\alpha = k \sin\theta$ and $\mathrm{Re}(\gamma)$, $\mathrm{Im}(\gamma) \geq 0$. The coefficients P and Q in each medium are related to those of the neighboring medium through a set of boundary conditions at the interface.

For a number of configurations of practical interest to contact and proximity printing, Heitman and van der Berg have numerically solved Eq. (2.271) for a normally incident plane wave ($\theta_0 = 0$). The most interesting case consists of a vacuum half-space (ϵ_0, μ_0), a glass layer ($\epsilon_1/\epsilon_0 = 2.25, \mu_1 = \mu_0$), a photoresist layer ($\epsilon_2/\epsilon_0 = 3.5, \mu_2 = \mu_0$), a silicon dioxide layer ($\epsilon_3/\epsilon_0 = 3.5, \mu_3 = \mu_0$), and an infinitely thick silicon substrate $\epsilon_4/\epsilon_0 = 21.17 + i0.466$, $\mu_4 = \mu_0$). The mask is at the interface between the glass layer and the photoresist. The wavelength in vacuum is $\lambda = 0.43$ μm. The interaction of the vacuum–glass interface with all other interfaces has been neglected since the glass layer has a thickness of about 1000λ. The amplitude transmission factor of the normally incident wave at the vacuum–glass interface equals 0.8 for E polarization and 1.2 for H polarization, respectively. Since the vacuum–glass interface has been removed to infinity (by neglecting interactions), it is assumed that the incident wave is generated by a source at infinity with amplitudes of either 0.8 or 1.2, respectively. The power flow density transmitted by the incident wave in the positive z direction S_z' is given by

$$S_z' = \begin{cases} \frac{1}{2}(\epsilon_1/\mu_1)^{1/2}(0.8)^2 = 0.48(\epsilon_0/\mu_0)^{1/2} & \text{for } E \text{ polarization,} \\ \frac{1}{2}(\mu_1/\epsilon_1)^{1/2}(1.2)^2 = 0.48(\mu_0/\epsilon_0)^{1/2} & \text{for } H \text{ polarization.} \end{cases} \qquad (2.274)$$

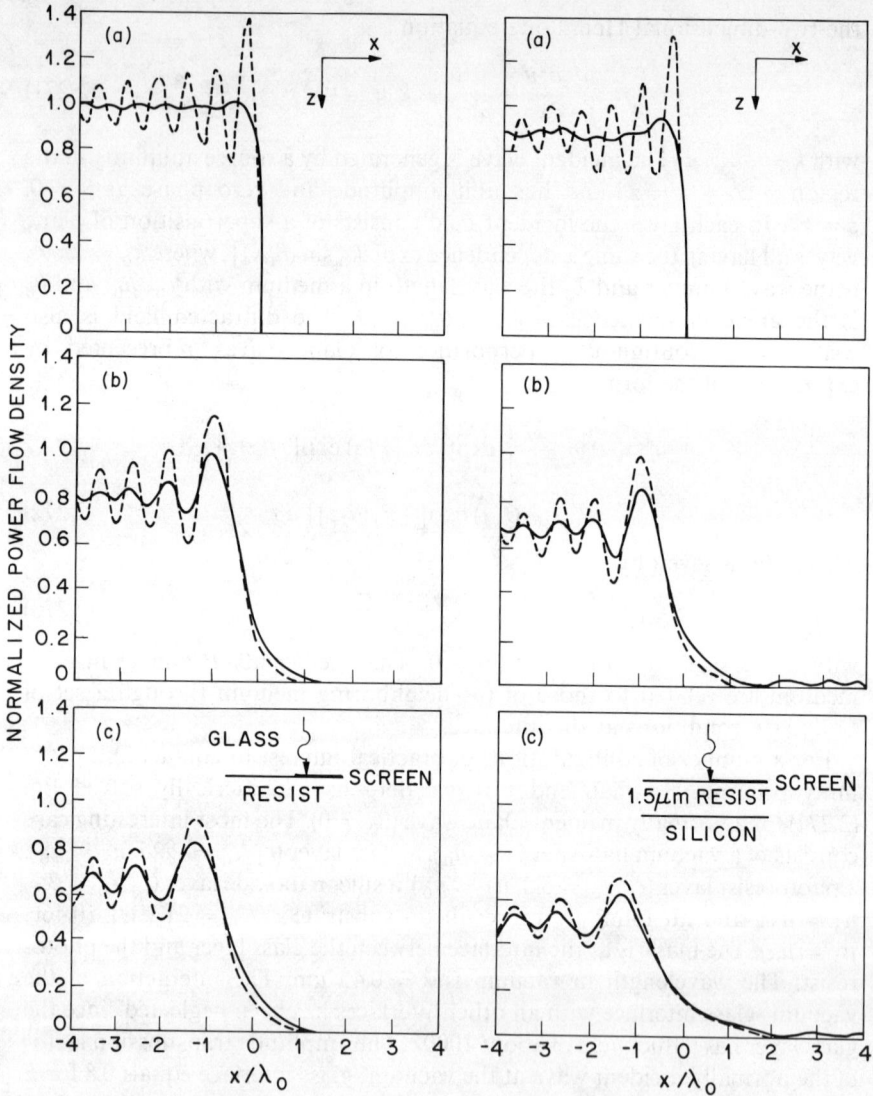

Fig. 74. Normalized power flow density $p(x, z)$ in the z direction as a function of x, while $z = 0$ (a), 0.75 (b), and 1.5 (c) μm, respectively. (——) E-polarization, (---) H-polarization. (From Heitman and van der Berg.[215])

Fig. 75. Same as Fig. 74, but for the glass–photoresist–silicon case. (From Heitman and van der Berg.[215])

The normalized power flow density, $\mathscr{P} = \mathscr{P}(x,z)$, is defined as

$$\mathscr{P} = S'_z/S_z, \qquad (2.275)$$

where $S_z = S_z(x,z)$ is the Poynting vector in the z direction

$$S_z = \tfrac{1}{2}\mathrm{Re}[(\mathbf{E}\times\mathbf{H}^*)\cdot\mathbf{i}_z] = \begin{cases}(2\omega\mu)\,\mathrm{Im}(u^*\partial u/\partial z) & \text{for } E \text{ polarization,}\\ (2\omega\epsilon)\,\mathrm{Im}(u^*\partial u/\partial z) & \text{for } H \text{ polarization.}\end{cases}$$

(2.276)

Figures 74 to 77 show $\mathscr{P}$ as a function of x and z for four different configurations. In Fig. 74, the simple configuration of a screen at the interface of two half-spaces, that is, glass–photoresist, is shown. The number of peaks decreases as the depth into the resist is increased. The average magnitude of the diffracted field remains unchanged, however, because of the assumption of a nonabsorptive photoresist. In Fig. 75 the results for the system glass–resist–silicon case are shown. The average magnitude of $\mathscr{P}$ decreases for increasing depth. The introduction of the silicon layer creates a standing wave in the photoresist.

In Fig. 76 the ideal contact printing situation of glass–photoresist–silicon-dioxide–silicon is considered. The transition layer between the photoresist and the silicon introduces higher level curves compared with those in Fig. 75. In Fig. 77 the influence of a thin air gap between the glass and the resist illustrates the typical nonideal contact printing situation.

Other results of exact solutions to Eqs. (2.260) and (2.261) have been published by Neerhof and Mur,[216] who investigated the diffraction pattern due to a plane H-polarized electromagnetic wave, when this wave is incident on an infinitely long slit of finite width in an opaque screen of nonvanishing thickness. A Green's function formulation of the problem leads to a system of four coupled integral equations in which the field distributions in the slits occur as unknowns. Numerical results are presented for the field just below the screen as well as the far field pattern.

Lin[217] has derived formulas for the near-field diffraction of an infinite slit, allowing TE, TM, and unpolarized illumination. He has solved the problem by synthesizing the exact Sommerfeld solution[218] of the Maxwell equations, Eq. (2.260) and (2.261), describing the diffraction of the slit as due to two interacting half planes. One incident wave and two "Sommerfeld" diffracted waves are superposed with proper phase adjustment. The interaction terms are evaluated in terms of the Sommerfeld solution, making the final result very convenient for a digital computer. Lin has given extensive results of the diffraction of a three-wavelength slit. Figures 78 and 79 illustrate the evolution of the TE component at zero to about six wavelengths from the same

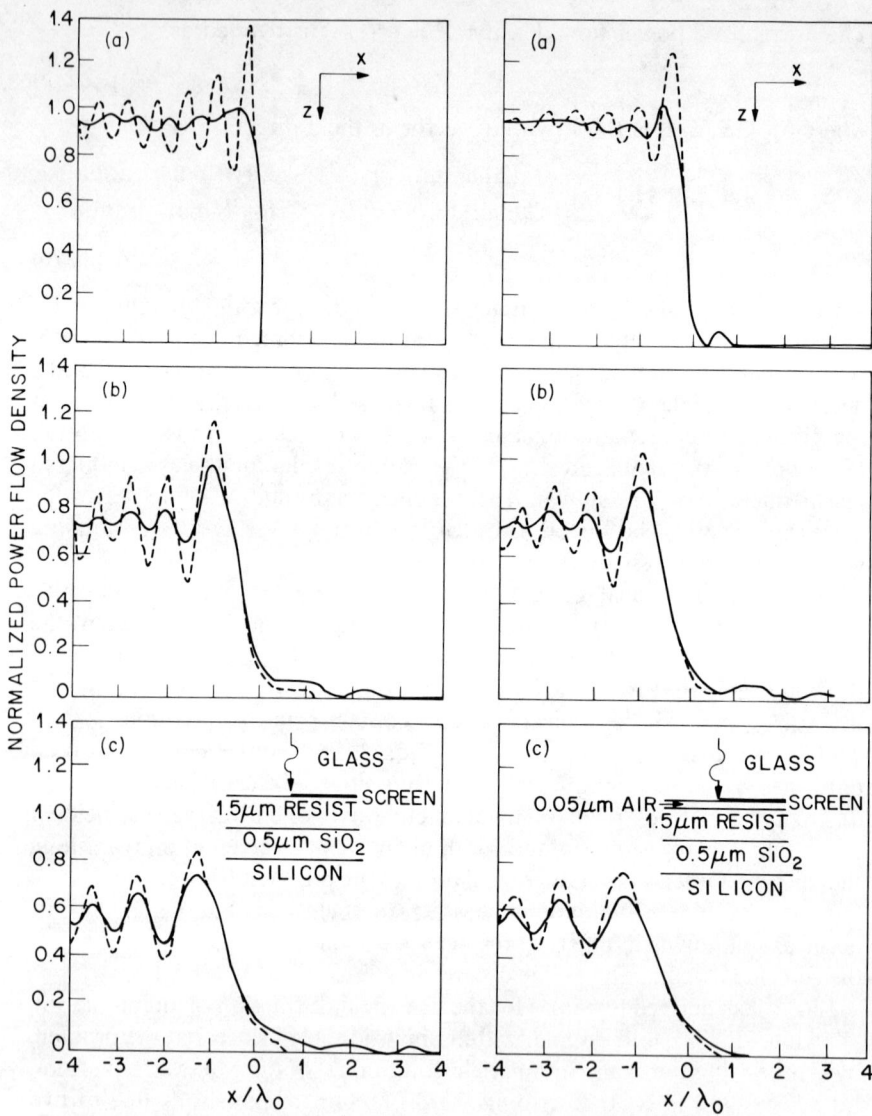

Fig. 76. Same as Fig. 74, but for the glass–photoresist–silicon-oxide–silicon case. $z = 0$ (a), 0.75 (b), 1.5 (c) μm. (From Heitman and van der Berg.[215])

Fig. 77. Same as Fig. 74, but with an air gap. $z = 0.05$ (a), 0.8 (b), and 1.55 μm. (From Heitman and van der Berg.[215])

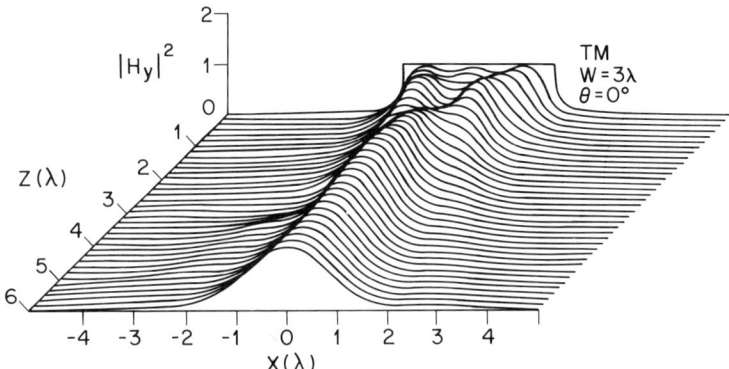

Fig. 78. Three-dimensional plot showing the evolution of the diffracted wave from the aperture plane for the TM case. The slit is located at $z = 0$ and is centered at $x = 0$. The width is 3λ and the incident angle is $0°$. (From Lin.[217])

three wavelength slit. TE and TM waves of normal incidence are used. Although the TE and TM waves start from entirely different shapes, they converge into similar ones. For normal incidence, the field of the TE wave has three peaks (Fig. 79) at $z = 0$, gradually becomes two-peaked and finally has one peak. The TM wave is uniform in the aperture (Fig. 78) but immediately leads into three peaks, and then goes through the same peak-decreasing process.

For general mask geometries, proximity printing constitutes a formidable problem for theoretical analysis and simulation. Exact solutions can only be found in a few simple cases such as the simple straight edge or slit situation just considered. In proximity printing, the ultimate resolution is determined

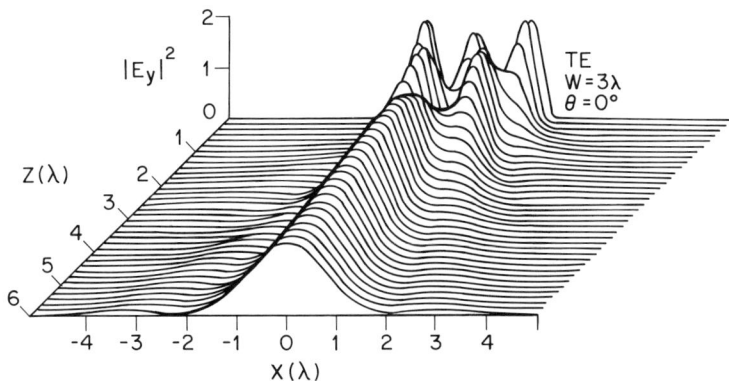

Fig. 79. Same as Fig. 78 for TE case. (From Lin.[217])

by the exposing wavelength and the gap between mask and wafer. In order for the theoretical analysis to have any "real" impact, two-dimensional mask patterns have to be simulated. Even with today's high-speed computers, this is an impossible task. In the following, several approximations are presented to evaluate diffraction from arbitrary two-dimensional objects. The Fresnel diffraction[219-221] is often used to understand diffraction phenomena qualitatively. Fraunhofer diffraction[220,221] is of great importance in the analysis of image-forming instruments. The physical-optics approximation, also known as Kirchhoff's approximation,[222] has been used to evaluate two-dimensional diffraction patterns quantitatively. All approximations can be derived from the exact Kirchhoff integral for the scalar wave function, u,

$$u(x, y, z) = -\frac{1}{4\pi} \int\int_S u(x_s, y_s, z_s) \frac{\partial G}{\partial n} dx_s \, dy_s, \qquad (2.277)$$

where $G = e^{ikr}/r$ is the Green's function for the Helmholtz equation,

$$\nabla^2 u + k^2 u = 0. \qquad (2.278)$$

In the case of proximity printing,[214] the surface is chosen to be the mask plane $z = 0$ and an infinitely large hemispherical surface. Since the amplitude of the wave function vanishes at infinity, Eq. (2.277) becomes

$$u(x, y, z) = \frac{z}{2\pi} \int\int_S u(x_s, y_s, 0)\left(ik - \frac{1}{r}\right)\frac{\exp(ikr)}{r^2} dx_s \, dy_s, \qquad (2.279)$$

where $r^2 = (x - x_s)^2 + (y - y_s)^2 + z^2$.

The physical-optics approximation is introduced by the following considerations. At the transparent part of the mask, except in the immediate neighborhood of the rim of the opening, u and $\partial u/\partial n$ will not differ appreciably from the values obtained in the absence of the screen, and if the opaque part of the mask is thin and very highly conducting, these quantities are zero. Mathematically, this can be expressed as

$$u = u_i, \qquad \partial u/\partial n = \partial u_i/\partial n \qquad (2.280)$$

for the transparent part, and

$$u = \partial u/\partial n = 0 \qquad (2.281)$$

for the opaque part of the mask with

$$u_i \sim \frac{e^{ikr}}{r}, \qquad \frac{\partial u_i}{\partial n} \sim \frac{e^{ikr}}{r}\left(ik - \frac{1}{r}\right)\cos(n, r). \qquad (2.282)$$

PHYSICS OF VLSI PROCESSING AND PROCESS SIMULATION 235

When z is large, then

$$e^{ikr} = \exp(ikz)\left\{1 + \frac{1}{2z^2}[(x - x_s)^2 + (y - y_s)^2]\right\} \quad (2.283)$$

and neglecting the $1/r$ term in Eq. (2.279) leads to the Fresnel approximation,

$$u(x, y, z) = \frac{ik}{4\pi z} e^{ikz} \int\int_s u(x_s, y_s, 0) \exp\left\{\frac{ik}{2z}[(x - x_s)^2 + (y - y_s)^2]\right\}. \quad (2.284)$$

When z is completely in the far-field region, the quadratic terms in Eq. (2.284) can also be neglected, and only the factor $\exp[ik/2z(xx_s + yy_s)]$ remains in the integral. If this condition for z holds, it is called the Fraunhofer approximation, which is essentially a Fourier transformation. Although Eq. (2.282) is widely used, the error becomes unnecessarily large for oblique illumination.[214]

Lin has critically compared the regions of validity of the different approximations.[214] Generally, the approximations become better as the width W of a feature decreases, as the mask-to-wafer distance z increases, or when the wavelength λ increases. Since a closed form expression of the error is not available, Lin has compared exact solutions with those generated by using the previously mentioned approximations. Figure 80 compares diffraction patterns for $v = W^2/z = 1.1, 11,$ and 50. In Fig. 80a, where $v = 1.1$, the zone is considered as a far-field. Though the diffraction patterns change shape from $W = 5\lambda$ to $W = \lambda$, the Fresnel approximation describes the actual form quite well. In the case of $v = 11$, Fig. 80b, significant changes of the diffraction patterns occur, even from the TE to the TM case. The Fresnel approximation is no longer adequate. When $v = 50$ (Fig. 80c), the patterns look similar at $W = 100\lambda$. For $W = 1$ and 10λ even the physical-optics result fails. The results have been summarized by Lin in the following way:

(1) For one-dimensional objects, the exact electromagnetic diffraction methods mentioned earlier should be used, because of precision and efficiency.

(2) For two-dimensional objects, the physical-optics approximation should be used. It gives results with errors better than 5% if W and z are larger than 2λ. Only when $W^6/z^5 < 12$, the Fresnel approximation has an error of less than 5%.

Lin[223] has calculated the intensity distribution of the diffracted images of masks with various sizes at 5 and 10 μm from the mask plane, using near-uv and deep-uv illuminations. The wavelengths of the illumination and the resist spectral intensity are shown in Table X.

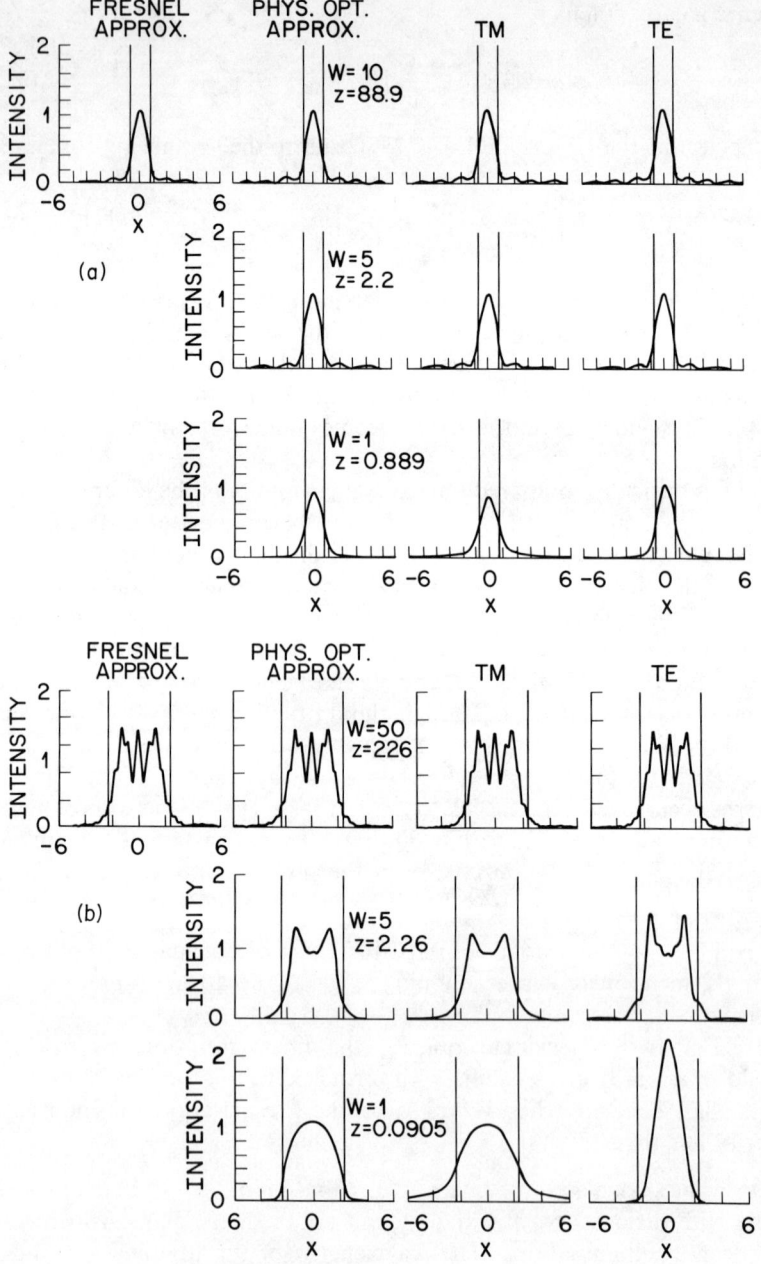

Fig. 80. Comparison of diffraction patterns at $v = 1.1$ (a), 11 (b), and 50 (c) for $\theta = 0°$. (From Lin.[214] Copyright North-Holland Physics Publishing, Amsterdam, 1980.)

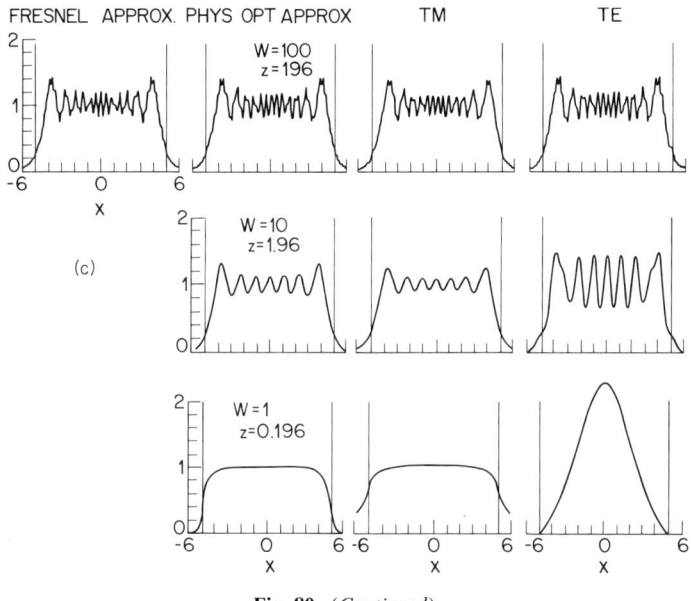

Fig. 80. (*Continued*)

Lin analyzed four mask patterns consisting of a $1 \times 1.2\text{-}\mu m^2$ contact hole, a $1 \times 4\text{-}\mu m^2$ isolated mask opening, the corresponding isolated opaque line, and a $1 \times 5\text{-}\mu m^2$ and $1 \times 4\text{-}\mu m^2$ line pair. The intensity distributions are shown in Figs. 81a–d, respectively. Each point in Fig. 81 is represented by a number that indicates the intensity level. There are a total of 20 levels ranging from 0, 1, 2, …, 1, 2, …, to 9. Each level is 1.5 dB from the next level. The level 0 indicates all intensities larger than or equal to unity, which is the

TABLE X

SPECTRAL PARAMETERS USED IN THE SIMULATION

Deep UV, Using PMMA						
Wavelength, Å	2050	2150	2250	2350	2450	2550
Relative illumination spectrum	2	8	13	22	30	29
Relative sensitivity spectrum	255	131	62.5	15.5	3.22	0.214
Near UV, Using AZ1350J						
Wavelength, Å	3132	3341	3650	4047	4358	
Relative illumination spectrum	84	60	192	136	169	
Relative sensitivity spectrum	920	1680	1910	1820	1240	

intensity required to expose completely a large uniform transparent area at a predetermined development condition. The boundaries between levels are constant intensity contours. When the resist is very thin, the constant intensity can be considered as the resist image. The distance between boundaries is an indication of image contrast. A shorter distance represents a higher contrast, which leads to a better linewidth control. The 1×1.2-μm^2 contact hole shown in Fig. 81a becomes ellipitical at the proximity distances because of loss of the higher spatial frequencies. Moreover, the long and short axes are opposite to those of the rectangular hole. This is plausible because the diffracted image becomes the Fourier transform of the original object at a reasonably large distance with respect to the dimension of the object. The deep-UV image at a 5-μm gap exhibits a high contrast when compared to the other images. If an image corresponding to the boundary of levels 6 and 7 is developed, a similar size image can be obtained at a 10-μm gap despite some

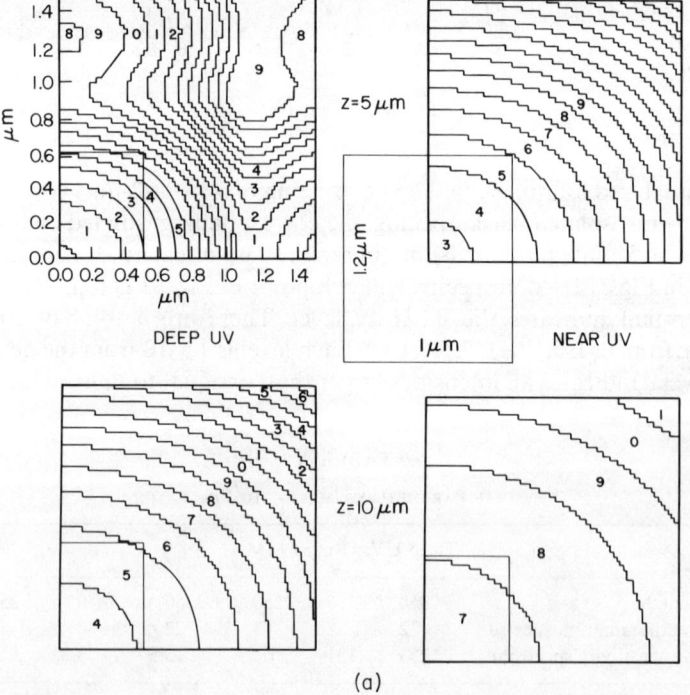

Fig. 81. (a) Intensity distribution of the proximity image of a contact hole at 5 and 10 μm from the mask, using near-UV and deep-UV illuminations. (b) Same as (a), but a rectangular opening was used instead of a contact hole. (c) Same as (b), but an opaque bar complementary to the rectangular opening was used. (d) Same as (a), but two rectangular openings of uneven lengths were used. (From Lin.[223])

contrast loss. Therefore, some gap error due to poor wafer flatness or mask wafer positioning can be tolerated. When near-UV exposure is used, the image has less contrast and is less tolerant to gap error.

Figure 81b shows the diffraction images of a $1 \times 4\text{-}\mu m^2$ opening. The deep-UV image at a 5-μm gap again shows good contrast and image fidelity. However, diffraction produces zigzagging line edges. Fortunately, when the periodicity of these zigzagging patterns is comparable to those of standing wave patterns, a standard process to remove the standing waves such as postexposure bake can also smoothen the line edges. When the linewidth is developed to the boundary between levels 2 and 3 approximating the mask line width closely, the line length is seen to be shorter. At a 10-μm gap, an

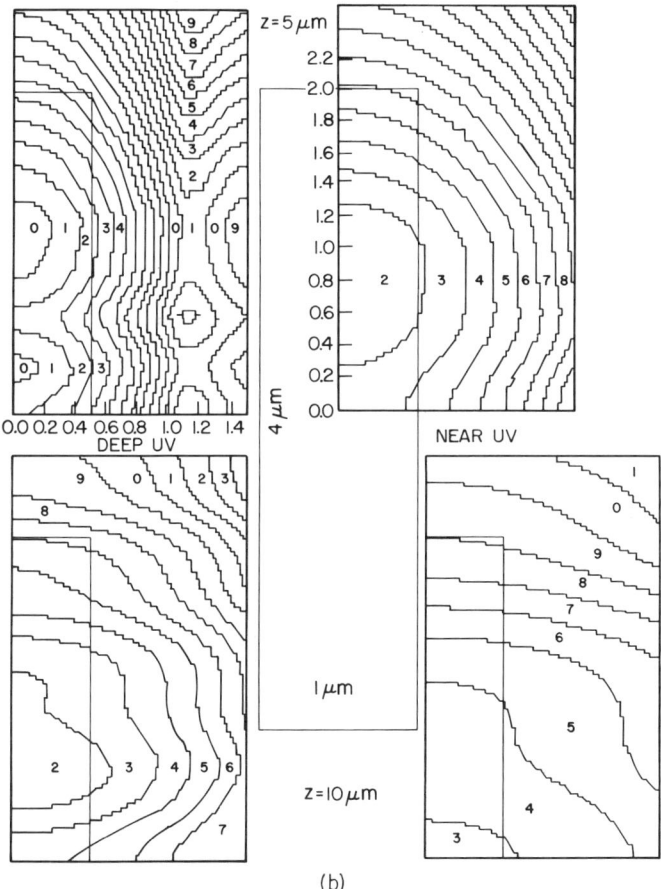

(b)

Fig. 81. (*Continued*)

8-shaped image is produced. The near-UV 5-μm gap situation is similar. At a 10-μm gap and with near-UV light, the long and short axes of the original rectangle start to reverse.

Figure 81c shows the diffracted images of a 1 × 4-μm^2 opaque rectangle in a transparent background, corresponding to the so-called light-field mask patterns. A particular problem associated with these light-field images is caused by the unwanted bright spots in the middle of the dark bars. For example, the boundary between levels 2 and 3 would produce an acceptable image but not for the bright spot at the vertical center line near the end of the rectangle.

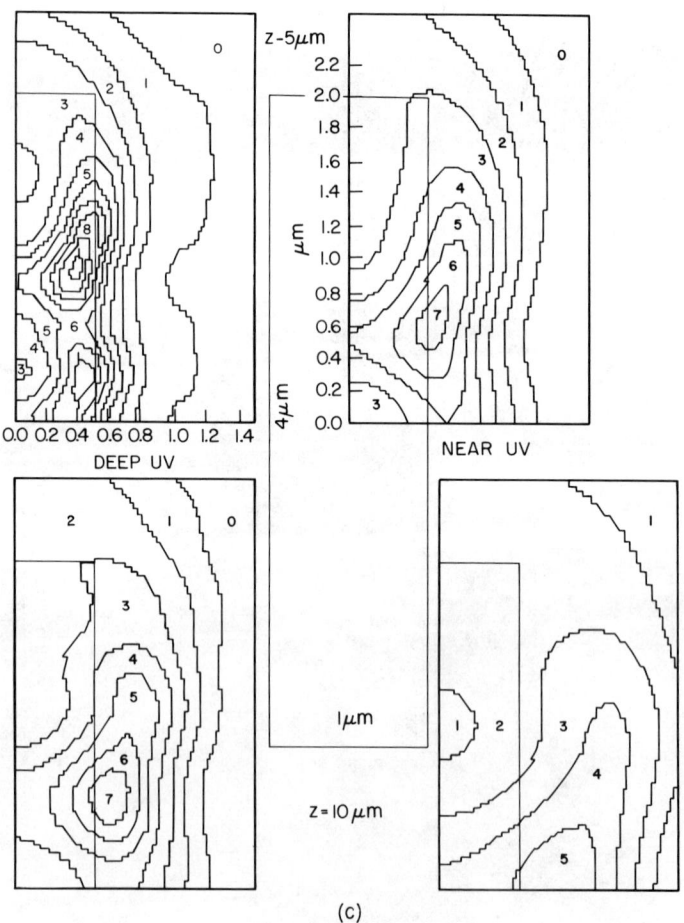

(c)

Fig. 81. (*Continued*)

Figure 81d shows the diffracted images from a $1 \times 5\text{-}\mu m^2$ opening placed 1 μm beside a $1 \times 4\text{-}\mu m^2$ opening. This is by far the most difficult image to print. In all cases except using deep-UV and 5-μm gap, the two bars are either jointed or become three bars.

(ii) *Projection Printing.* In projection printing, the image of the object—the mask—is projected onto the wafer through a high-resolution optical lens, whose demagnification can vary between 1 and 20 times. The complete optical system is usually diffraction limited.

Figure 82 describes the basic components of a projection printer. Contrary to proximity printers, there is an imaging lens between the mask and the wafer, which means that the mask-to-lens and lens-to-wafer distances have to be controlled together with the orientation of the mask, the wafer and the lens.

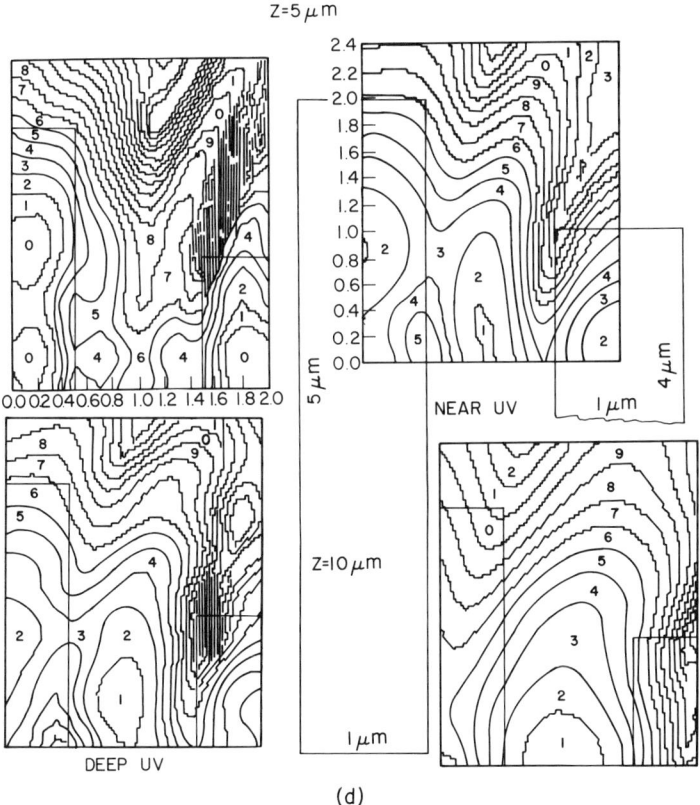

(d)

Fig. 81. (*Continued*)

The deviations from the ideal diffraction-limited system are measured by the modulation transfer function (MTF) of the lens. The MTF includes all aberrations such as spherical aberration, coma, astigmatism and field curvature. For a mask consisting of lines and spaces of spatial frequency ξ_P, the MTF is defined as the ratio of the mask modulation to the image modulation

$$\text{MTF}(\xi_P) = \frac{M_{\text{image}}(\xi_P)}{M_{\text{mask}}(\xi_P)}, \qquad (2.285\text{a})$$

where

$$M_{\text{mask}} = \frac{I_{\text{max}} - I_{\text{min}}}{I_{\text{max}} + I_{\text{min}}} \qquad (2.285\text{b})$$

and an equivalent expression for M_{image}. The factors I_{max} and I_{min} are the maximum and minimum intensities, respectively. For an idealized optical system as in Fig. 83, the angle θ between the maximum diameter of the exit pupil and the image plane determines the resolution. This is described by the numerical aperture NA, defined by

$$\text{NA} = n \sin \theta \qquad (2.286)$$

or the effective f/number,

$$f/\text{number} = 1/2\,\text{NA}, \qquad (2.287)$$

where n is the refractive index of the surrounding medium. The quality of the

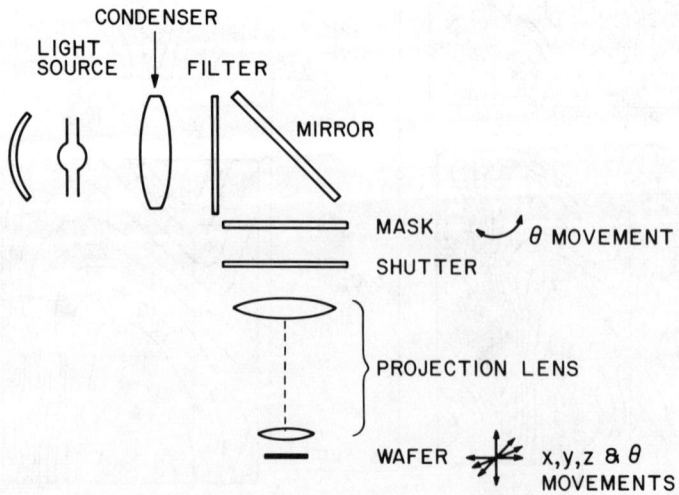

Fig. 82. Elements of a typical refractive-type projection mask aligner. (From Lin.[214] Copyright North-Holland Physics Publishing, Amsterdam, 1980.)

image depends critically on the illumination conditions. The illumination condition of most practical interest is partially coherent illumination. Partial coherence is quantified by the ratio

$$\sigma = NA_c/NA_o \tag{2.288}$$

where NA_c and NA_o are the numerical apertures of the condenser and objective lens, respectively. The parameter σ describes the degree of filling of the entrance pupil of the imaging lens by the source. It is the ratio of the imaged source at the entrance pupil to the pupil diameter. An incoherent source is a source of infinite dimension ($\sigma = \infty$), whereas a coherent source ($\sigma = 0$) is a point source. Figure 84 shows graphically coherent and partially coherent ($\sigma = 1$) illumination conditions for a periodic object of period P.[224] In the fully coherent case (a) the entrance pupil accepts only diffraction orders 0 and ± 1. When these diffraction orders reach the image plane, the image is reconstructed from orders 0 and ± 1. With an infinitesimal decrease of the mask period P, no image would be formed because only order 0 reaches the image plane. Figure 84a corresponds to the case of a diffraction-limited lens illuminated with a coherent plane wave. The fundamental spatial frequency cuts off abruptly at $1/P = NA/\lambda$. It can be seen from Fig. 84b that for $\sigma \geq 1$, some of the diffracted light of order ± 2 just enters the pupil and hence the cutoff frequency is twice the coherent cutoff. The MTF of a partially coherent system is shown in Fig. 85 for three σ-values.[224] The size parameter β is related to the half-pitch through $P/2 = \beta\lambda/NA$. For each value of σ, a best exposure value exists that for $\beta = 1$ allows a reproduction of all mask features in the resist. We shall consider other cases that illustrate the influence of partial coherence.

The concept of the MTF is a useful parameter to specify the performance of optical projection lenses. However, because images of interest in microfabrication are not sinusoidal, the concept of the MTF is difficult to generalize. Over the last few years, the direct calculation of intensity distributions

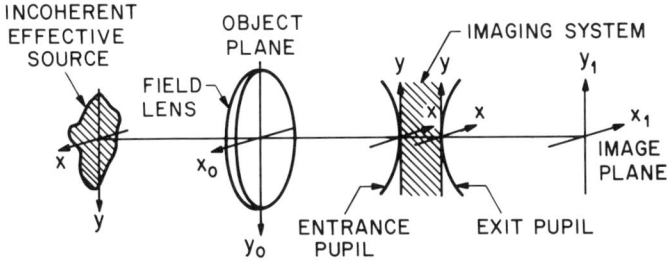

Fig. 83. Optical system showing the entrance and exit pupils of the imaging system. (From Kinzly.[227])

of lens images has advanced considerably. Contrary to the near-field diffraction problems in the case of proximity printing, which have only been solved for a few simple cases, a rigorous theory is available to calculate the imaging properties of lenses for arbitrary objects.

The basic theory of imaging with partially coherent light was developed by Hopkins[225] and Wolf[226] in the early 1950s. Several authors, especially Kinzly,[227] Watrasiewicz,[228] Kintner,[229] Offner and Meiron,[230] Hershel,[231]

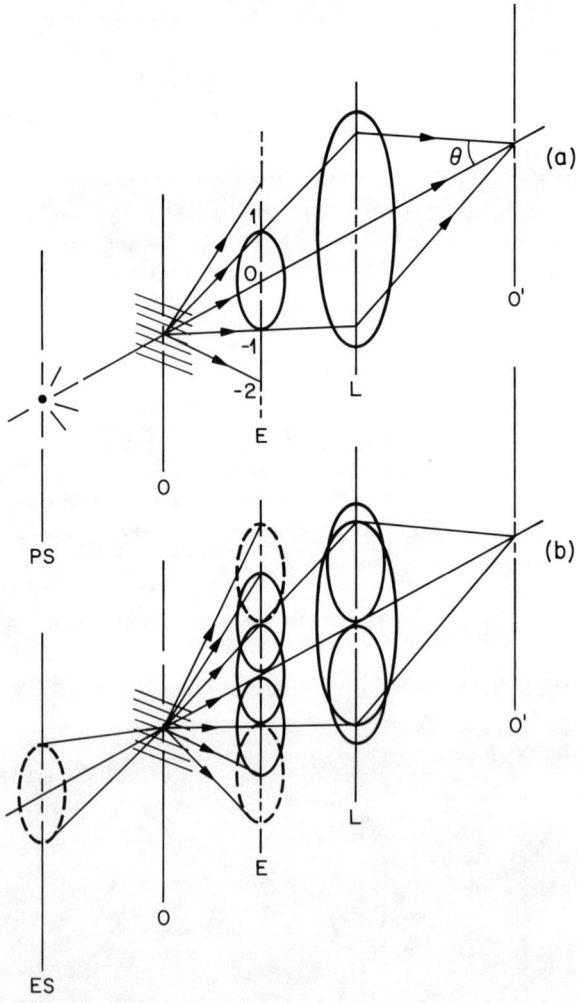

Fig. 84. Interpretation of partially coherent illumination: (a) point source, coherent illumination, (b) extended source, partial coherence. (From Bruning.[224] This figure was originally presented at the Spring 1982 Meeting of the Electrochemical Society, Inc. held in Montreal, Quebec, Canada.)

Nyssonen,[232] and Considine,[233] simplified the original theory and applied it to projection printing. Lacombat and Dubroeucq[234] and Tigreat[235] have studied step-and-repeat systems illuminated with partially coherent light. O'Toole and Neureuther[236] have incorporated the work of Kintner[229] into the program SAMPLE.[237] In the following, we give a short introduction to the theory of image formation under partially coherent illumination. Our analysis follows Born and Wolf[219] and Kintner.[229] Other treatments can be found in other books on this subject.[220,238,239] We shall use dimensionless Cartesian coordinates x_o and y_o, where

$$x_o = (NA_o/\lambda)\bar{x}_o, \tag{2.289}$$

$$y_o = (NA_o/\lambda)\bar{y}_o. \tag{2.290}$$

The geometric coordinates in the object plane are given by $\bar{x}_o$ and $\bar{y}_o$; NA_o is the numerical aperture of the objective, and λ is the mean wavelength of the quasi-monochromatic illumination. Points in the image plane conjugate to (x_o, y_o) in the object plane are described by the same coordinates (x_o, y_o). Points in the pupil plane are described by the coordinates

$$\xi = \bar{\xi}/NA_o, \tag{2.291}$$

$$\eta = \bar{\eta}/NA_o, \tag{2.292}$$

where $\bar{\xi}$ and $\bar{\eta}$ are the geometrical coordinates.

The imaging properties of the system may be described by the transmission function $K(x_o, y_o; x_1, y_1)$, which is defined as the complex amplitude per

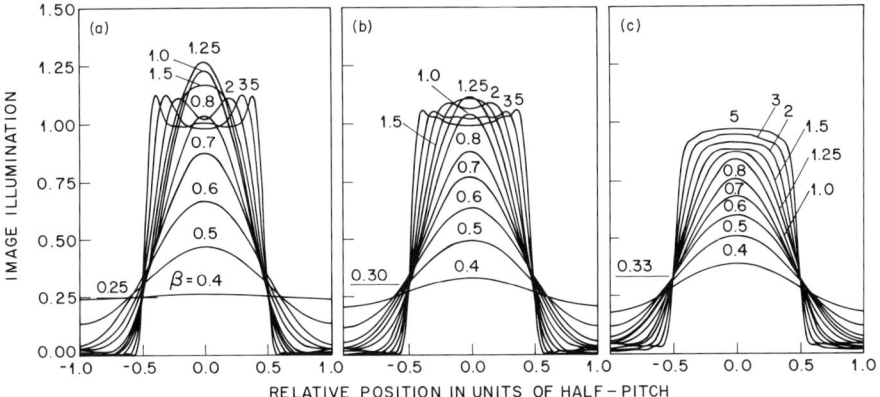

Fig. 85. Image intensity profile of a square wave target of period or pitch p. The horizontal axis is in units of p and each profile is characterized by a dimensionless sign parameter β where $p = 2\beta\lambda/NA$. (a) $\sigma = 0.5$, (b) $\sigma = 0.7$, (c) $\sigma = 1.0$. For 4000 Å and 0.4NA, β is the linewidth in micrometers. (From Bruning.[224] This figure was originally presented at the Spring 1982 Meeting of the Electrochemical Society, Inc. held in Montreal, Quebec, Canada.)

unit area in the x_o-y_o plane, at the point (x_1, y_1) in the image plane. In a well-corrected system, $K(x_o, y_o; x_1, y_1)$ can be written as

$$K(x_o, y_o; x_1, y_1) = K_A(x_1 - x_o, y_1 - y_o). \tag{2.293}$$

A region "A" with this property is called an *isoplanatic* region of the system. In what follows, we restrict our discussion to objects that are so small that they fall within such a region, and we shall use K instead of K_A. For coherent illumination, the transmission function is equal to the pupil function.

Let $J_o(x_o, y_o; x'_o, y'_o)$ be the mutual intensity for points (x_o, y_o), (x'_o, y'_o) in the object plane. The intensity in the image plane is given by

$$J_1(x_1, y_1; x'_1, y'_1) = \int\!\!\int\!\!\int\!\!\int_{-\infty}^{+\infty} J_o(x_o, y_o; x'_o, y'_o) K(x_o, y_o; x_1, y_1)$$

$$\times K^*(x'_o, y'_o; x'_1, y'_1) \, dx_o \, dy_o \, dx'_o \, dy'_o. \tag{2.294}$$

Suppose that a portion of the object plane is occupied by a transparent or semitransparent object that is illuminated with partially coherent quasi-monochromatic light, which originates in a primary source and reaches the object plane after the passage through the condenser. We specify the object by an appropriate transmission function $F(x_o, y_o)$. Then we can write

$$J_o(x_o, y_o; x'_o, y'_o) = F(x_o, y_o; x'_o, y'_o) F^*(x_o, y_o; x'_o, y'_o) J_o^-(x_o, y_o; x'_o, y'_o), \tag{2.295}$$

where J_o^- is the mutual intensity of the light incident on the object. When J_o^- is of the form

$$J_o^-(x_o, y_o; x'_o, y'_o) = J_o^-(x_o - x'_o, y_o - y'_o), \tag{2.296}$$

which is the case for the Köhler illumination of interest here (see Born and Wolf,[219] 10.5.2); and if the object is so small that it forms an isoplanatic area, then the intensity $I_1(x_1, y_1) = J_1(x_1, y_1; x_1, y_1)$ in the image plane is given by

$$I_1(x_1, y_1) = \int\!\!\int\!\!\int\!\!\int_{-\infty}^{+\infty} J_o^-(x_o - x'_o, y_o - y'_o) F(x_o, y_o) F^*(x'_o, y'_o)$$

$$\times K(x_1 - x_o, y_1 - y_o) K^*(x_1 - x'_o, y_1 - y'_o) \, dx_o \, dy_o \, dx'_o \, dy'_o. \tag{2.297}$$

By expressing F and J_o^- in two-dimensional Fourier integrals of the form

$$F(x, y) = \int\!\!\int_{-\infty}^{+\infty} \mathscr{F}(\xi, \eta) \exp[-2\pi i(\xi x + \eta y)] \, d\xi \, d\eta, \tag{2.298}$$

we can rewrite Eq. (2.297) to

$$I_1(x_1, y_1) = \int\int\int\int_{-\infty}^{+\infty} \mathcal{T}(\xi', \eta'; \xi'', \eta'') \mathcal{F}(\xi', \eta') \mathcal{F}^*(\xi'', \eta'')$$
$$\times \exp\{-2\pi i[(\xi' - \xi'')x_1 + (\eta' - \eta'')y_1]\} d\xi' d\xi'' d\eta' d\eta'', \quad (2.299)$$

where $\mathcal{T}(\xi', \eta'; \xi'', \eta'')$, the transmission cross coefficient of the system, is given by

$$\mathcal{T}(\xi', \eta'; \xi'', \eta'') = \int\int_{-\infty}^{+\infty} \mathcal{J}_0^-(\xi, \eta) \mathcal{K}(\xi + \xi', \eta + \eta') \mathcal{K}^*(\xi + \xi'', \eta + \eta'') d\xi d\eta. \quad (2.300)$$

The variables $\mathcal{K}$ and $\mathcal{J}_0^-$ are Fourier transformations of Eq. (2.293) and (2.296). The function $\mathcal{T}$ characterizes the complete optical system, including condenser and objective. When the illumination is incoherent, the function $\mathcal{T}$ becomes the optical transfer function,[238] and, for coherent illumination, it becomes the product of the transmission functions $\mathcal{K}$ and $\mathcal{K}^*$.

Kintner[229] has shown that imaging calculations can be greatly simplified under the assumption that the functions representing the optical system, $\mathcal{K}(\xi, \eta)$ and $\mathcal{J}_0^-(\xi, \eta)$, are symmetric around the optical axis. The following subsection follows his work on image calculations.

(iii) *Image Calculations.*

One-Dimensional Calculations. If the object function varies in one dimension only, Eqs. (2.299) and (2.300) can be simplified to

$$I(\xi) = \int\int_{-\infty}^{+\infty} \mathcal{T}(\xi + \xi', \eta + \eta') \mathcal{F}(\xi + \xi') \mathcal{F}^*(\xi') d\xi' d\eta' \quad (2.301)$$

with

$$\mathcal{T}(\xi_1, \xi_2) = \int\int_{-\infty}^{+\infty} \mathcal{J}(\xi'', \eta'') \mathcal{K}(\xi_1 + \xi'', \eta'') \mathcal{K}^*(\xi_2 + \xi'', \eta'') d\xi'' d\eta''. \quad (2.302)$$

In the case of a one-dimensional object of even periodicity, the object can be represented by a Fourier series consisting of cosine terms only. An even periodic function with period $P = 1/\xi_P$

$$F(x) = \sum_{n=-\infty}^{+\infty} a_n \cos(2\pi n \xi_P x) \qquad (2.303)$$

with $a_n = -a_n$ has the Fourier transform

$$\mathscr{F}(\xi) = \sum_{n=-\infty}^{+\infty} a_n \delta(n\xi_P - \xi). \qquad (2.304)$$

The coefficients $\{a_n\}$ are complex when the object function includes phase variations. Inserting Eq. (2.304) into $I(\xi)$ gives

$$I(\xi) = \sum_{n=-\infty}^{+\infty} \left(a_n a_0^* \mathscr{T}(n\xi_P; 0) + \sum_{m=1}^{\infty} \left\{ a_{n+m} a_m^* \mathscr{T}[(n+m)\xi_P; m\xi_P] \right. \right.$$
$$\left. \left. + a_{n-m} a_{-m}^* \mathscr{T}[(n-m)\xi_P; -m\xi_P] \right\} \right) \delta(n\xi_P - \xi)$$
$$= \sum_{n=-\infty}^{+\infty} c_n \delta(n\xi_P - \xi), \qquad (c_n = -c_n) \qquad (2.305)$$

and

$$I(x) = \sum_{n=-\infty}^{-\infty} c_n \cos(2\pi n \xi_P x). \qquad (2.306)$$

Because the intensity distribution in the image is real and because the functions describing the object and the optical system are assumed to be real, the coefficients $\{c_n\}$ are real.

Even in the one-dimensional case, the integration for $\mathscr{T}$ is a tedious task. Where the source and objective apertures are circular and the objective pupil includes aberrations, this calculation has no simple solution. Two approximations exist that simplify the calculations. First, in the cases where the source aperture is small, the integration can be reduced to one dimension and numerical methods become attractive. Equation (2.302) becomes

$$\mathscr{T}(\xi_1; \xi_2) = \int_{-\infty}^{+\infty} \mathscr{J}(\xi'') \mathscr{K}(\xi_1 + \xi'') \mathscr{K}^*(\xi_2 + \xi'') \, d\xi'', \qquad (2.307)$$

where

$$\mathscr{K}(\xi) = \begin{cases} \exp[2\pi i \phi(\xi)] & |\xi| \leq 1 \\ 0 & |\xi| > 1 \end{cases} \qquad (2.308)$$

and $\phi(\xi)$ describes phase variations in $\mathscr{K}$. Within the framework of this model, effects of aberrations such as defocusing and spherical aberration can be investigated. In the coherent limit, where the effective source is a point, $\mathcal{J}(\xi) = \delta(\xi)$ and this model describes the imaging system exactly. For finite source sizes the model becomes less accurate.[232] Second, in the cases of an aberration-free system, the problem reduces to finding the area of mutual intersection of the source aperture and the shifted objective pupils (see Fig. 86). Kintner has published an efficient algorithm for the calculation of $\mathcal{T}$, which may also be applied to systems with annular sources, such as systems with dark-field illumination.

For simple objects in one dimension, analytical solutions can be found for the intensity distribution in the limits of coherent and incoherent illumination.[235,240] When the object function is an odd periodic function of position with equal lines and spaces of period P and spatial frequency $\xi_P = 1/P$, it can be represented by

$$F(x) = a_0 + \sum_{n=1}^{\infty} a_n \sin(2\pi n \xi_P x) \qquad (2.309)$$

analogous to Eq. (2.303). In the case of incoherent illumination, the optical system is linear in intensity. Under the assumption that only the fundamental spatial frequency of the periodic pattern is significant, the image intensity for incoherent illumination is given by

$$I(x) = \frac{I_{max}}{2} \left\{ 1 + \frac{4}{\pi} \left[1 - \frac{4}{\pi} \sin(\xi_P \lambda f) \right] \sin(2\pi \xi_P x) \right\}, \qquad (2.310)$$

where I_{max} is the illumination intensity for an unpatterned area. In the case

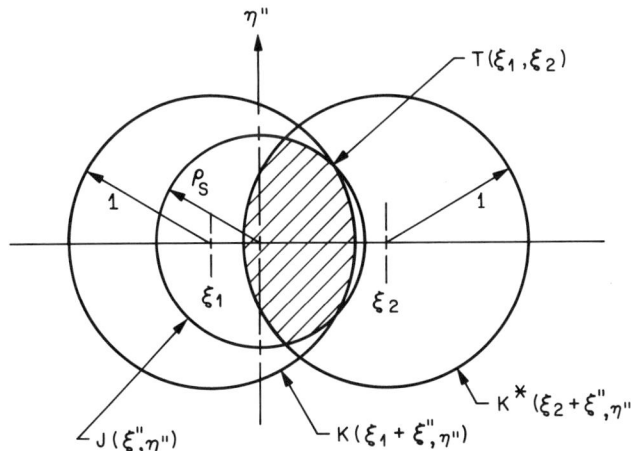

Fig. 86. Computation of the cross-coefficient function $T(\xi_1, \xi_2)$. (From Kintner.[229])

of coherent illumination, the image amplitude is given by[240]

$$A(x, z) = \frac{A_{max}}{2}\left\{1 + \frac{4}{\pi}\exp[i\phi(z)]\sin(2\pi\xi_P x)\right\}. \quad (2.311)$$

The phase angle ϕ describes the various aberrations within the optical system. For a nearly perfect system the phase term depends only on the focus condition and can be approximated by $\phi(z) \simeq \pi z \lambda / \xi_P^2$. The variable z is the distance from the focal plane. The intensity of the coherent image is given by $I(x) = AA^*$,

$$I(x) = \frac{I_{max}}{4}\left[1 + \frac{8}{\pi}\cos\phi\sin(2\pi\xi_P x) + \frac{16}{\pi^2}\sin^2(\pi\xi_P x)\right]. \quad (2.312)$$

In the case of partially coherent illumination, Eq. (2.301) is evaluated numerically.

A comprehensive computer program called SAMPLE[237] is available that has been used to explore the effects of spatially coherent illumination of a periodic mask pattern. Partial coherence affects both the image intensity incident on the resist and the resulting resist profiles after development.

The basic effects of imaging with partially coherent light can be seen in Fig. 87, which shows the calculated image intensity near the edge of a mask

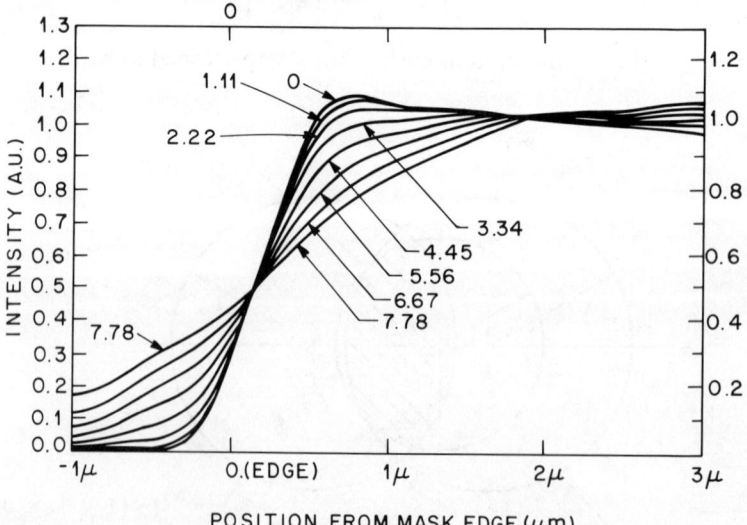

Fig. 87. Effect of focus error for $\sigma = 0.6$ on the image of a mask pattern of 2 μm lines and 6 μm spaces. The mask is imaged at $\lambda = 4357$ Å through a lens with NA = 0.28. (From O'Toole and Neureuther.[236])

pattern consisting of 2-μm lines and 6-μm spaces.[236] Since the mask is periodic, it is reflexive around both the $x = -1$-μm and $x = -3$-μm axis. The partial coherence factor σ has been defined in Eq. (2.288).

The numerical aperture of the lens is 0.28 and the wavelength is 0.436 μm. The focus error for the curves is taken in units of 0.4 Rayleigh units; one Rayleigh unit is 2.78 μm = $\lambda/(2NA_o)^2$. The focus error d is defined as the distance in micrometers between the resist surface and the plane of perfect focus. An increase of the focus error causes a decrease in the edge slope S:

$$S = \frac{dI}{dx}\bigg|_{x=\text{mask edge}} \tag{2.313}$$

and an increase in the low-intensity portion of the image.

The image slope near the edge does not vary too much from the slope at the edge. Figure 88 shows the edge slope of the same pattern (2-μm line, 6-μm space) for different σ's. The vertical lines represent focus errors at 1, 2, 3, and 4 Rayleigh units. The edge slope varies considerably with focus error, but it is fairly constant as a function of σ. This indicates that the edge slope is not significantly improved with the use of partial coherence.

The low portion of the image intensity distribution—the toe intensity—depends strongly on σ. It is defined as the intensity of the image at a distance

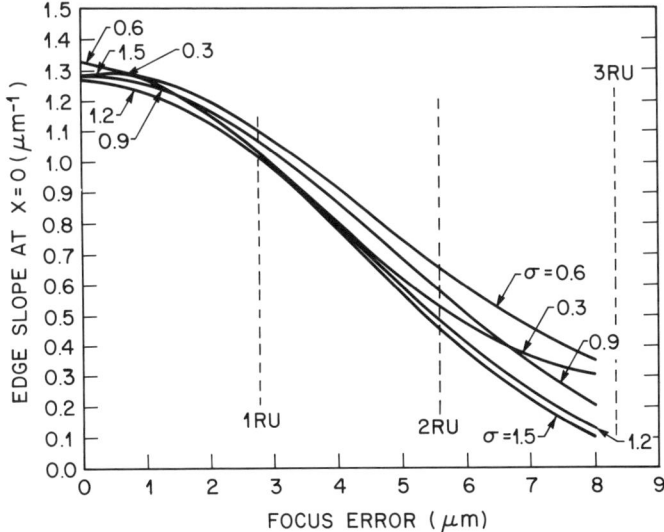

Fig. 88. Edge slope versus focus error for various σ's. The vertical dashed lines indicate the Rayleigh units. Illumination and lens parameters as in Fig. 87. (From O'Toole and Neureuther.[236])

Δx from the mask edge

$$I_T = I|_{x = \pm \Delta x}, \qquad (2.314)$$

where

$$\Delta x = \pm \tfrac{1}{2}(1.22\lambda/2\mathrm{NA}_o) \qquad (2.315)$$

and $\pm$ refers to the low-intensity side of the mask edge (toward the resist line). In Eq. (2.315), Δx is chosen as half the Rayleigh resolution limit for incoherent light.

Figure 89 shows the toe intensity of the same pattern for various σ and $\Delta x = -0.5$ μm. I_T rises rapidly with focus error and depends strongly on σ. Toe intensities for $\sigma \geq 0.9$ are significantly greater than those for smaller σ for all cases of focus error. I_T's for $\sigma \leq 0.7$ and $d \leq 4$ μm are very close to each other, indicating that nothing is gained by reducing σ below 0.7. Reduction of σ would only result in less light output from the condenser system and therefore would increase the exposure time.

Two-Dimensional Calculations. Lin[241] has studied nonperiodic diffraction-limited images in and out of focus with different partially coherent illumination conditions. He has assumed circular pupils, nonreflective substrates, absorption-free photoresists, and quasi-monochromatic illumination. His calculation starts from Eq. (2.297) for the mutual intensity

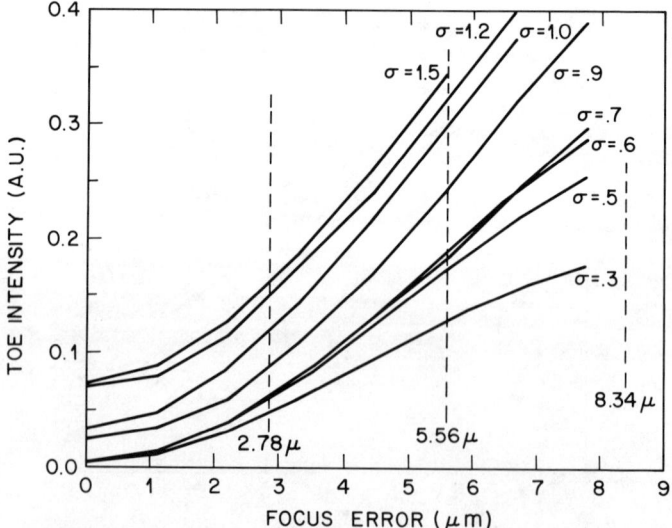

Fig. 89. Toe intensity versus focus error for various σ's and $\Delta x = 0.4$ μm. The vertical dashed lines indicate the Rayleigh units. Illumination and lens parameters as in Fig. 87. (From O'Toole and Neureuther.[236])

distribution of illumination. Assuming a uniform circular source, the mutual intensity of the light incident on the object is given by[219]

$$J_o^-(x_o - x_o', y_o - y_o') = [2J_1(\sigma)/v]I_o^-, \qquad (2.316)$$

where J_1 is the Bessel function of the first kind of order 1 and

$$v = (2\pi/\lambda)\sqrt{(x_o' - x_o)^2 + (y_o' - y_o)^2} n_c \sin \theta_c. \qquad (2.317)$$

The factor I_o^- is the intensity (assumed uniform) of the incident light, and $n_c \sin \theta_c$ is the numerical aperture of the condenser at the side of the imaging lens.

The factor I in Eq. (2.316) is taken to be one at the transparent part of the mask and zero at the opaque parts. The transfer function $\mathcal{K}$ is expressed by an integral and evaluated numerically. It is given by[219]

$$\mathcal{K}(x_o - x_o', y_o - y_o') = \frac{i}{\lambda^2} \int\int G\left\{\frac{\exp[ik(s-r)]}{sr}\right\} d\xi \, d\eta, \qquad (2.318)$$

where $k = 2\pi/\lambda$, G is the aberration function of the imaging lens, r and s are the distances from the object and image points, $(x_o, y_o, 0)$ and (x_o', y_o', z_o'), to the point (ξ, η, ζ) in the pupil plane of the imaging lens as shown in Fig. 90. In the absence of aberrations, G is set to 1. Defocus is incorporated in s. The expression for $\mathcal{K}$ in Eq. (2.318) can be simplified using the paraxial approximation and integration

$$\mathcal{K}(x_o - x_o', y_o - y_o') = \frac{n \sin \theta}{2\pi\lambda^2} \exp\left[\frac{iv}{(n \sin \theta)^2}\right] \int_0^1 J_0(u\rho) \exp\left(-\frac{iv\rho^2}{2}\right) \rho \, d\rho, \qquad (2.319)$$

where $J_0(u\rho)$ is the Bessel function of the first kind of order zero, and $n \sin \theta$ is

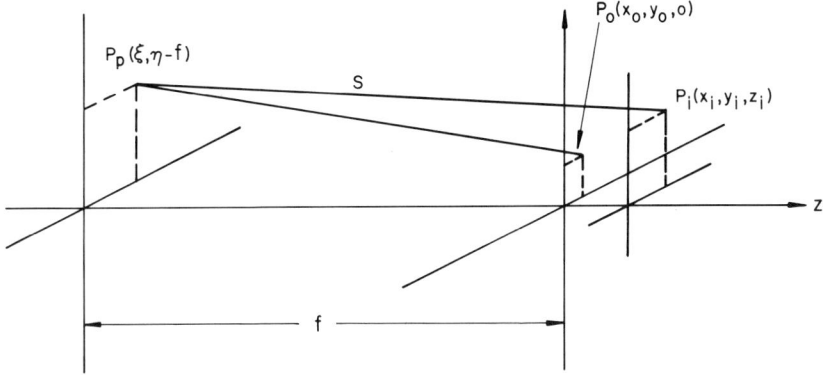

Fig. 90. Coordinate system used in the evaluation of the coherent transfer function. (From B. J. Lin.[241] © 1980 IEEE.)

the numerical aperture of the imaging lens. The variables u and v are normalized dimensionless coordinates for x and z of the form

$$u = (2\pi/\lambda)n \sin \theta [(x_o - x'_o)^2 + (y_o - y'_o)^2]^{1/2}, \quad (2.320)$$

$$v = (2\pi/\lambda)(n \sin \theta)^2 z. \quad (2.321)$$

The four-dimensional integral in Eq. (2.97) has been solved using Simpson's

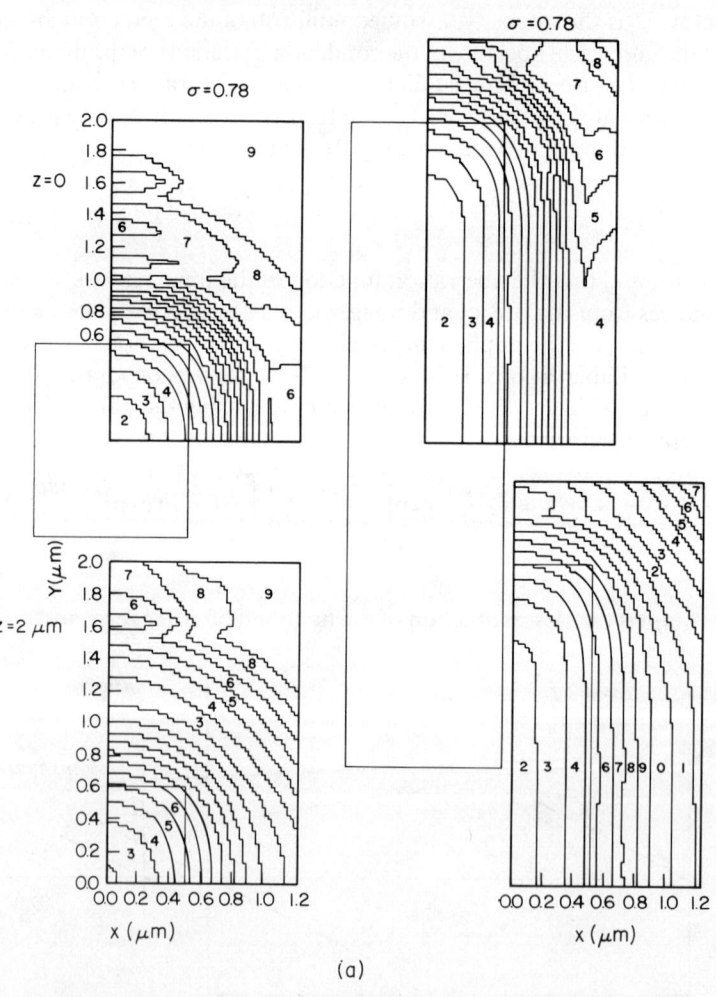

Fig. 91. (a) Image in focus ($z = 0$) and out of focus ($z = 2$ μm) of a 1×1.2-μm contact hole and a 1×4-μm rectangular opening. The imaging lens has a numerical aperture of 0.32. The wavelength of the light is 4045 Å. (b) Image of same pattern as in (a), except $\sigma = \infty$ and $\sigma = 0$, corresponding to total incoherence and total coherence, respectively. (c) Image from a rectangu-

and Newton's rules. Several simplifications have been employed by Lin to reduce the amount of computation. First, the matrices J_o and $\mathcal{K}$ in Eq. (2.97) are computed outside the integral, replacing time-consuming computations by simple indexing operations. Second, since J_1 has to be a real function, the computation of imaginary parts of $\mathcal{K}\mathcal{K}^*$ can be dropped in the two outer

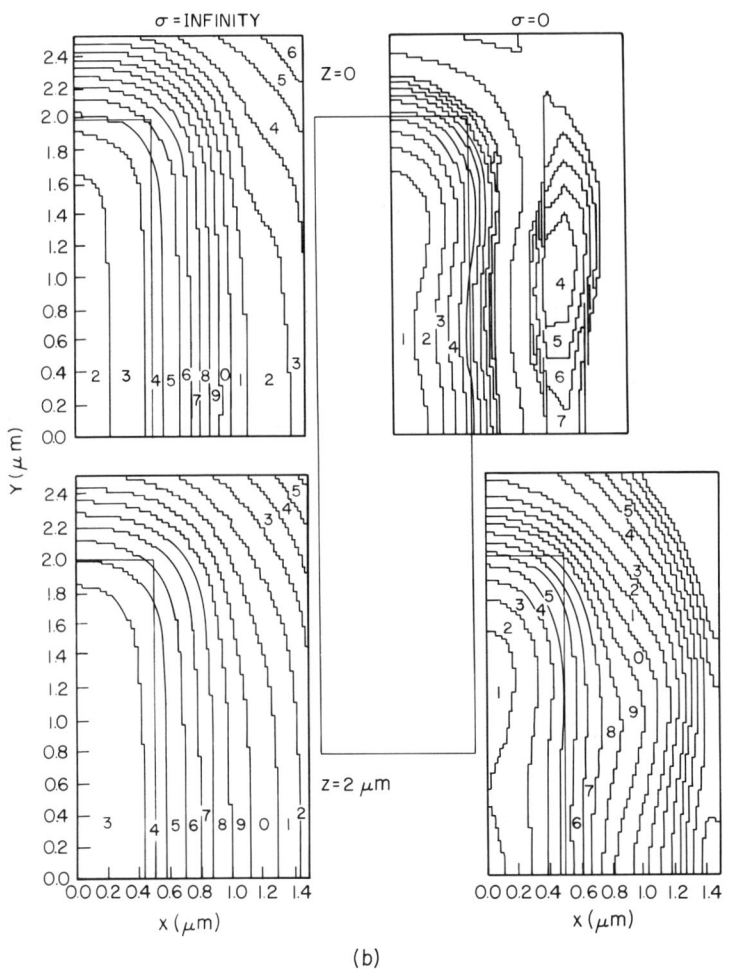

(b)

Fig. 91. (*Continued*)

lar bar at $\sigma = 0.44$. Illuminations using single and dual wavelengths are compared. (d) Image of two parallel rectangular openings of uneven lengths, for NA = 0.32 at $\lambda = 4045$ Å and NA = 0.16 at $\lambda = 2200$ Å. (From Lin.[223])

integration loops. Third, when only the image in the focus plane is of interest, all matrix elements in $\mathscr{K}$ are real.

Figure 91a shows calculated diffraction images of a 1-μm × 1.2-μm contact hole and a 1 × 4-μm^2 rectangular opening. An aberration-free lens with NA = 0.32 has been used for the calculations. The illumination is treated as monochromatic at λ = 4047 Å and a partial coherence factor of σ = 0.78 is used. The focal plane is specified by $z = 0$, which means that $z = 2$ μm means a 2-μm focus error from either side of the focal plane. Each point in the figure is represented by a number that indicates the intensity level. Similar to the preceding proximity printing results, there are a total of 20 levels ranging from 0, 1, 2, ..., 1, 2, ..., to 9. Each level is 1.5 db from the next level. Again, the level 0 indicates all intensities larger or equal to unity, which is the intensity required to expose completely a large uniform transparent area at a predetermined development condition.

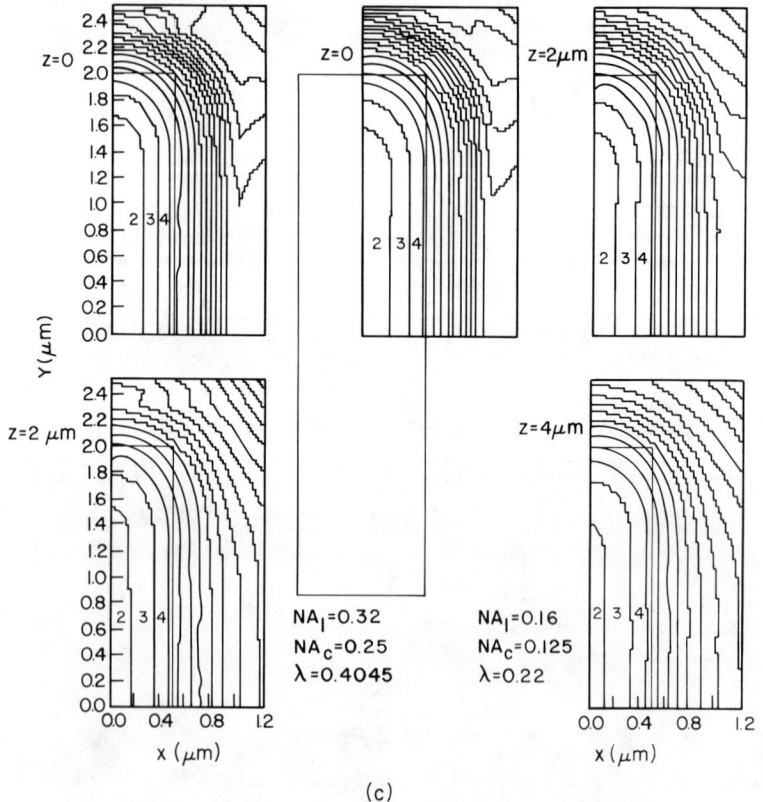

Fig. 91. (*Continued*)

Figure 91b compares the effect of illumination coherence on the image, for the same pattern. In the limit of complete incoherence, $\sigma = \infty$, the lines are less wavy but the image contrast is lower than that in Fig. 91a.

The situation of single versus dual wavelength exposure is shown in Fig. 91c. The spectral parameters used in the calculation are given in Table X.

Figure 91d compares calculated diffraction images for a lens with NA = 0.32 at $\lambda = 4047$ Å to a lens with NA = 0.16 at $\lambda = 2200$ Å. For both cases the coherence factor is $\sigma = 0.78$. The result for the shorter wavelength and

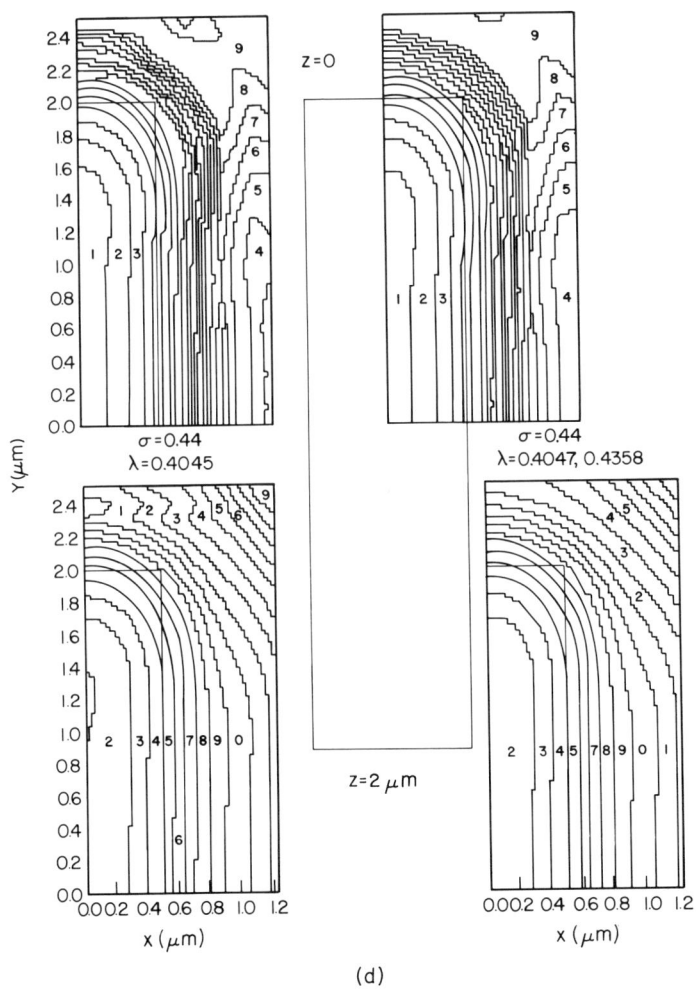

(d)

Fig. 91. *(Continued)*

smaller aperture at a defocus of $z = 4$ μm is similar to the other case at a defocus of $z = 2$ μm, indicating that wavelength reduction is more preferable than a higher NA.

(iv) *Resist Exposure.* The photoresist process is separated into two distinct parts: exposure and development. Exposure is the optical process that chemically alters the photoresist; development is a surface dissolution process that removes the resist at a rate proportional to the exposure. In the following we discuss a model that is capable of describing accurately the exposure process of positive photoresists.[242,243] Positive photoresists are composed of a photoactive compound, a base resin, and a suitable organic solvent system.[244] The basic role of the photoactive compound is to inhibit dissolution of the photoresist. Illumination destroys the inhibitor and results in an increased dissolution rate. Negative resists do not follow this model. Exposure of a negative resist results in cross-linking of large molecules. This makes exposed areas unsoluble. The modeling of these materials has not been attempted. For positive resists, a Lambert–Beer law is used to describe the optical absorption

$$\partial I(z, t)/\partial z = -I(z, t)[a_1 m_1(z, t) + a_2 m_2(z, t) + a_3 m_3(z, t)], \quad (2.322)$$

where $I(z, t)$ is the light intensity at a given depth z in the film at time t, a_1 to a_3 are the molar absorption coefficients of the inhibitor, the base resin and reaction products, and m_1 to m_3 are the corresponding molar concentrations. The destruction of inhibitor by the light intensity is modeled by

$$\partial m_1(z, t)/\partial t = -m_1(z, t)I(z, t)C, \quad (2.323)$$

where C is the fractional decay rate per unit area. Inserting the initial conditions and assumptions

$$I(0, t) = I_0 \quad (2.324)$$

$$m_1(z, 0) = m_{10} \quad (2.325)$$

$$M_2(z, 0) = m_{20} \quad (2.326)$$

$$m_3(z, t) = m_{10} - m_1(z, t) \quad (2.327)$$

into Eq. (2.322), we obtain

$$\partial I(z, t)/\partial z = -I(z, t)[AM(z, t) + B] \quad (2.328)$$

$$\partial M(z, t)/\partial t = -I(z, t)M(z, t)C \quad (2.329)$$

subject to the initial conditions

$$M(z, t) = 1 \quad (2.330)$$

$$I(z, 0) = I_0 \exp[-(A + B)z] \quad (2.331)$$

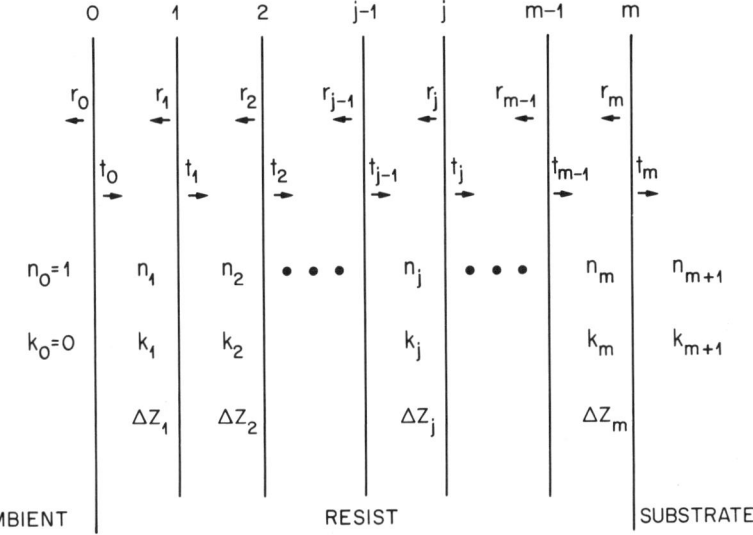

Fig. 92. Cross section of substrate and resist. The resist is subdivided into layers of thickness Δz_j. Various terms needed in the computation are defined in this figure. (From Dill et al.[243] Copyright 1975 by International Business Machines Corporation; reprinted with permission.)

and boundary conditions

$$I(0, t) = I_0 \tag{2.332}$$

$$M(0, t) = \exp(-I_0 Ct). \tag{2.333}$$

The variable $M(z, t) = m_1(z, t)/m_{10}$ is the fractional inhibitor concentration, and A, B, and C are constants that depend on the resist parameters and the exposure conditions. The two-dimensional inhibitor distribution within the resist after exposure is obtained by interpolation of the $M(z)$-function in the lateral x-direction.

The exposure parameters A, B, and C describe the exposure-dependent optical properties of the photoresist. The parameters A and B describe the absorption constant α of the resist[243]

$$\alpha = AM(z, t) + B \tag{2.334}$$

In actual calculations it is preferable to use the complex index of refraction of the photoresist

$$\tilde{n} = n - ik, \tag{2.335}$$

where n is the real part of the index and k is the extinction coefficient

$$k = \alpha\lambda/4\pi, \tag{2.336}$$

where λ is the exposing wavelength. This leads to

$$\tilde{n} = n - i\{\lambda[AM(z,t) + B]\}/4\pi. \quad (2.337)$$

To calculate the exposure distribution within the resist, thin-film optical computation techniques are used.[243,245] The optical properties of the resist vary during exposure as a function of depth. Therefore, the resist is subdivided into sublayers thin enough to be treated as if they had isotropic properties. Interference effects (standing waves[246] have a periodicity of $\lambda/2n$, so the sublayers must be thin compared to this value. Another criterion for the sublayer thickness is that the change in inhibitor concentration between neighboring sublayers must be small. The sublayer thickness Δz_j is usually in the range of 10–500 Å.

Berning[245] has published a procedure to perform thin-film optical computations on a computer. Figure 92 shows schematically the subdivided resist layer and the substrate. Complex reflection r_j and transmission coefficients t_j are calculated for each interface, starting from the substrate. The terms r_j and k_j are the refractive indices and extinction coefficients of the sublayers of thickness Δz_j.

One calculates the reflection and transmission coefficient

$$r_{j-1} = \frac{[\exp(-2i\phi_j)](F_j - r_j) - F_j(1 - F_j r_j)}{F_j \exp(-2i\phi_j)(F_j - r_j) - (1 - F_j r_j)} \quad (2.338)$$

$$t_{j-1} = \frac{(F_j^2 - 1)t_j \exp(-i\phi_j)}{F_j(F_j - r_j)[\exp(-2i\phi_j)] - (1 - F_j r_j)} \quad (2.339)$$

with the boundary conditions at the substrate

$$r_m = F_{m+1} \quad (2.340)$$

$$t_m = 2\sqrt{n_0 \operatorname{Re}(n_{m+1})}, \quad (2.341)$$

where

$$F_j = (n_0 - n_j)/(n_0 + n_j) \quad (2.342)$$

is the classical Fresnel coefficient with respect to air and

$$\phi_j = (2\pi/\lambda) n_j \Delta z_j \quad (2.343)$$

is the optical phase thickness of the layer.

The calculation starts at the substrate $j = m$ and progresses toward the resist surface. The reflectance and transmittance of the substrate coated with dielectrics and resist are given by

$$R = |r_0|^2, \quad (2.344)$$

$$T = |t_0|^2. \quad (2.345)$$

The absorbtance $\mathscr{A}_j$ of each sublayer is calculated from

$$\mathscr{A}_j = (1 - R)\left(1 - \frac{P_j}{P_{j-1}}\right) \prod_{1 \leq q \leq j-1} \left(\frac{P_q}{P_{q-1}}\right) \qquad (2.346)$$

where P_j is the magnitude of the Poynting vector, given by the recursive relation

$$\frac{P_j}{P_{j-1}} = \frac{|t_{j-1}|^2(1 - |r_j|^2)}{|t_j|^2(1 - |r_{j-1}|^2)}. \qquad (2.347)$$

The calculation for the absorptance starts at the surface ($j = 0$) and proceeds inward toward the resist film. The values for $\mathscr{A}_j$ are used to calculate the optical intensity I_j that exposes the resist in the jth sublayer

$$I_j = I_0 \mathscr{A}_j / (AM_j + B)\Delta z_j, \qquad (2.348)$$

where M_j is the inhibitor concentration in the same sublayer, given by Eq. (2.239),

$$\partial M_j / \partial t = I_j M_j C. \qquad (2.349)$$

The destruction of inhibitor changes the absorption in each layer. This change is treated by dividing the exposure time into small substeps, so that the changes in the inhibitor concentration are small. The calculation starts by evaluating I_j with M_j constant, incrementing the exposure time by Δt, and calculating a new value of M_j, and so on. The variable M_j is altered according to Eq. (2.333)

$$M_j|_{t+\Delta t} = M_j|_t \exp(-I_j(\Delta t)) \qquad (2.350)$$

with the initial condition, Eq. (2.330),

$$M_j|_{t=0} = 1. \qquad (2.351)$$

Figure 93 shows the intensity distribution $I(z)$ within a 0.584-μm photoresist film with a 6000 Å SiO_2 layer at the beginning of exposure by uniform incident illumination at $\lambda = 4358$ Å. Also included in the figure is the resulting inhibitor concentration after an exposure of 57 mJ/cm^2.

The final success of the calculation of resist exposure depends strongly on the choice of A, B, and C in the preceding equations. The constants A, B, and C have been measured by a variety of workers for different resists under various exposure and development conditions.[231,243,247-250] Table XI summarizes the most recent data on common resists. The values for A, B, and C are strongly dependent on the prebake conditions. This is in Fig. 94, which shows A, B, C for two resists. In Fig. 94a the dependence of A shows a masked decrease for prebake temperatures in excess of 90°C, whereas B has a

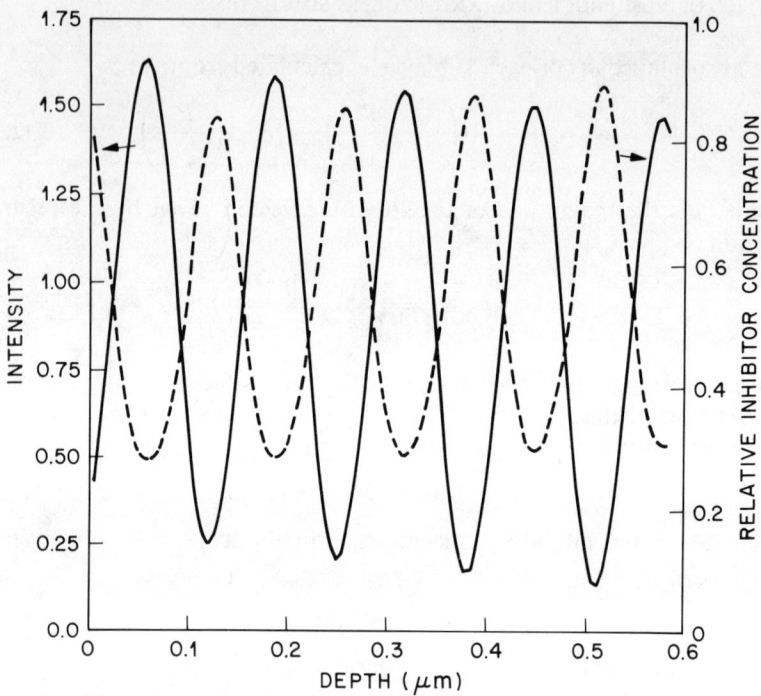

Fig. 93. Simulated intensity and inhibitor profile within 0.584 μm AZ1350 photoresist on 600 Å of SiO_2 on Si. (—) Intensity of exposing light, (---) inhibitor concentration after exposure to 57 mJ cm^2 at $\lambda = 4358$ Å. (From Dill et al.[243] Copyright 1975 by International Business Machines Corporation; reprinted with permission.)

TABLE XI

SUMMARY OF ABC DATA

Resist type	Wavelength (Å)	A (μm^{-1})	B (μm^{-1})	C	Reference, prebake conditions
AZ1350	3650	0.801	0.316	0.011	[248], not given
	4047	0.926	0.045	0.0139	[250], 80°C, 15 min
	4358	0.532	0.101	0.010	[248]
AZ1470	3650	0.860	0.381	0.0136	[250], 80°C, 15 min
	4047	0.852	0.086	0.0135	[250]
	4358	0.557	0.072	0.0110	[250]
AZ2400	3650	0.269	0.057	0.012	[249], 85°C
	4047	0.258	0.012	0.0122	[250], 80°C, 15 min
	4358	0.025	0.016	0.001	[249], 85°C
Kodak 809	3650	0.545	0.138	0.013	[250], 80°C, 15 min
	4047	0.680	0.075	0.0137	[250], 80°C, 15 min
	4358	0.422	0.073	0.010	[249], 85°C
Polychrome	3650	0.269	0.057	0.012	[249], 85°C
PC129	4047	0.199	0.066	0.011	[249]
	4358	0.025	0.016	0.011	[249]
AZ111	4047	0.569	0.042	0.008	[248], 85°C, 15 min

moderate increase and C is not affected at all. For one of the resists, Fig. 94b shows the values for A, B, and C versus time for one temperature $T = 110°C$.

To model the development of exposed resist, an analytical relationship between inhibitor concentration and etch rate has been suggested,[242,243] based on a least-squares fit to experimental data

$$R = \exp(E_1 + E_2 M + E_3 M^2). \qquad (2.352)$$

The parameters E_i give good agreement to experimental curves for high

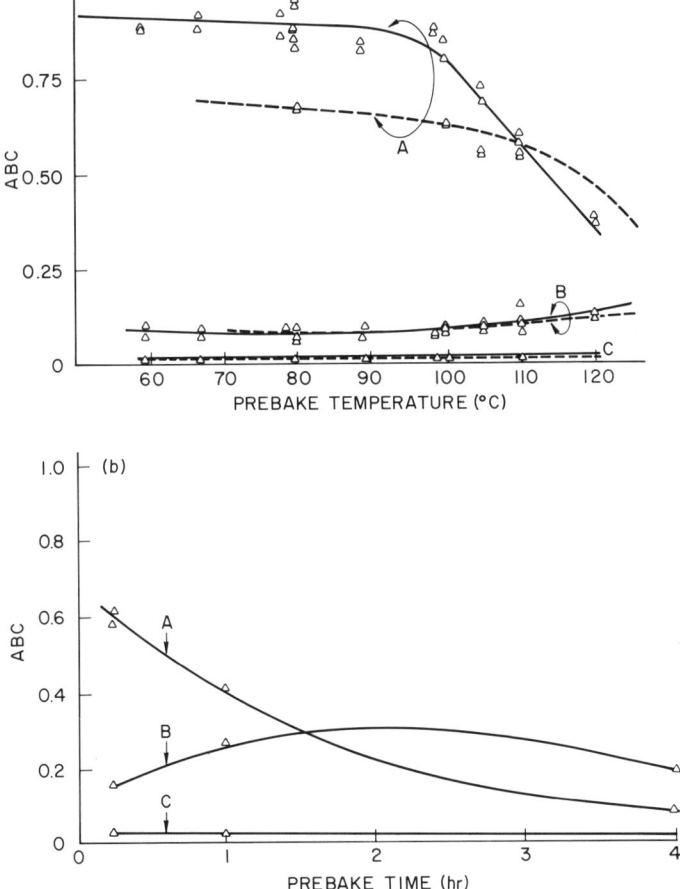

Fig. 94. Dependence of the values of the exposure parameters A, B, and C on the prebake conditions. (a) Dependence of A, B, and C on the prebake temperature for AZ1470 (—) and Kodak 809 (-··-) photoresist. (b) Dependence of A, B, and C on the prebake time for AZ1470 photoresist at a fixed temperature $T = 110°C$. In all cases the exposing light has a wavelength $\lambda = 4047$ Å. (From Neureuther and Wong.[250])

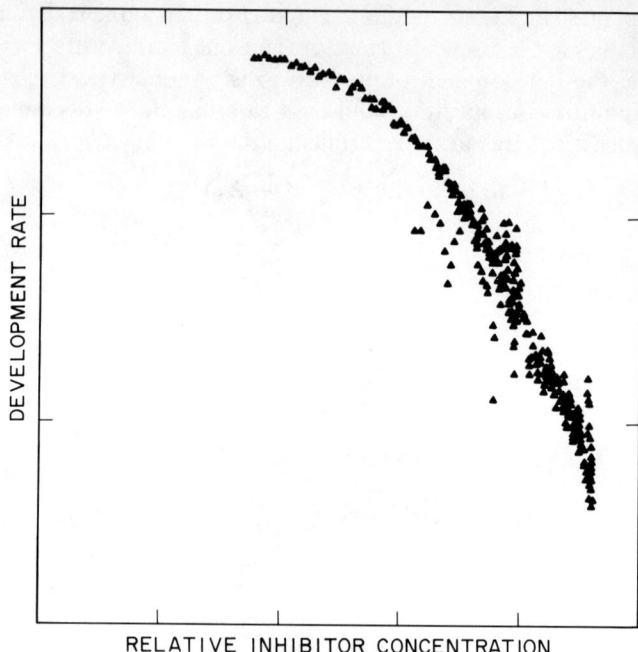

Fig. 95. Development rate for AZ1350J photoresist in a mixture of 1:1 AZ developer:H_2O at 20°C (Prebake at 70°C). (From Dill et al.[243] Copyright 1975 by International Business Machines Corporation; reprinted with permission.)

TABLE XII

SUMMARY OF ETCH RATE DATA

Resist type	E_1	E_2	E_3	Reference, development conditions
AZ1350J	5.63	7.43	−12.6	[246], 1:1 AZ developer: H_2O, 21°C (70C prebake)
	2.24	16.92	−15.0	[246], AZ developer (conc.), 21°C (70°C prebake)
	4.39	5.69	−9.0	[246], 1:1AZ developer: H_2O, 21°C (100°C prebake)
	15,269	−11.366	1.007	[251], AZ developer (conc.) 23°C (85°C, 15-min prebake)
AZ2400	10.7	−5.897	1.75	[251], 1:4, 23°C, 30-min prebake
Kodak 809	6.46	2, 19	−6.11	[249], 1:1 809 developer: H_2O, 20°C
Polychrome PC 129	8.96	−1.97	−4.51	[231,249], conc. 20°C
AZ111	5.96	−1.19	−2.27	[246], 1:1 AZ developer: H_2O, 21°C (70°C prebake)
	6.46	2.19	−6.11	[249], conc. 20°C (85°C, 30-min prebake)

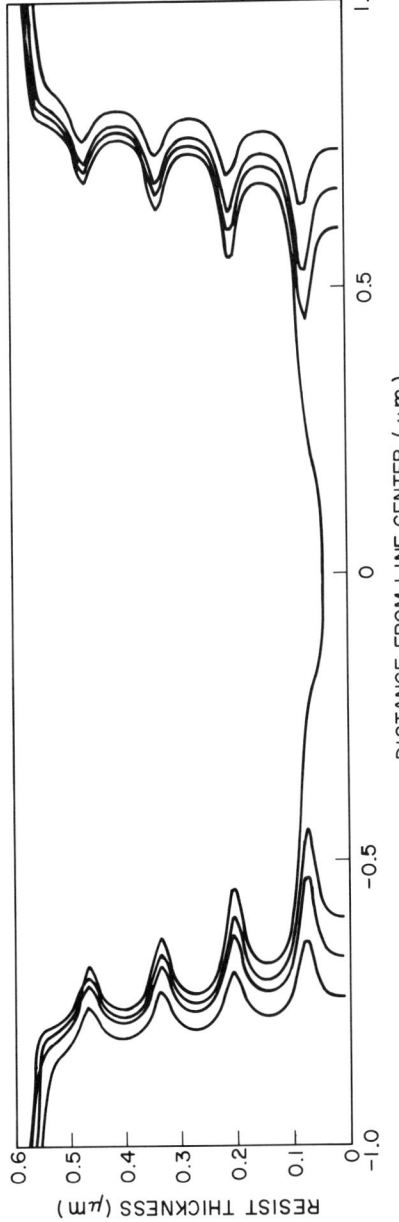

Fig. 96. Edge profiles for a 1 μm line in AZ1350J resist developed for 70, 80, 90, and 120 sec in a mixture of 1:1 AZ developer:H$_2$O. (From Dill et al.[243] Copyright 1975 by International Business Machines Corporation; reprinted with permission.)

values of inhibitor concentration ($M > 0.4$). For smaller values the fit using Eq. (2.352) can become unreasonable. The development rate of unexposed resist can be measured in a separate experiment to get the value for $M = 1$.

Figure 95 shows an experimentally determined development rate curve for AZ1350J photoresist. Table XII contains a summary of etch rate data for a variety of resists at different development conditions. With Eq. (2.352) specifying a result such as the one in Fig. 95, development rates can be obtained from the two-dimensional inhibitor concentrations yielding a matrix $R(x, z)$.

Subsection 5 presents the different techniques used to model resist development as a function of time. Typically, simulation starts at the top surface "developing" the resist according to Eq. (2.352). A typical example of a development simulation is given in Fig. 96, which shows the edge profiles of a nominal 1-μm line for development times of 70, 80, 90, and 120 sec.

b. *Electron-Beam Lithography*

Electron-beam (e-beam) lithography provides an elegant way to achieve the resolution necessary for further progress in microelectronics. Whereas optical lithography is based on diffraction phenomena, e-beam lithography

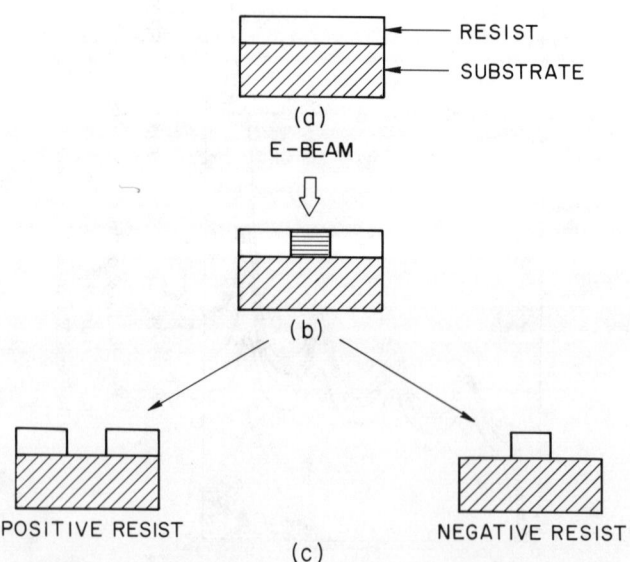

Fig. 97. Schematics of e-beam lithography processes. (a) Basic resist–substrate configuration, (b) exposure by e beam, and (c) development removes either exposed (positive resist) or unexposed areas (negative resist).

utilizes the scattering of energetic electrons for structure definition. Although fundamental limits exist for the electron-beam patterning process as well, they become dominant at considerably smaller dimensions (500 Å).

The use of focused electron beams to expose polymeric resist layers is the standard technique used in modern IC processing to create complex patterns.[252-254]

Figure 97 illustrates schematically the processes involved. After the resist is deposited on the wafer (a), the pattern is defined by the electron beam (b). Depending on whether the resist is positive or negative, the energy deposited in the resist by the electrons can break or cross-link the resist macromolecules. During development, exposed or nonexposed areas are then removed (c). The final resist pattern serves as a mask in subsequent processing steps (e.g., etching, ion implantation, or deposition).

Simulations of steps b and c in Fig. 97 include the actual design of the electron-beam exposure system and the calculations concerned with the exposure and development process. For this chapter we are not concerned with system design; interested readers are referred to Herriot and Brewer[255] and Thornton.[256]

For exposure simulations the quantity of major interest is the spatial distribution of energy dissipated by electrons in the resist due to interaction phenomena in the resist and the target. This stored energy forms an actual image of the desired pattern. The successful application of e-beam lithography is closely coupled to the understanding of the actual physical and chemical processes that govern interaction between beam electrons and the target. The ultimate resolution in e-beam lithography is not limited by the actual beam size but rather by electron scattering effects.

The final shape of the developed pattern in Fig. 97c depends on a variety of factors, including the charge density deposited, the accelerating voltage, the beam shape, the resist and the target properties, the proximity of other exposed regions, and the development conditions. The properties of a resist can be characterized by the energy D_c dissipated per unit area. This threshold energy is related to the deposited energy distribution $I(r, z)$[257]

$$qi\tau I(r, z) = D_c, \qquad (2.353)$$

where q is the electron charge, i is the beam current, and τ is the dwell time of the beam. The radially symmetric function $I(r, z)$ is the energy deposited per unit volume per electron and it allows the calculation of the energy at any point $r = \sqrt{x^2 + y^2}$ and depth z in the resist. For proper resist exposure, the total energy dissipated per unit volume has to fulfill the relation

$$\int_{-\infty}^{+\infty} qi I(r, z)\, dt = D_c. \qquad (2.354)$$

For a beam moving with velocity v, the linear charge density Q_L defines a relation analogous to Eq. (2.353)

$$E(y, z)Q_L = D_c, \qquad (2.355)$$

where

$$E(y, z) = \int_{-\infty}^{+\infty} qI(r, z)\, dt. \qquad (2.356)$$

The quantity $E(y, z)$ is the energy dissipated per unit volume per coulomb per centimeter.

In this section we present the physical foundations of electron scattering in solids, followed by a discussion of numerical and analytical techniques used to calculate the energy dissipation functions $I(r, z)$ and $E(y, z)$.

(i) *Electron Scattering in Solids.*

Nuclear Scattering. Fast electrons interact with electrically charged positive target nuclei and negative target valence electrons. As a result of these interactions, the incident electrons change both their direction and their velocity. Whereas the electron–electron collisions cause both an angular change and an energy loss, the electron–nuclear collisions are only responsible for a change in the direction of the incident electrons. The complete scattering history of an electron may be the accumulated result of a number of small deflections produced by different nuclei, or it may be due to one single deflection produced by one nucleus.

In the case of very thin resist layers on light substrates or thin foils, the probability that one electron will suffer more than one deflection is very small. This type of scattering is commonly referred to as *single* scattering. The other extremum occurs if an electron experiences a large number of collisions of similar size. This *multiple* scattering happens in the case of small angular deflections in thick targets. Multiple scattering exists as long as the resulting angular distribution is Gaussian.

For very thick films and/or lower beam energy, the number P_{elas} of scattering events increases, and one speaks of *electron diffusion*. In the opposite case, the number of scattering events becomes smaller, and the term *plural* scattering is used.

The theories of multiple scattering were initially developed by Bothe,[258] Williams,[259] Bethe et al.,[260] Lewis,[261] Meister,[262] Weymouth,[263] Moliere,[264] and Lenz.[265] Plural scattering theories were published by Bethe et al.,[260] Wentzel,[266] Moliere,[264] and Lenz.[265]

Multiple- and plural-scattering theories are first order in the scattering angle and they are valid only for total deflections up to 10 or 20°. For e-beam calculations they are of little value for simulations where scattering becomes so strong that the most probable angle of deflection exceeds 20° even for thin resist films.

The initial approximation for single small-angle scattering events is the Rutherford expression for the scattering of a charged particle by a nucleus of charge $Z_i q$,

$$\left.\frac{d\sigma}{dE}\right|_{\text{elas}} = \frac{(Z_i q^2)^2}{4m^2 v^4 [\sin^2(\theta/4) + \alpha_i^2]^2}, \quad (2.357)$$

where the subscript i indicates the ith atomic species, m and v are the mass and the velocity of the incoming electron and $\alpha_i = 2.33 Z_i^{1/3} E^{-1/2}$ is the atomic screening parameter given by Nigam et al.[267] In Eq. (2.357), any effect of inelastic scattering by an atomic electron is ignored. For small angles the intensity of inelastic scattering is given by an expression similar to Eq. (2.357), with a lower limit at an angle of the order of $\lambda/2\pi a$, where λ is the de Broglie wavelength of the scattered electron and a is the Bohr radius. In real cases the effect of electronic scattering can be incorporated in Eq. (2.357) by simply replacing Z by $(Z^2 + Z)^{1/2}$ [259]

$$\left.\frac{d\sigma}{dE}\right|_{\text{elas}} = \frac{Z_i(Z_i + 1)q^4}{4m^2 v^4 [\sin^2(\theta/2) + \alpha_i^2]^2}. \quad (2.358)$$

It is generally assumed that electrons scatter off individual atoms, neglecting any phenomena due to molecular binding. The probability for an electron with velocity v to collide with an atom of species i is given by

$$w_i = N_i v \sigma_i, \quad (2.359)$$

$$\sum_i w_i = 1, \quad (2.360)$$

where N_i is the density of atoms i and σ_i is the total scattering cross section obtained by integrating Eq. (2.358) over the full solid angle

$$\sigma_i = \frac{Z_i(Z_i + 1)\pi q^4}{m^2 v^4 \alpha_i^2 (\alpha_i^2 + 1)}. \quad (2.361)$$

An improvement over the Rutherford cross section in Eq. (2.358) is provided by the Mott cross section,[268] which is obtained through an exact solution of the Dirac equation describing the scattering of electron by an unscreened nuclear charge. Krefting and Reimer[269] and Ichimura et al.[270] have shown that Eq. (2.358) is a poor approximation for the description of nuclear scattering events, especially for electrons of lower energy ($E < 10$ keV) and for targets composed of heavy nuclei. For lithography calculations one is usually concerned with light substrates where Eq. (2.358) is still a good approximation. Nevertheless, it is quite common to define patterns on heavier substrates such as silicides (e.g. in the gate lithography step of a MOS process).

Krefting and Reimer[269] used the Mott cross section as derived from the Dirac equation by utilizing a partial wave expansion technique, whereas Ichimura et al.[270] applied Bühring's method[271] to generate the cross section. Figure 98 shows the ratio of Mott to Rutherford cross sections as a function of scattering angle θ and electron energy E for Al (a) and Ge (b).[272] The results were obtained using Byatt potentials,[273] which include the Coulomb potentials expressed by exponential terms. The influence of another approximation for a screened Coulomb potential according to Tietz[274] is shown in Fig. 98b for Ge at 10 and 100 keV. Kotera et al.[275] have investigated the applicability of the screened Rutherford and the Mott cross section for scattering in the 1–10-keV range for both Al and Au. Figure 99 compares results as a function of energy.

The results of Fig. 99 show that the classical Rutherford result increases more rapidly at lower energies but has a more rapid falloff for higher energies. Adesida et al.[276] have used this fact by multiplying the screening parameter

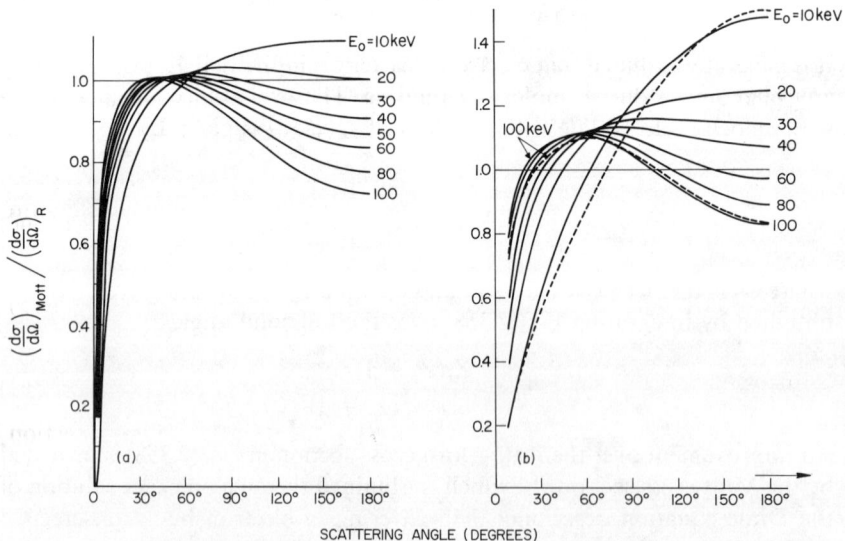

Fig. 98. Ratio of Mott to Rutherford cross section as a function of the scattering angle and the electron energy for (a) Al and (b) Ge. These results are obtained using Byatt potentials.[273,274] The dashed line in (b) is obtained from Tietz' approximation to the screened Coulomb potential[274] at 10 and 100 keV. (From Reimer and Krefting.[272])

in Eq. (2.358) by 0.48 to fit the Rutherford formula to the Mott result. No systematic study has been performed to analyze the impact of the inherent inaccuracies of Eq. (2.358) in lithography calculations.

The number P_{elas} of elastic scattering events in a resist film of thickness t is obtained by multiplying the elastic single-scattering cross section by the number of scattering centers. Using Eq. (2.361), we obtain

$$P_{elas} = \frac{Z_i(Z_i + 1)\pi q^4 N_A t}{m^2 v^4 \alpha_i^2 (\alpha_i^2 + 1) A}, \quad (2.362)$$

where N_A and A are Avogadro's number and the atomic weight of the solid,

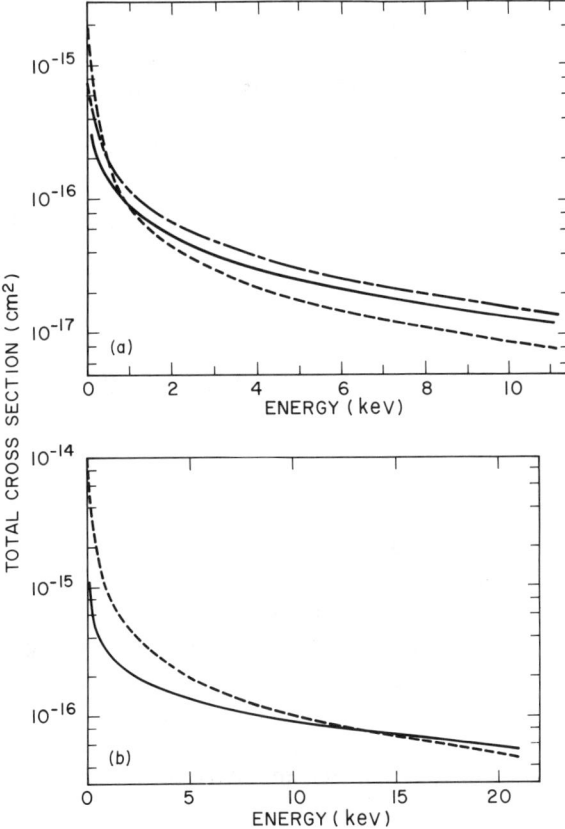

Fig. 99. Comparison of Mott and Rutherford cross sections for (a) Al and (b) Au. (a) — — —, Mott (HF); —, Mott (MHF); - - - -, screened Rutherford. (b) —, Mott; - - - -, screened Rutherford. (From Kotera et al.[275])

respectively. For typical electron resists, Greeneich[277] approximates Eq. (2.362) as

$$P_{\text{elas}} = 400t/E \qquad (2.363)$$

with the resist thickness t in micrometers and E in kilo-electron-volts.

Equation (2.363) can serve as an indication of whether plural or multiple scattering dominates for a given situation. Cosslett and Thomas[278-280] have established that for incident energies E between 10 and 30 keV, the transition from plural to multiple scattering occurs around $P_{\text{elas}} \cong 25$. Plural scattering is thus confined to $1 < P_{\text{elas}} < 25$, and multiple scattering involves a larger number of events until a diffuse condition is reached around $P_{\text{elas}} = 50$.

For typical situations, $t \cong 0.5$ μm and $E = 20$ keV, so $P_{\text{elas}} \cong 10$ and plural scattering prevails. The theory of plural scattering was pioneered by Wentzel[266] and Bothe.[258] Bothe suggested integrating the following expression for a known differential cross section $d\sigma/d\Omega$:

$$\psi(\chi) = \int_0^\infty \frac{d\sigma}{d\Omega} [J_0(\chi\theta) - 1]\theta \, d\theta, \qquad (2.364)$$

which can be used to calculate the angular distribution $f(\theta)$ by an integral transformation

$$f(\theta) = \int_0^\infty \exp\left[\frac{2\pi\rho t N_A \psi(\chi)}{A}\right] J_0(\chi\theta)\chi \, d\chi. \qquad (2.365)$$

In Eq. (2.364) and (2.365), J_0 is the Bessel function with index 0. The angular distribution is normalized as

$$\int_0^\infty f(\theta)\theta \, d\theta = 1. \qquad (2.366)$$

The evaluation of the integrals in Eqs. (2.364) and (2.365) depends on the form of the differential cross section. Moliere[264] has solved Eq. (2.365) for a simple cross section ignoring inelastic scattering events. Lentz[265] has included inelastic events by using a cross section of the form

$$\frac{d\sigma}{dE} = \frac{\bar{Z}q^4}{E^2}\left[\frac{\bar{Z}+1}{(\theta^2+\theta_0^2)^2} + \frac{2\theta_0^2}{\theta^2(\theta^2+\theta_0^2)}\right], \qquad (2.367)$$

where θ_0 is a screening parameter and $\bar{Z}$ is an averaged atomic number. Inserting this equation into Eq. (2.365) and integrating, one obtains

$$\theta_0^2 f(\theta) = \int_0^\infty \exp\left(\frac{2P_{\text{elas}}}{\bar{Z}}\left\{\frac{\bar{Z}+1}{\bar{Z}}[yK_1(y) - 1] + 2L(y)\right\}\right) J_0(xy) y \, dy, \qquad (2.368)$$

where $x = \theta/\theta_0$, $y = \theta_0 \chi$, K_1 is the normalized Hankel function of index 1, and

$$L(y) = -\tfrac{1}{2} y K_1(y) - K_0(y) - \ln y - \ln(\gamma/2) \tag{2.369}$$

and γ is Euler's constant.

Cosslett and Thomas[280] have compared the results of Lenz's calculations for the Bothe theory with experimental data for C, Al, Cu, and Au. Their findings confirm the Bothe theory.

In the region of multiple scattering, the angular distribution can be described by a Gaussian function.[258] The fraction of incident electrons scattered into direction θ is

$$f(\theta) = (2\pi\bar{\theta})^{-1} \exp(\theta^2/2\bar{\theta}^2), \tag{2.370}$$

where $\bar{\theta}$ is the most probable angle of scattering. Moliere[264] has expanded the original treatment by Bothe using a Thomas–Fermi expression for the atomic potential, where Bethe et al.[260] started from the Boltzmann equation.

Goudsmit and Saunderson[281] published a theory for multiple scattering that is valid for any scattering angle. They derived the angular distribution as a Legendre series

$$f(\theta)\theta\,d\theta = \sum_{l=0}^{\infty} (l + \tfrac{1}{2}) \exp\left[-\int_0^s G_l(s')\,ds'\right] P_{\text{elas}}(\cos\theta) \sin\theta\,d\theta, \tag{2.371}$$

where

$$G_l(s) = 2\pi l \int_0^\pi \frac{d\sigma}{d\Omega}(\theta, s)[1 - P_{\text{elas}}(\cos\theta)] \sin\theta\,d\theta \tag{2.372}$$

and s is the pathlength. The evaluation of the series in Eq. (2.371) proceeds using convenient recursion relations suggested by Spencer.[282–284]

Moliere's theory[264] takes into account occasional large-angle scattering events. His distribution has the form

$$f(\theta)\theta\,d\theta = \phi\,d\phi [2 \exp(-\phi^2) + f^{(1)}/B + f^{(2)}/B^2 + \cdots] \tag{2.373}$$

with the variable $\phi = \chi/(\chi_c \sqrt{B})$, where χ_c and B depend on the pathlength and energy of the electron.[283] The function $f^{(i)}$ in Eq. (2.271) has been tabulated by Moliere[264] and Bethe.[285] Extensions of Moliere's original work have been published by Nigam and Mathur.[286] The relationship between the Goudsmit–Sanderson and the Moliere theories has been pointed out by Lewis[261] and has also been discussed by Bethe.[285] Mott and Massey[287] give a comprehensive summery of the work on multiple scattering.

For a cylindrically symmetric scattering law, the transport equation can be simplified and has been solved by Fermi as indicated by Rossi and Greisen[288] for the spatial probability distribution function[289]

$$H(r, z) = (3\lambda/4\pi z^3) \exp(-3\lambda r^2/4z^3) \qquad (2.374)$$

with the mean free path

$$\lambda = \pi N \int_0^\pi \frac{d\sigma}{d\Omega} [1 - P_{\text{elas}}(\cos\theta)] \sin\theta \, d\theta, \qquad (2.375)$$

where N is the number of atoms in the resist. The radial distance r is measured

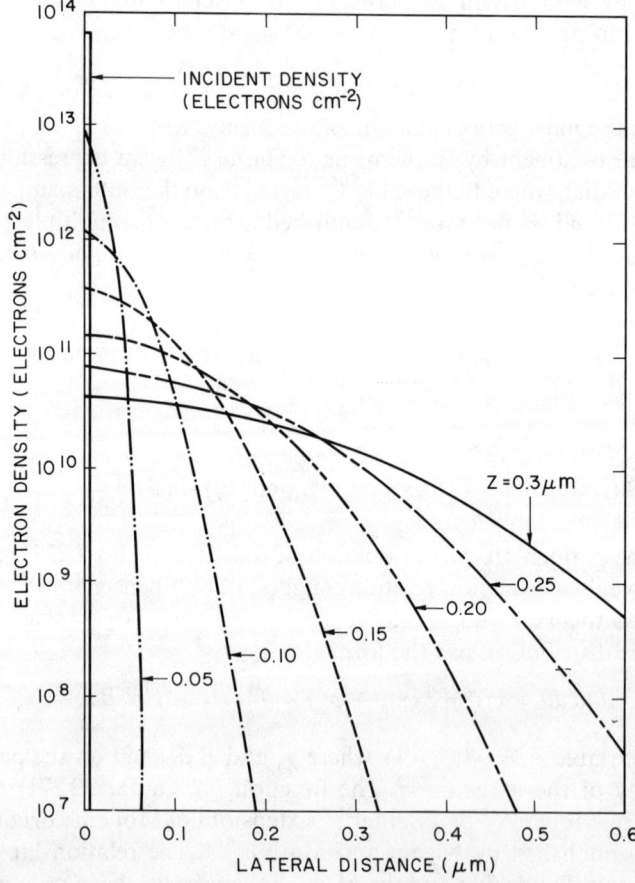

Fig. 100. Delta function response for forward-scattered electrons ($E_0 = 15$ keV) in a thin film of Ge. The curves show the electron density as a function of the lateral distance for various depths inside the Ge substrate. The total incident electron density of 10^{-5} C/cm^2 is uniformly distributed over 75 Å. (From Nosker.[290])

from the axis pointing in direction of the electron entering the target and z is the vertical distance along the axis of penetration.

Nosker[290] has utilized Eq. (2.375) to study the scattering of electrons in solids. Figure 100 shows the radial distribution for 15 keV electrons into Ge as a function of depth. A plot of this type is called δ-function response because all electrons enter the target at the same point.

Electron–Electron Scattering and Energy Loss. The first theory for the stopping power of matter for fast particles was given by Bohr.[35] Although this theory gave the general structure of cross-section formulas correctly, certain aspects of scattering remained unclear because of the lack of quantum mechanics. In 1930 Bethe[64] published a quantum-mechanical treatment based on the Born approximation, deriving a number of important results for the collision cross section and stopping power.

The energy transferred to an electron in the stopping medium is given classically by Bethe and Ashkin[291] and Birkhoff[292]

$$W = q^4/Ep^2, \tag{2.376}$$

where p is the impact parameter, defined in Fig. 13. For a range of impact parameters between $p_{\min}$ and $p_{\max}$, the stopping power, which is the energy lost per unit pathlength, is

$$-\frac{dE}{ds} = \frac{4\pi N_A q^4 \rho Z}{m_0 v^2 A} \int_{p_{\min}}^{p_{\max}} \frac{dp}{p} \tag{2.377}$$

with the target mass density ρ. In Bethe's original treatment this equation becomes

$$-\frac{dE}{ds} = \frac{4\pi N_A q^4 \rho Z}{m_0 v^2 A} \ln \frac{m_0 v^2}{I} \tag{2.378}$$

with the average ionization I of an atom. Expressions for I have been published[293] and can be approximated by Eqs.(2.55). A more complete quantum-mechanical treatment using the Mott electron–electron scattering formula gives

$$-\frac{dE}{ds} = \frac{4\pi N_A q^4 \rho Z}{m_0 v^2 A} \ln \left[\frac{m_0 v^2}{2I} \left(\frac{e}{2}\right)^{1/2} \right]. \tag{2.379}$$

This equation is used in most simulations of electron transport, and it is commonly referred to as the continuous-slowing-down-approximation (CSDA).

Although the average rate of energy loss can be obtained from Eq. (2.379), it does not apply to the process where an electron occasionally loses a large fraction of its energy in a single collision. Furthermore, the generation of secondary electrons by energy loss processes are ignored.

Several approaches to treat the statictical nature of the energy loss process have been published. The effect of energy *straggling* becomes more important in the later stages of electron travel. One result of straggling is that ionization can occur at increased depths because of the increased range of some electrons.[294] Landau[295] has given an expression for the distribution of energy loss by assuming that W is small compared to the initial energy E_0. Blunck and Leisegang[296] refined Landau's result by including the binding of atomic electrons in more detail.

Shimizu et al.[297,298] have proposed a more comprehensive approach to treat energy loss by incorporating differential cross sections for each of the elementary excitation processes such as electron–conduction-electron scattering, electron–plasmon scattering, and electron–core-electron scattering. The drawback of this rigorous technique is that it requires very accurate knowledge of the excitation functions of all individual processes. These functions, however, are only available for a small number of elements (e.g., Al).

Reimer and Krefting[272] have assumed that the primary electron loses part of its energy through core–electron excitation associated with the generation of secondary electrons and the rest through continous energy loss [Eq. (2.379)]. They used Gryzinski's core–electron cross section[299–301]

$$\left.\frac{d\sigma}{dW_c}\right|_{core} = \frac{q^4 E_{nl}}{W_c^3 E_0}\left(\frac{E_0}{E_0 + E_{nl}}\right)^{3/2}\left(1 - \frac{W}{E}\right)^{E_{nl}/(E_{nl}+W_c)}$$
$$\times \left\{\frac{W_c}{E_{nl}}\left(1 - \frac{E_{nl}}{E_0}\right) + \frac{4}{3}\ln\left(2.7 + \left(\frac{E - W_c}{E_{nl}}\right)^{1/2}\right)\right\}, \quad (2.380)$$

which approaches $\pi q^4/(EW_c^2)$ for free electrons, E_{nl} is the ionization energy for the atomic shell with quantum numbers n and l, and W_c is the energy loss through core-electron excitation. For valence–electron excitation, the difference between Bethe's stopping power and the above core–electron loss is used, that is,

$$-\left.\frac{dE}{ds}\right|_{valence} = -\left.\frac{dE}{ds}\right|_{CSDA} - \left.\frac{dE}{ds}\right|_{core}. \quad (2.381)$$

Ichimura et al.[270] have applied Krefting and Reimer's approach with considerable success to study electron scattering in Al targets. They have also compared the use of Eq. (2.381) to the direct technique of Shimizu et al.,[298] finding good agreement between the two methods.

Shimizu and Everhart[302] and Adesida et al.[276] have used a slightly different approach by extending Gryzinski's core excitation function to valence electrons by incorporating an appropriate value for the mean binding energy $\bar{E}_{nl}$ which satisfies

$$-\int W \left(\frac{d\sigma}{dW}\right)_{\bar{E}_{nl}} dW = -\left.\frac{dE}{ds}\right|_{\text{CSDA}} - \left.\frac{dE}{ds}\right|_{\text{core}}. \quad (2.382)$$

Although this approach seems straightforward, its application to organic resists is limited because of the inherent complexity of the integral in Eq. (2.382).

The transfer of energy from the primary electron to the atomic target electron can result in secondary electrons whose kinetic energy can be as large as half the primary energy. For example, 20-keV electrons can easily

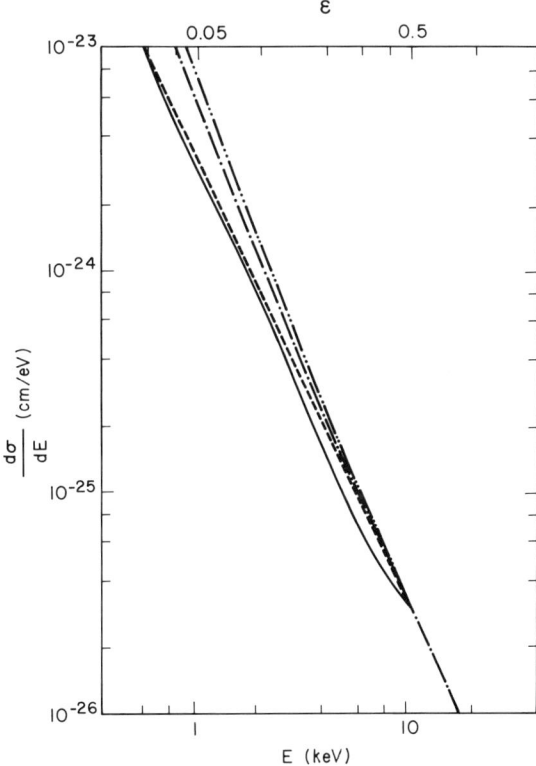

Fig. 101. Comparison among various differential cross sections for inelastic scattering ($E_0 = 20$ keV). (—) Møller cross section,[305] (-·-·-) Gryzinsky ($E_{nl} = 284$ eV),[299] (-··-··) Gryzinski ($E_{nl} = 532$ eV), (---) free electron limit. (From Murata et al.[306])

generate 2-keV secondaries along their path. According to the Bethe formula, however, these secondaries are nearly 10 times as efficient as the primary in depositing energy per unit pathlength.

The choice of the differential cross section determines the energy distribution and number of secondary electrons. Several choices for the cross section exist that have been used in simulations. Figure 101 compares several of these cross sections as a function of energy. The classical treatment of Coulomb interaction between free electrons gives for the differential cross section of one electron to produce secondary electrons[303-306]

$$\frac{d\sigma}{d\epsilon} = \frac{2\pi q^4}{mv^2 E}\left[\frac{1}{\epsilon^2} + \frac{1}{(1-\epsilon)^2}\right], \tag{2.383}$$

where ϵ is the normalized energy transfer W/E. The quantum-mechanical treatment adds an additional term and yields

$$\frac{d\sigma}{d\epsilon} = \frac{2\pi q^4}{mv^2 E}\left[\frac{1}{\epsilon^2} + \frac{1}{(1-\epsilon)^2} - \frac{1}{\epsilon(1-\epsilon)}\right]. \tag{2.384}$$

Gryzinski's result in Fig. 101 has been obtained from Eq. (2.380). As can be seen, the quantum-mechanical result in Eq. (2.384) gives the smaller result. The Gryzinski cross section is given for two different ionization energies E_{nl} corresponding to the ls state of O and C, respectively. It shows a larger value than the quantum-mechanical expression over the complete energy range.

(ii) *Calculation of Energy Deposition Functions.*

Numerical Methods—Monte Carlo. The complexity of the physics underlying electron interaction in solids makes it easy understand to why Monte Carlo (MC) methods have been *the* preferred choice. Contrary to ion implantation, where Boltzmann transport equation methods have had some impact for one-dimensional calculations, e-beam exposure simulations have to be at least two-dimensional to be applicable to real situations.

Since the availability of computers, MC calculations have been applied to study phenomena related to electron scattering in solids. Reviews of this early work can be found in the proceedings of the 1965 X-ray, Optics and Microanalysis conference,[307] and in several reviews.[283,308]

Green[309] used MC to calculate x-ray production in Cu using experimentally obtained scattering data. He used a simplified model by splitting the complete particle trajectories into a fixed number of linear elements of 0.1-μm length. Energy loss was included via Eq. (2.379).

Based on the multiple-scattering theory of Goudsmit and Saunderson,[281] Bishop,[310] and Shimizu *et al.*[311] replaced experimental scattering data by theoretical scattering expressions. Whereas Bishop's approach was based on Green's fixed steplength technique, Shimizu scaled the steplength with energy, thus accounting for the fact that electrons suffer increased scattering at lower energies. A number of authors have used this multiple-scattering approach to study microanalysis problems.[312,313]

Murata et al.[314] presented a model where the steplength is chosen equal to the elastic mean free path

$$\lambda = \left(\sum_i N_i \sigma_i\right)^{-1} \quad (2.385)$$

where N_i is the density of atoms of species i and σ_i is the corresponding cross section in Eq. (2.361). This is equivalent to the use of a fixed steplength and does not constitute a true single-scattering approach. However, the authors have shown that it is more appropriate for heavy-element targets than a full multiple-scattering approach.

Reimer[315] published the first "real" single-scattering MC calculation. He used two expressions of the cross sections for elastic and inelastic scattering to obtain angular distributions. For the elastic part, he used

$$\sigma_{elas} = \frac{\pi q^4 Z^2}{4E_0^2}\left[\left(\sin^2 \frac{\epsilon}{2}\right)^{-1} - 1\right] \quad (2.386)$$

for scattering angles $\theta \geq \epsilon = 10°$. For large scattering angles θ such that $\epsilon \leq \theta \leq 90°$, inelastic scattering is given by

$$\sigma_{inelas} = \frac{2\pi q^4 Z^2}{3E^2}\left(\frac{1}{\sin^3 \epsilon} - 1\right). \quad (2.387)$$

The mean free path is obtained from

$$\lambda = -\ln R/(\sigma_{elas} + \sigma_{inelas}) \quad (2.388)$$

with a random number $0 \leq R \leq 1$.

Shimizu et al.[311] have compared the results of Reimer's single-scattering and the approach based on Lewis' multiple-scattering theory. They found that although the single-scattering model needs a far less complicated program, the calculation times required are typically one to two orders of magnitude larger. A typical example is shown in Fig. 102, which compares electron trajectories in a Cu target for $E_0 = 30$ keV.

Saitou,[316] Kyser and Murata,[317] and Hawryluk et al.[257] were the first to apply the single-scattering concept to study the exposure of polymeric resists with energetic electrons. The major difference between the various single-scattering approaches lies in the calculation of the pathlength between elastic scattering events.

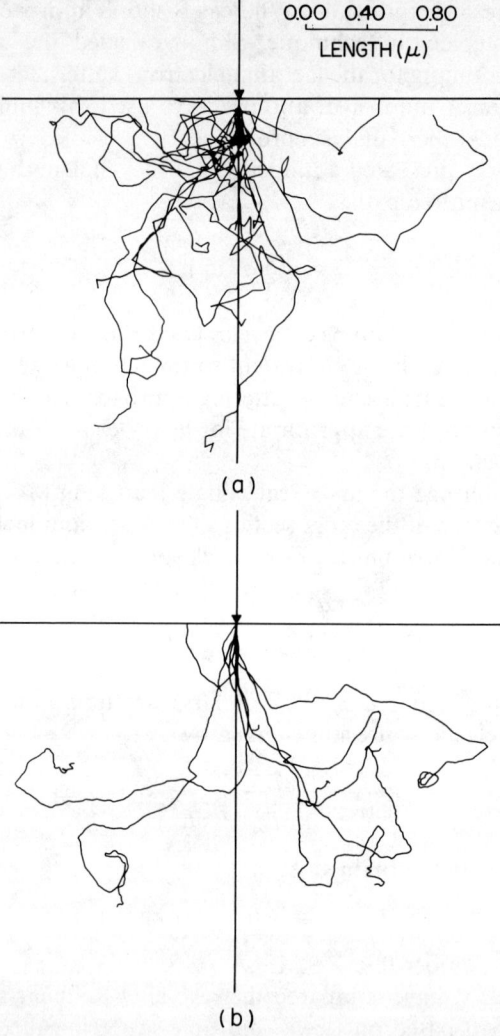

Fig. 102. Electron trajectories in a Cu target for $E_0 = 30$ keV (a) multiple-scattering result; (b) single-scattering result. (From Shimizu et al.[311])

Kyser and Murata[317] assume that the distance along a particle trajectory between successive scattering events is given by the mean free path in Eq. (2.385)

$$\lambda = \left(\sum_i \sigma_i n_i\right)^{-1} = \sum_i \frac{A_i}{N_A \rho \sigma_i C_i}, \qquad (2.389)$$

where A_i is the atomic weight of the scattering atom in grams per mole and C_i is the weight fraction for a specific element i.

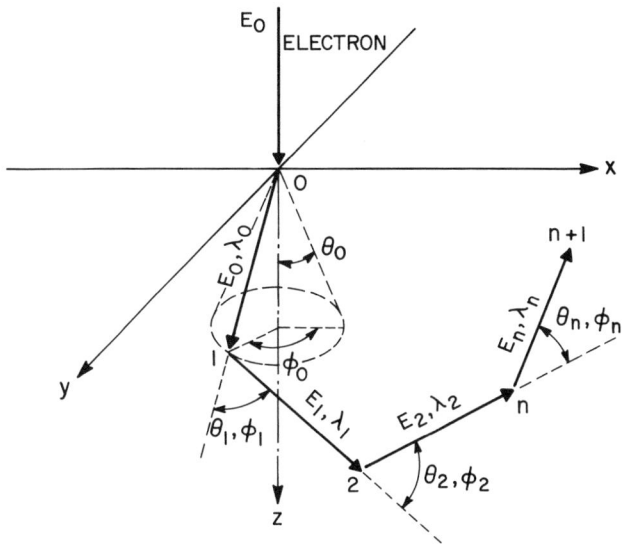

Fig. 103. Geometry and step sequence for Monte Carlo calculations in thick resists. (From Kyser and Murata.[317] This figure was originally presented at the Spring 1974 Meeting of The Electrochemical Society, Inc. held in San Francisco, California.)

Equation (2.389) is called the *mean free path approximation*. Hawryluk et al.[257] use a free path derived from Eq. (2.389)

$$s = \lambda \ln R \tag{2.390}$$

with a random number $0 \leq R \leq 1$. A second random number is used to determine the species i in the preceding equation, which acts as scattering center in the next elastic collision. The probability for atom i to act as a scattering center is given by

$$P_i = n_i \sigma_i / \sum_i n_i \sigma_i. \tag{2.391}$$

Figure 103 illustrates schematically the step sequence in the single-scattering model. Suppose an electron enters the surface at the origin, where it experiences the first elastic scattering event. The first pathlength s_0 is obtained either from Eq. (2.389) or (2.390), depending on the pathlength model used. The angle of scattering θ_0 is calculated from

$$\frac{\int_0^{\theta_0} (d\sigma/d\Omega)_i \sin x \, dx}{\int_\pi^0 (d\sigma/d\Omega)_i \sin x \, dx} = R_3 \tag{2.392}$$

and, with Eq. (2.358),

$$\theta_0 = \arccos\left[\frac{R_3(1 + 2\alpha_i^2) - \alpha_i^2}{R_3 + \alpha_i^2}\right] \tag{2.393}$$

with a third random number R_3. Furthermore, the axial symmetry of the scattering potential is accounted for by a random azimuthal angle Φ_0

$$\Phi_0 = 2\pi R_4. \tag{2.394}$$

The triple (s_0, θ_0, Φ_0) fixes the position of the next elastic scattering center. This procedure is repeated until the particle has stopped.

A more detailed sequence is shown in Fig. 104. Using (s, θ, Φ) and the coordinates of the two preceding scattering centers $\mathbf{r}_{i-1}$ and $\mathbf{r}_i$, the coordinates of the next event $\mathbf{r}_{i+1}$ are given by

$$\mathbf{r}_{i+1} = \mathbf{I}_{i+1} + \mathbf{r}_i, \tag{2.395}$$

where $|\mathbf{I}_{i+1}| = s$, and

$$\mathbf{r}_{i+1} = \mathbf{r}_i + s \sin\theta \cos\Phi \left(\frac{-|\mathbf{I}_i|^2 \hat{\mathbf{z}} + |\mathbf{I}_i \cdot \hat{\mathbf{z}}| \mathbf{I}_i}{|\mathbf{I}_i \times \hat{\mathbf{z}}||\mathbf{I}_i|} + \frac{\mathbf{I}_i \times \hat{\mathbf{z}}}{|\mathbf{I}_i \times \hat{\mathbf{z}}|} \right) + \frac{\mathbf{I}_i}{|\mathbf{I}_i|} s \cos\theta. \tag{2.396}$$

In order to calculate the energy loss between elastic scattering events, Hawryluk et al.[257] and Kyser and Murata[317] have used Bethe's CSDA. From Eq. (2.379), the threshold energy has to be larger than $I/\sqrt{(2/e)}$. For low energies the steplength becomes very small and the number of steps increases significantly.

The main concern in Monte Carlo calculations is to obtain a reasonably small standard error in the final result. The standard error is inversely proportional to the square root of the number of electrons. A reduction of the error by a factor k can be achieved only by increasing this number by a factor k^2. Typical exposure calculations are performed using 20,000–50,000 trajectories for one particular target–exposure situation.

The outline scheme must also account for multilayered targets. Figure 105 illustrates the sequence of events for a target composed of a resist layer on a substrate. Both the mean free path and the free path model described earlier determine the pathlength only from the scattering parameters appropriate to the layer containing the initial scattering point. However, electron scattering across layers will result in a pathlength different from an actual pathlength.

Horiguchi et al.[318] calculated the free pathlength in a different way. For a two-layer target, their result is given by

$$\bar{\lambda} = \lambda_2 - (\lambda_1 - \lambda_2) \exp[-(x/\lambda_1)], \tag{2.397}$$

where x is the distance from the interface of the two layers. Although their model is equivalent to the technique of Hawryluk et al.,[257] it is inherently faster because it requires only one function call to the random number generator.[319]

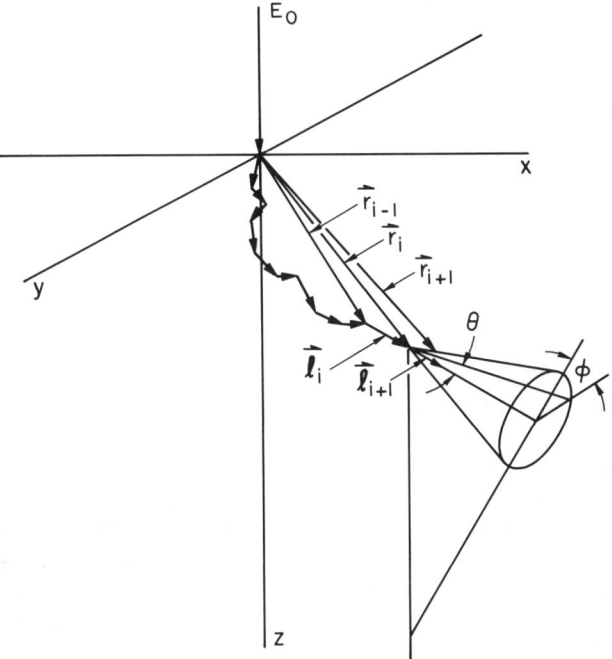

Fig. 104. Electron path from r_{i-1} through r_i to r_{i-1} and definition of variables. (From Hawryluk et al.[257]

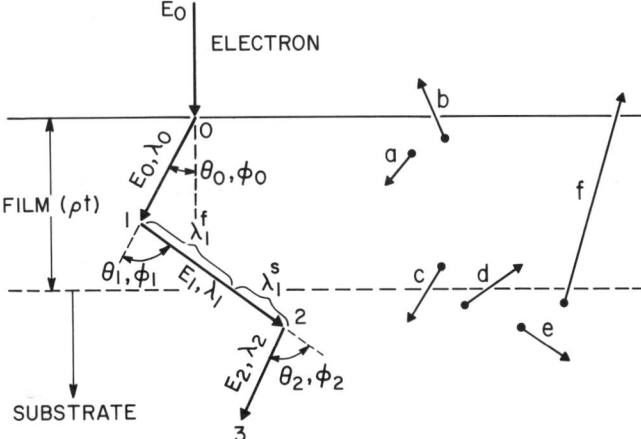

Fig. 105. Geometry and step sequence for Monte Carlo calculation in thin films. The letters a–f refer to the various scattering processes in the layers. (From Kyser and Murata.[317] This figure was originally presented at the Spring 1974 Meeting of The Electrochemical Society, Inc. held in San Francisco, California.)

Figure 106 shows calculated results for the pathlengths for a three-layer target consisting of 0.29-μm CMS resist, 0.3-μm Mo on Si. The difference between λ_{exact} and λ decreases for larger x because the probability that an electron will scatter into the Mo film decreases for larger x.

Chung and Tai[320] have performed single-scattering MC calculations of energy dissipation in polymeric resist where they compared Bethe's CSDA and Landau's energy straggling model.[295] In the straggling model the energy loss W is sampled from the distribution function f_{LD} via

$$R = \int_0^W f_{\text{LD}}(W', s)\, dW', \qquad (2.398)$$

where R is a random number, $0 \leq R \leq 1$.

The authors found excellent agreement both between the two energy loss models and the experimental results.

The single-scattering model has formed the basis for most of the energy deposition calculations. Figure 107 shows simulated particle trajectories of 50 electrons in 0.5-μm GMC resist on Si for energies of 3, 10, 20, and 50 keV. These trajectories were obtained from a MC simulation assuming a free pathlength model [Eq. (2.390)] using the CSDA for energy loss. The trajectories have been projected onto the $(x-z)$-plane, and the y component is not shown. These figures provide qualitative results for the movement of electrons inside the target. Basically, one can distinguish between two different scattering components: (1) a lateral forward scattering component and (2) a backscattering component that can extend over large lateral distances.

MC calculations are usually performed for idealized point and line sources. In the case of a point source, the half-space is divided into donut-shaped

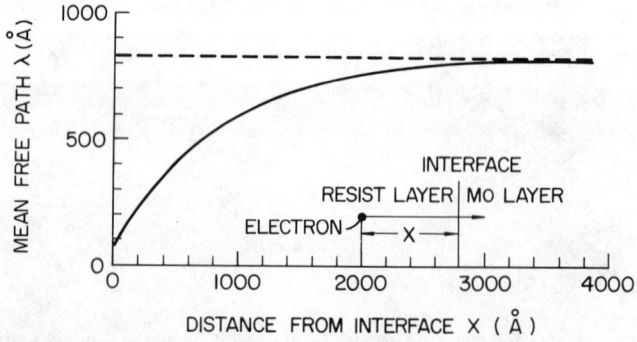

Fig. 106. Calculated mean free paths from different models as a function of the distance x from the interface between the resist and the Mo substrate. —, present model; - - -, conventional model. (From Horiguchi et al.[318])

volumes such as in Fig. 108a. The energy loss of all electrons traversing or stopping in this volume is added up and stored as a histogram $I(r, z)$ [Eq. (2.353)]. A typical example is shown in Fig. 109a for a 0.4-μm PMMA film on an Al substrate. In a line exposure the half-space is divided into infinite parallelepipeds as shown in Fig. 108a. All energy is added up to calculated $E(y, z)$ given in units of electron-volts per coulomb per square centimeter. Figure 109b shows the exposure result for the same target as above.

Murata *et al.* [306] have extended the single-scattering MC method in order to include the effects of secondary electrons. Their pathlength model is

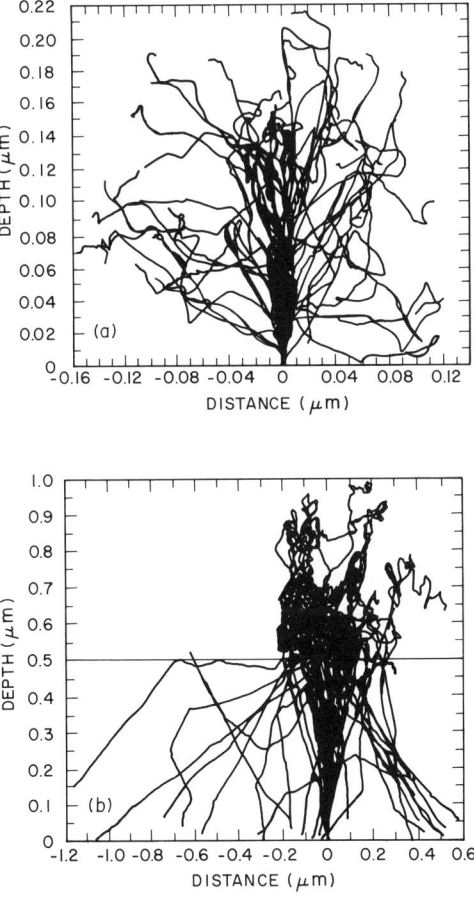

Fig. 107. Simulated particle trajectories for 50 electrons in a target with 0.5 μm of resist for four different energies of 3 (a), 10 (b), 20 (c) and 50 (d) keV. Note the different depth scales in the four cases. (From Fichtner *et al.*[7] © 1984 IEEE.)

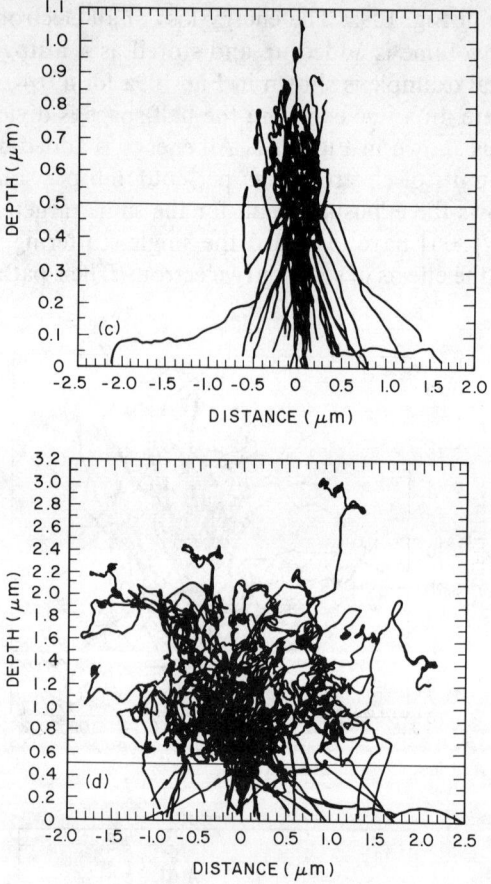

Fig. 107. (*Continued*)

based on the original work of Schneider and McCormack.[321] Figure 110 shows the relation between the energy lost and the pathlength for this *knock-on* process. The dashed line is the energy spectrum as obtained from the original CSDA model. In the knock-on model, primary electrons lose energy continuously until secondaries are generated.

The determination of elastic or inelastic collision is obtained from the scattering mean free path λ_T given by

$$1/\lambda_T = \sum_i n_i \sigma_i^{elas} + \sum_i n_i Z_i \sigma_i^{inelas} = \sigma_T = 1/\lambda_{elas} + 1/\lambda_{inelas}. \quad (2.399)$$

The steplength s is obtained analogous to Eq. (2.390)

$$s = -\lambda_T \ln R. \quad (2.400)$$

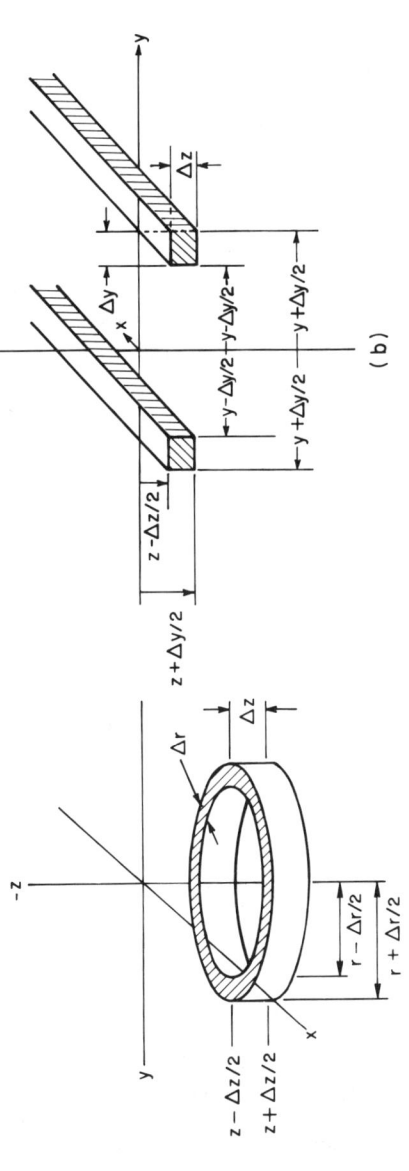

Fig. 108. (a) Donut-shaped geometry used to calculate $I(r, z)$. (b) Parallelepipeds used to calculate $E(y, z)$. (From Hawryluk et al.[257])

Defining the probabilities for elastic and inelastic scattering

$$P_{elas} = \lambda_T/\lambda_{elas} \qquad (2.401a)$$

$$P_{inelas} = \lambda_T/\lambda_{inelas} \qquad (2.401b)$$

it follows that $P_{elas} - P_{inelas} = 1$. If $P_{elas} < (>) R$, the scattering will be elastic (inelastic).

A newly generated electron will travel in a similar fashion, continuously losing energy along its way. When an inelastic collision occurs, the energy of the secondary is determined by an integrated expression of Eq. (2.383) or (2.384).

Figure 111 shows trajectories of 50 electrons and the corresponding secondaries generated for a thick poly(methyl methacrylate) (PMMA) resist substrate exposed at $E_0 = 20$ keV. The energy loss process can be visualized from Fig. 111b. Some electrons will lose energy abruptly to a secondary and

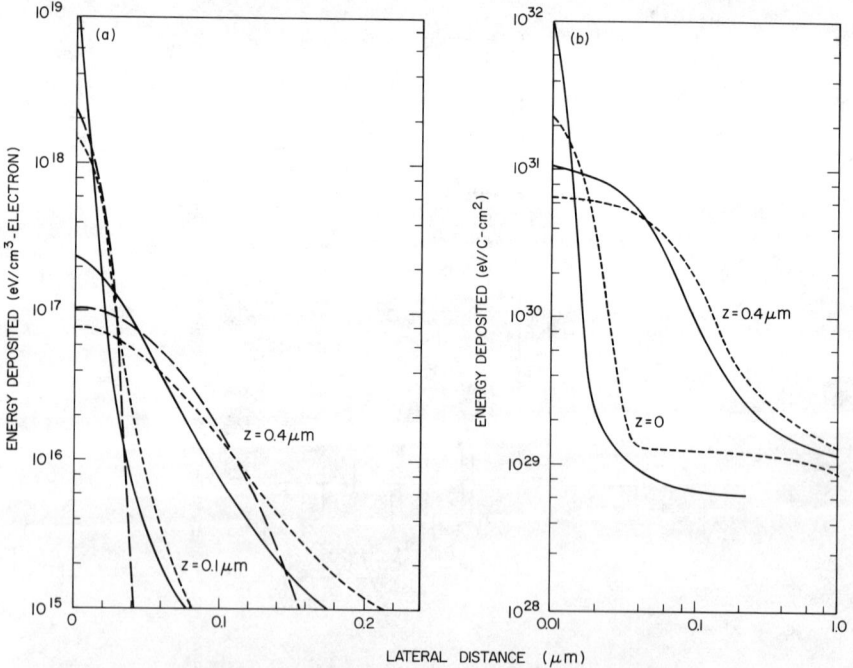

Fig. 109. (a) Energy density profiles for delta-function point source (20-keV electrons into 0.4-μm PMMA). MC results for Al (—) are compared with analytical profiles from Hawryluk et al.[257] (---) and Greeneich and van Duzer.[323] (-----) (b) Energy density profiles for a delta line source. MC results for Si (—) are compared with the plural-scattering profile of Hawryluk et al.[257] (----) (From Kyser and Murata.[317] This figure was originally presented at the Spring 1974 Meeting of The Electrochemical Society, Inc. held in San Francisco, California.)

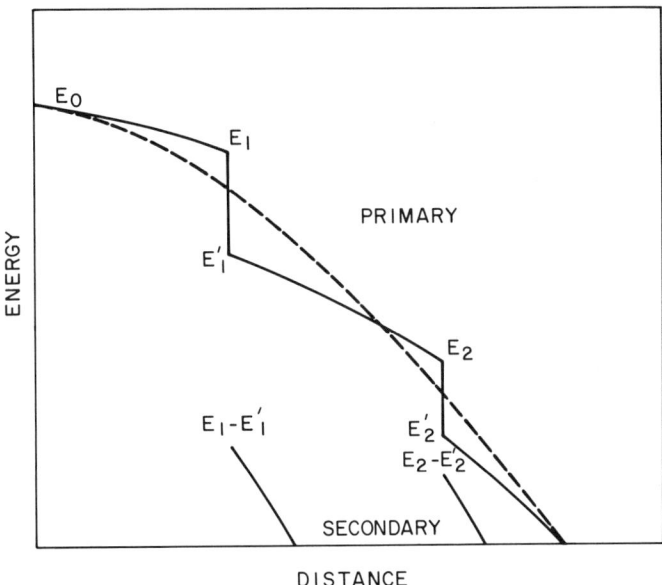

Fig. 110. Schematic illustration of the electron energy versus path length in the hybrid model for the knock-on process. (From Murata et al.[306])

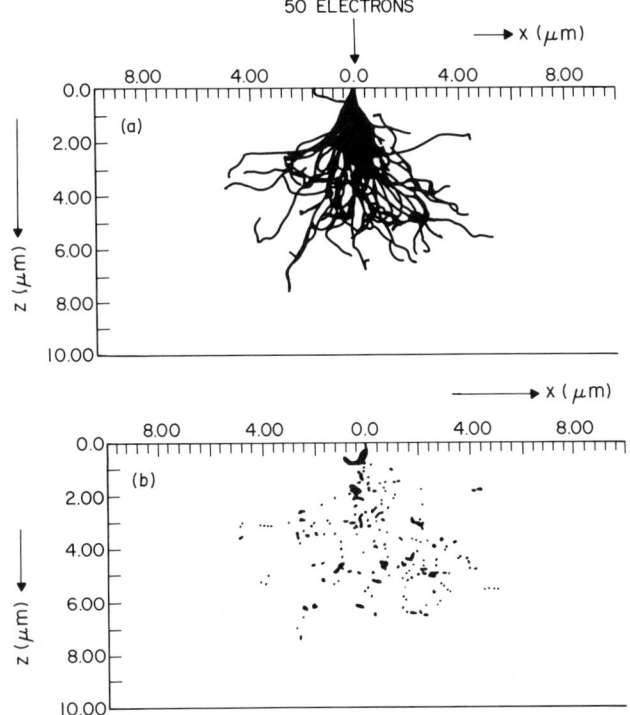

Fig. 111. Trajectories of the (a) primary and (b) secondary electrons for thick PMMA resist at 20 keV. (From Murata et al.[306])

stop in an early stage. Others will travel a long distance without a major loss along the way.

The quantity of major interest in beam simulations is the deposited energy. Murata et al.[306] have performed calculations for a thin PMMA film with and without a Si substrate for $E_0 = 20$ keV. A typical result is shown in Fig. 112 for the top and the bottom interface of the resist. Curves 1–4 and 5–8 represent results with and without the Si substrate. The production of secondary electrons predicts a significantly larger intensity around the center. At the bottom (3, 4), the intensity increases by 50% in the range 0.1–0.4 μm. For the

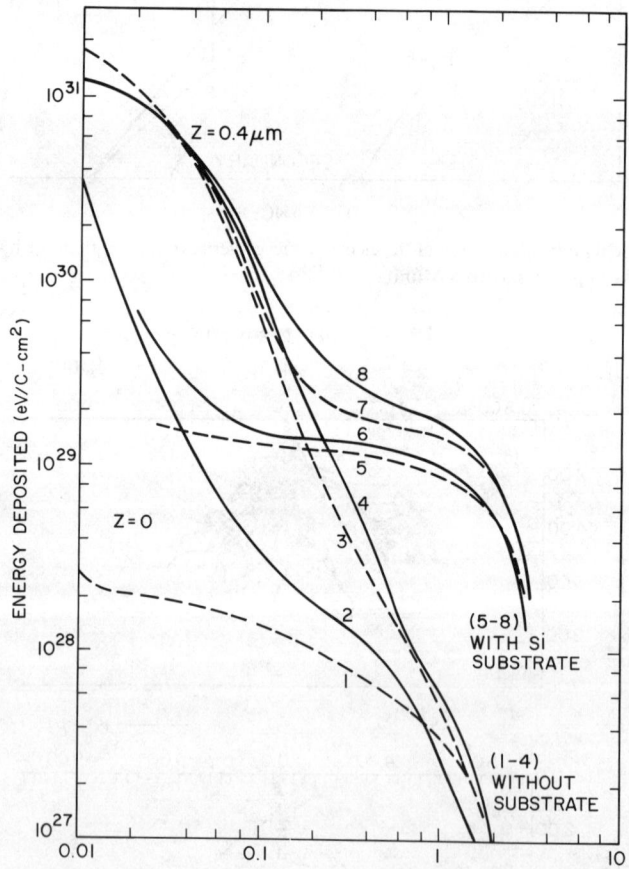

Fig. 112. Lateral distribution of the absorbed energy density for a point source exposure at 20 keV into 0.4-μm PMMA without (curves 1–4) and with (curves 5–8) substrate. Results are compared with conventional results (old model, no secondaries) for both an isolated thin PMMA film and a film on Si. —, Knock-on, – –, old model. (From Murata et al.[306])

substrate, contributions from secondary electrons are generated in the resist by highly energetic backscattered primaries, or backscattered secondaries that were generated in the substrate.

Analytical Methods. The energy distribution function $I(r, z)$ is the sum of three components (Fig. 113): (1) $I(r, z)|_{FR}$, the energy loss of forward-scattered electrons in the resist, (2) $I(r, z)|_{BR}$, the energy loss of electrons backscattered in the resist, and (3) $I(r, z)|_{BS}$, the equivalent substrate component[322]

$$I(r, z) = I(r, z)|_{FR} + I(r, z)|_{BR} + I(r, z)|_{BS}. \quad (2.402)$$

Both the plural- and the multiple-scattering model have been used to calculate the energy density for the forward-scattering component $I(r, z)|_{FR}$. In both models the flux of electrons is simulated by replacing the resist between the surface and the plane of observation by an effective scattering source at the top. The pathlength is the distance between the observation point and the effective source.

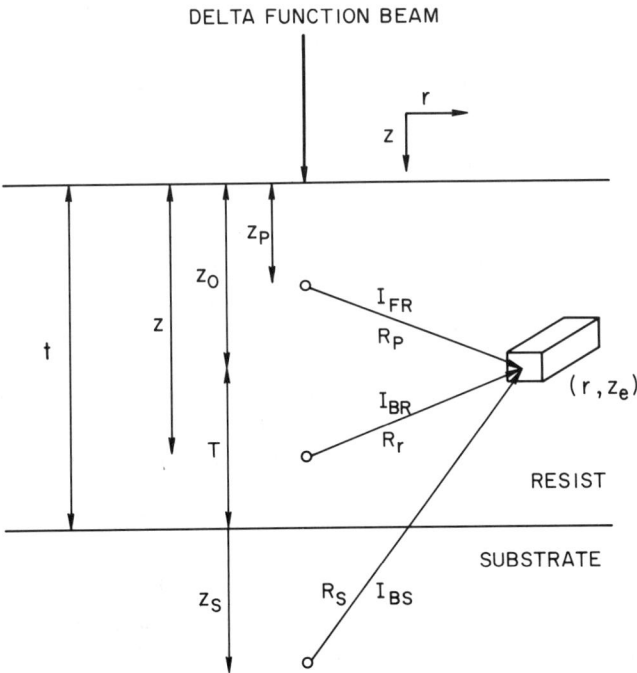

Fig. 113. Geometry for computing the delta function scattering response of the resist and the substrate. (Adapted from Greeneich and van Duzer.[322] © 1977 IEEE.)

In Section II.4.b.1, the differences between plural and multiple scattering are noted. The plural-scattering model is more complicated, although more accurate, if applied to typical e-beam exposures [Eq. (2.363)].

Greeneich and van Duzer[322,323] used Lenz's plural-scattering model in Eq. (2.368) to calculate $I(r, z)|_{FR}$. For a delta function response, they write the forward-scattering contribution as the product of the flux of electrons passing through an incremental volume around the observation point at (r, z_0) times the energy loss rate in this volume.

$$I(r, z)|_{FR} = F(r, z)E(s, E_0), \quad (2.403)$$

where $F(r, z) = [f(\theta)/2\pi R^2](\theta/\sin \theta)^{1/2}$ is the electron flux and $\theta = \arctan(r/2)$. The energy loss rate E is an approximation to the CSDA equation for an electron of initial energy E_0 after traveling the distance $s = (r^2 + z^2)^{1/2}$.

The contribution of backscattered electrons is based on Everhart's approach[324] to calculate the backscattering coefficient η, which is defined as the ratio of backscattered to incident electrons. As can be seen in Fig. 113, two different backscattering components exist.

For the backscattering in the resist film, Greeneich and van Duzer obtain

$$I(r, z)|_{BR} = \frac{0.045\bar{Z}}{\pi R_{Be}} \int_z^{z\,\max} \left(\frac{E(z')}{E_0}\right)^{2\beta - 2} \frac{E(s)}{(R + z' - z_0)^2} dz', \quad (2.404)$$

where $R = [(z' - z)^2 + r^2]^{1/2}$, $\beta^2 = 0.045\bar{Z}$, $\bar{Z}$ is an average atomic number, and R_{Be} is the Bethe range, defined as the pathlength for which Eq. (2.379) goes to zero.[325] An equivalent expression has been used for the backscattering contribution from the substrate. Results from the plural scattering model are shown in Fig. 114 for the energy density $I(r, z)$ as a function of radial distance from the origin. The exposure conditions of the simulation are for 0.4-μm PMMA resist film on Al exposed with 20-keV electrons.

Hawryluk et al.[257] have used the multiple-scattering model of Nosker[290] to calculate the forward-scattering contribution. They obtain the density distribution of electrons in the resist by a convolution of Eq. (2.374)

$$D(r, z) = \int_0^{2\pi} \int_0^\infty H(r_0, z)N(X)X\, dX\, d\phi, \quad (2.405)$$

where $N(X) = N_0 a/\pi \exp(-aX^2)$ is the Gaussian distribution of electrons in the resist and N_0 is the total number of incident electrons. The forward-scattering component is obtained in analogous form to Eq. (2.403)

$$I(r, z) = \frac{D(r, z)}{N_0} \left\{\frac{2\pi q^4 N}{E_0} \ln\left[\frac{E_0}{I}\left(\frac{e}{2}\right)^{1/2}\right]\right\}. \quad (2.406)$$

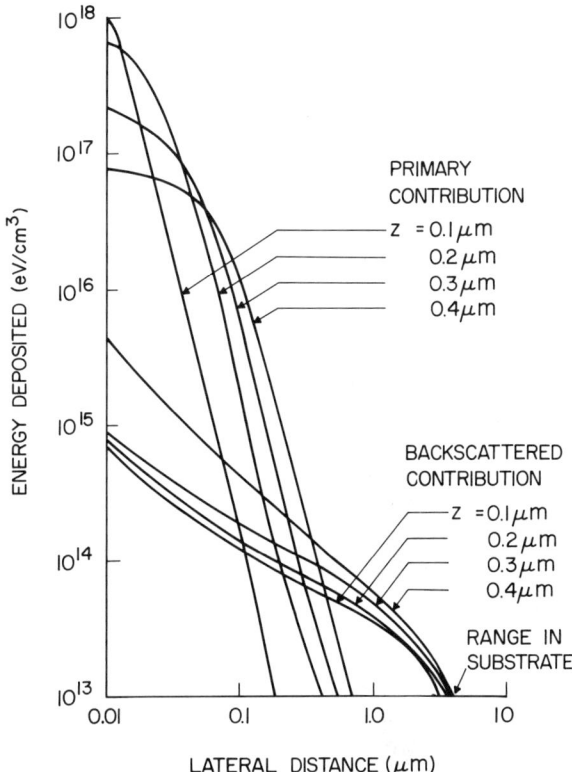

Fig. 114. Calculated absorbed energy density for primary and backscattered contributions at selected depths in a 0.4-μm PMMA film on Al exposed at 20 keV. (From Greeneich and van Duzer.[323])

Hawryluk et al.[257] also used the Everhart model to calculate the backscattering contributions from both the resist and the substrate. They used the original Thomas–Whiddington law for the energy–range relationship

$$v^4 = v_0^4 - C_T \rho s = v_0^4(1 - s/R_{TW}), \qquad (2.407)$$

where v_0 is the initial electron velocity, R_{TW} is the Thomas–Whiddington electron range in the material of density ρ, and $C_T = 5.05 \times 10^{42}$ cm^2/g·sec^4. For the substrate component they obtain

$$I(r,z)|_{BS} = \frac{\beta}{\pi R_{TW}^2} \int_0^{\xi_{max}} \frac{(1-\xi)^{\beta-1}}{\{(\xi+\delta) + [\kappa^2 + (\xi+\delta)^2]^{1/2}\}^2} \left(-\frac{dE}{ds}\right) d\xi, \qquad (2.408)$$

where

$$-\frac{dE}{dS} = \frac{2\pi q^4 N}{E_0 g(\xi, \kappa, \delta)} \ln\left[\frac{E_0}{I}\left(\frac{e}{2}\right)^{1/2} g(\xi, \kappa, \delta)\right] \qquad (2.409)$$

and

$$g(\xi, \kappa, \delta) = \left\{1 - \xi - \frac{\xi}{\xi + \delta}[\kappa^2 + (\xi + \delta)^2]^{1/2} \right.$$
$$\left. - \frac{\rho}{\rho_s}\frac{\delta}{\xi + \delta}[\kappa^2 + (\xi + \delta)^2]^{1/2}\right\}^{1/2}, \qquad (2.410)$$

where $\xi = z_S/R_{TW}$, $\kappa = r/R_{TW}$, $\delta = T/R_{TW}$, and ρ and ρ_S are the densities of the resist and substrate, respectively.

Figure 109a contains a comparison between both analytical models and the MC calculations of Kyser and Murata[317] for 0.4-μm PMMA on Si exposed at $E_0 = 20$ keV. The MC results are consistently higher than the other materials at $r = 0$. The plural scattering result for an Al substrate is also compared for a line source exposure. While the MC data show a considerably higher energy at the origin, good agreement between the models is obtained at larger lateral distances.

(iii) *Proximity Functions and Proximity Effect Correction (PEC) Techniques.* The effect of electron scattering in the resist and the underlying substrate can lead to undesired "cross-talk" between adjacent regions. This phenomenon is called *proximity effect*.

Proximity effects can occur in two ways in the exposure of dense patterns:

(1) *Intrashape proximity effect*: Isolated shapes with small dimensions receive less exposure from electron scattering than large features. As a result of this underexposure, small shapes develop more slowly and less completely than larger shapes. If the development is adjusted to satisfy the smallest feature, larger shapes will overdevelop and grow.

(2) *Intershape proximity effect*: The large-angle backscattering contributions can lead to an appreciable exposure increase between closely adjacent shapes. Increased packing density will therefore lead to an overexposure or even pattern loss.

Three techniques have been developed to minimize proximity effects. In the first, the exposure of each shape or part of a shape is adjusted. Parikh[326-329] has been a pioneer in this field. Others[330-332] have used this technique successfully. For some exposure systems such as raster-scan or projection machines, exposure control is difficult and pattern size control has to be preferred. In this method the exposure is chosen high enough to ensure the proper development of the smallest, most isolated feature. Pattern dimensions are then adjusted to ensure the critical dose at the pattern edges. This second technique relies on changes in the actual feature dimension to be exposed. Several papers have been published describing this technique in detail.[333-338] The third and most general technique involves a change of both exposure and pattern parameters. Wittels[253] gives several references to his own work in this area.

All correction algorithms subdivide larger patterns to save computer time. Subdivision is not a trivial problem; the interested reader should see Grobman et al.,[330] Parikh and Schreiber,[339] Kikuchi et al.,[340] Kratschmer,[341] and Molzen and Grobman.[342] A thorough discussion of proximity effect correction techniques is contained in Wittels.[253]

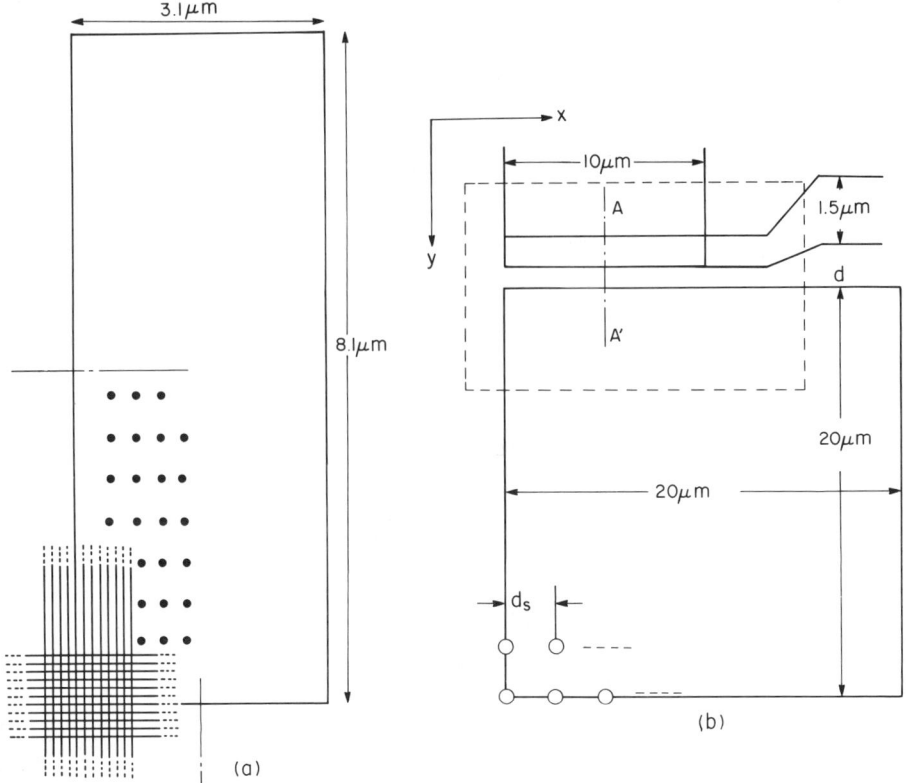

Fig. 115. (a) Pattern shape and configuration layout an the resist surface for exposure calculations. The dots in the pattern are the origins for the point sources. (From Murata et al.[343]) (b) Schematic of investigated pattern consisting of two shapes. The interelectrode spacings are $d = 0.5$, 1.0, and 1.5 μm. (From Nomura et al.[344])

At a first glance, it would seem obvious that the results of MC calculations should be used in any proximity effect correction (PEC) scheme. Murata et al.[343] and Nomura et al.[344] have presented results of three-dimensional MC simulations that take the pattern geometry into account. They have calculated the absorbed energy density for both the single pattern and two adjacent patterns in Fig. 115. The resist film is subdivided into small cubes with 0.1-μm sides. At any point, all contributions from each dot exposure in the pattern (see Fig. 115a) are added up to obtain the total absorbed energy density in the resists. Figure 116 shows the deposited energy for pattern (b) in the form of isoenergy contours at the resist surface (Fig. 116a) and the resist–substrate interface (Fig. 116b). The contours are labeled in percentages where the contour closest to the feature boundary has been set to 100% ($= 1.12 \times 10^{26}$ eV/cm·C).

Figure 117 presents the same results redrawn along the line A–A'. The results clearly display both intra- and interproximity effects and their influence on the energy distribution. At the top a sharp reduction from 100 to

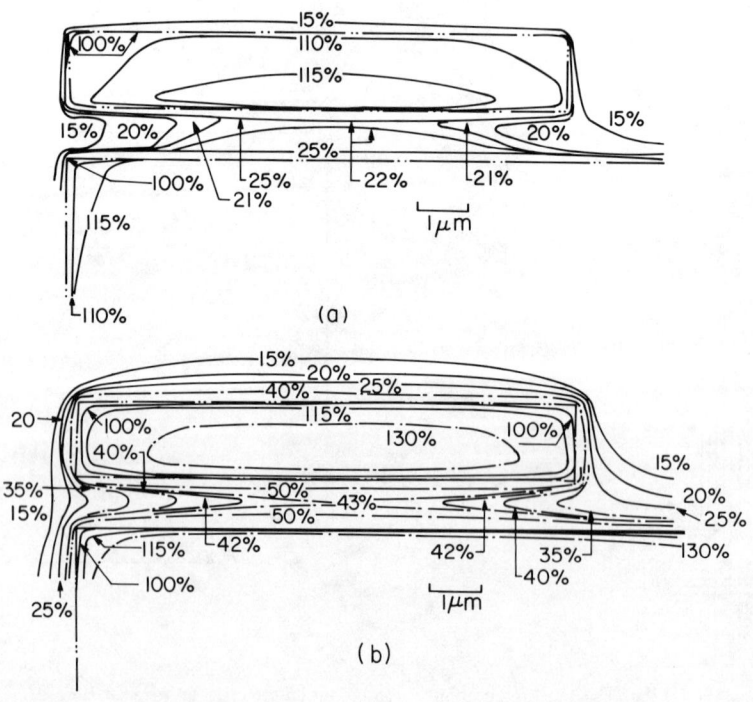

Fig. 116. Calculated isoenergy contours at the (a) top and (b) bottom of the resist. The region enclosed by the dashed lines in Fig. 115b is shown. For contour labeling see text. (From Nomura et al.[344])

25% occurs with 0.2 μm from the shape boundary because of the dominance of forward scattering. At the bottom of the resist, however, contours are smeared out between the features because of backscattering components.

While MC calculations offer considerable insight into the physical nature of proximity phenomena, they are too expensive and thus impractical to be applied to dense patterns.

Chang[345] was the first to propose the use of an analytical radial exposure intensity distribution (EID) equivalent to the delta point source response. A

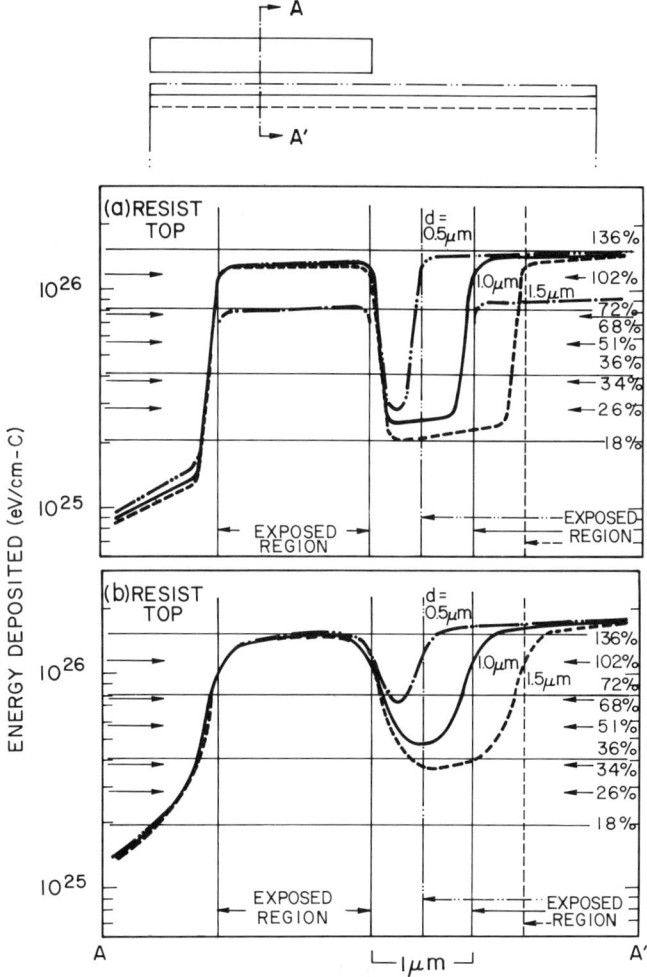

Fig. 117. Variation of the deposited energy density along the line $A-A'$. Interelement spacings are 0.5, 1.0, and 1.5 μm. For labelling see text. (From Nomura et al.[344])

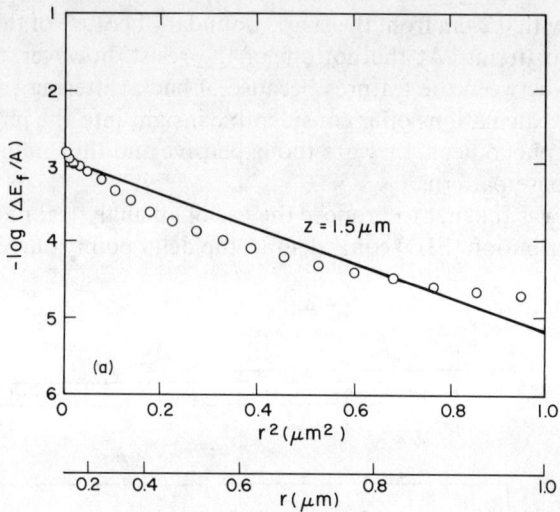

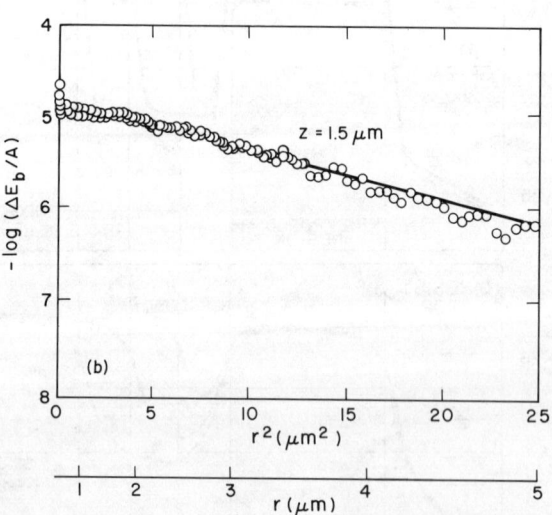

Fig. 118. Plot of the radial distribution of the energy distribution due to (a) forward scattering and (b) backward scattering of 25-keV electrons in 1.5-μm PMMA on Si. The symbols are the results of MC calculations and the solid lines are a least-squares approximation to Eq. (2.411). Note the two abscissas r and r^2, the change in the ordinate scales and the constant $A = (8/\pi) \times 10^{18}$ eV/cm^3 electron. (From Parikh and Kyser.[346])

typical example for an EID function is shown in Fig. 109a. An analytical expression not only provides a simple formula to be used in PEC techniques, but also contains several macroscopic parameters that can be experimentally obtained for a particular situation.

Distinguishing between forward- and backward-scattered components, Parikh and Kyser[346] approximate the EID by a two Gaussian expression

$$f(r) = k\left[\exp\left(\frac{r^2}{\beta_f^2}\right) + \frac{\eta_E \beta_f^2}{\eta_b^2}\exp\left(-\frac{r^2}{\beta_f^2}\right)\right], \qquad (2.411)$$

where β_f and β_b are the "characteristic widths" of the two components, k is a constant and $\eta_E = \eta K_E$ with Everhart's backscattering coefficient η. The factor K_E is the ratio of the average energy deposited by backward-scattered electrons to the corresponding forward contribution.

For a real Gaussian beam of the form

$$g(r) = \exp(-r^2/\beta^{*2}), \qquad (2.412)$$

an equation analogous to Eq. (2.411) with generalized parameters can be obtained

$$\beta_f^* = (\beta_f^2 + \beta^{*2})^{1/2}, \qquad (2.413a)$$

$$\beta_b^* = (\beta_b^2 + \beta^{*2})^{1/2}. \qquad (2.413b)$$

Several techniques are available to obtain the EID function parameters. To fit Eq. (2.411) to MC results, Parikh and Kyser[346] used least-squares procedures. Figure 118 shows a typical result of a linear least-squares fit of EID data for forward-scattered electrons (a) and backward-scattered (b) electrons for PMMA on Si. The result in Fig. 118b is a much better fit than that for forward scattering. This result is not unexpected since the contribution from forward scattering has a stronger than exponential dependence. However, since for most cases $\beta^* > \beta_f$, the result will be dominated by the beam parameters and not by β_f.

For heavy substrates, the backward-scattered component of the EID function exhibits a strong nonlinearity. In these cases, large-angle backscattering is favored. This leads to an electron distribution that is narrower than that for lighter substrates. Furthermore, the higher stopping power makes large-angle scattering normal to the incidence axis very unlikely. A combination of these effects leads to a backward-scattering component that is peaked around the origin but that decreases quickly for larger distances laterally. A typical example is given in Fig. 119, which shows the backscattering component for a PMMA–Au resist–substrate situation. To model the nonlinearity, Parikh and Kyser suggest the following two Gaussian expressions:

$$f(r) = k\left\{\exp\left(-\frac{r^2}{\beta_F^2}\right) + \frac{\eta_E \beta_F^2}{(1+\eta_{ds})\beta_{bs}^2}\left[\exp\left(-\frac{r^2}{\beta_{bs}^2}\right) + \frac{\eta_{ds}\beta_{bs}^2}{\beta_{bd}^2}\exp\left(-\frac{r^2}{\beta_{bd}^2}\right)\right]\right\},$$

(2.414)

where η_{ds} and η_{bs} describe the importance of the diffuse- and the peaked backward-scattering components, respectively, with similar definitions for β_{bs} and β_{ds}.

Kyser et al.[347] have used the SPECTRE program developed by Parikh[327] to determine unique sets of Gaussian parameters β_f, β_b, and η_E. They in-

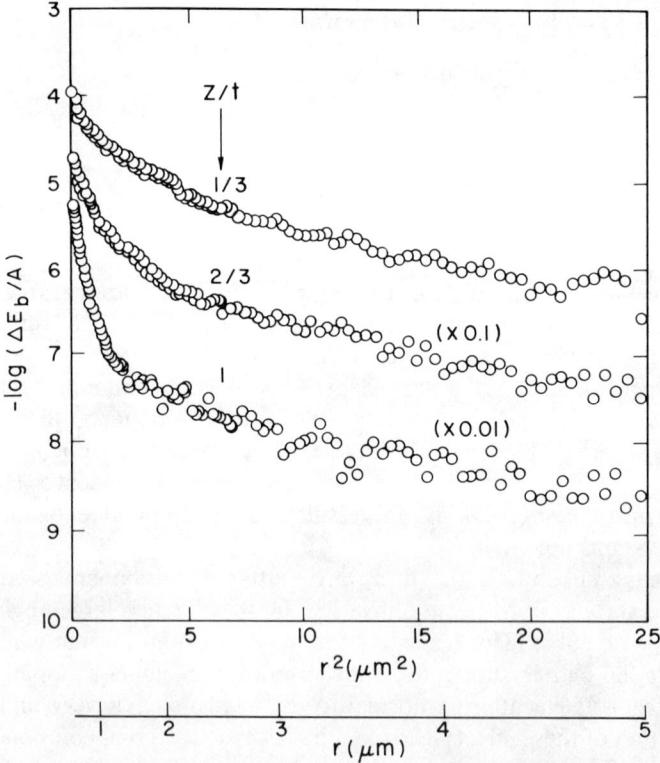

Fig. 119. Energy deposition due to the backward-scattered electrons in 1.5-μm PMMA on Au for different depths (z/t) in the resist. Note the two abscissas r and r^2, the change in the ordinate scales, and the constant $A = (8/\pi) \times 10^{18}$ eV/cm^3 electron. (From Parikh and Kyser.[346])

cluded developer effects[348] and obtained good agreement with experimental data.

Aizaki[349] has used MC results to investigate the influence of the substrate atomic weight on the proximity effect function. Kyser and Ting[350] have studied the dependence of proximity effects on the beam voltage.

Shaw,[351] Wittels,[253] and Grobman et al.[330] developed various techniques to obtain the parameters in Eq. (2.411) for a particular experimental condition. Parikh[326] compared experimental results with calculated data for PMMA resist on Si substrates.

PEC techniques rely strongly on the use of Eq. (2.411). Greeneich[352] has used the two Gaussian approximation to study linewidth control. He models the forward-scattering term β_f in terms of the number P_{elas} of elastic scattering events [Eq. (2.362)]

$$\beta_f = 34.73 P_{\text{elas}} + 5.47 P_{\text{elas}}^2. \quad (2.415)$$

The backward-scattering contribution β_b is fitted to the MC data of Parikh and Kyser[346] as a function of incident energy. The parameter η_E is obtained from Everhart's backscattering theory. Using this simple model, Greeneich[353] developed a PEC algorithm based on a dose compensation curve that links dose adjustment factors, developed resist images, and calculated values of

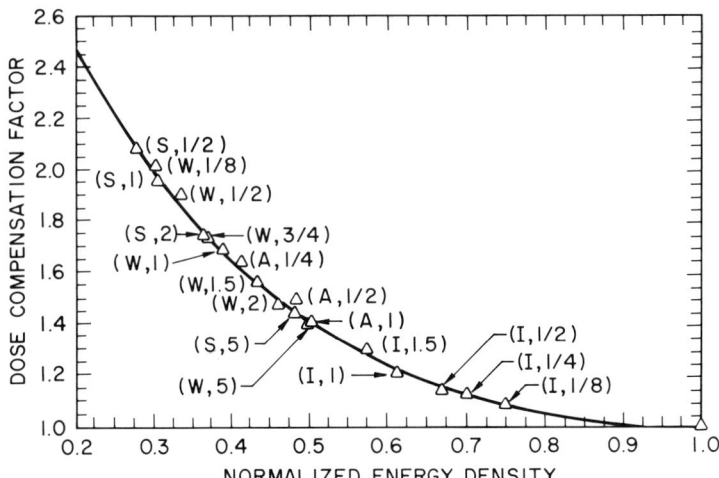

Fig. 120. Dose compensation curve for the two Gaussian model at 20 keV for 0.5-μm PMMA on Si and 120-sec development in 1:1 MIPK:IPA. Data points refer to the pattern type and dimensions: S = square, W = window, A = array of lines and spaces, I = resist island between large window exposures. (From Greeneich.[353])

energy density at the pattern edges. This dose compensation curve is insensitive to changes in the choice of the model, the substrate parameters, the exposure conditions, resist thickness, and development conditions. An extension to multilayered substrates is found in Greeneich.[354]

In Fig. 120, data from SEM measurements on pattern quality for a variety of different shapes and different experiments are compared with the theoretical prediction of the two Gaussian model.[353] The data include sizes of 0.5 μm and larger. A 0.5-μm film of PMMA on Si is exposed with a 20-keV beam and later developed for 110 sec in 1:1 MIBK:IPA developer. The energy is normalized to the calculated critical energy–dose combination necessary to develop the center of a large area. The variety of shapes includes isolated windows, resist islands between large-area windows, arrays of lines and spaces, and small windows of varying sizes. Further results can be found in Greeneich.[353]

The successful application of a PEC algorithm depends strongly on the computing resources required. Considerable work has been done in this field; interested readers are referred to the literature.[340–342,355,356]

(iv) *Development Simulations.* The least understood and probably most critical part of the complete e-beam simulation process is the exposure–development model. In this subsection, we discuss some of the approaches taken together with representative results. The presentation of actual numerical methods used in the development simulation is given in Subsection 5.

The energy absorbed in an organic resist alters the chemical structure of the polymer, thus affecting its physical and chemical properties. Depending on the resist type, the interaction with the electrons results either in bond breaking and chain fission (positive resists) or in cross-linking (negative resists). During the development process, exposed or unexposed areas are then removed as illustrated in Fig. 97. Greeneich[277] has given a good summary of resist properties.

Most of the work on development simulation has been focusing on positive resists. The processes that occur during the development of negative resists are considerably more complicated and even less understood. Heidenreich et al.[357,358] and Lin[359] have published a simple exposure model for negative resists. Nakata et al.[360] and Nakayama et al.[361] have presented some results on PEC for negative resists.

Positive resists are developed in a chemical solution that acts as a good solvent for exposed areas and as a bad solvent elsewhere. The increase in solubility is caused by the lower molecular weight in the irradiated regions.

Greeneich and co-workers[362-364] have shown that the degraded average number molecular weight $\overline{M}_f$ is given by

$$\overline{M}_f = \overline{M}_n/(1 + N_{sc}), \qquad (2.416)$$

where $\overline{M}_n$ is the initial number average molecular weight and N_{sc} is the number of scission events per molecule given by

$$N_{sc} = g_s I \overline{M}_n / \rho N_A \qquad (2.417)$$

with the absorbed energy density I and the radiation chemical yield for scission events g_{sc}. Combining Eqs. (2.416) and (2.417) yields

$$\overline{M}_f = \overline{M}_f/(1 + K\overline{M}_n) \qquad (2.418)$$

with $K = g_{sc} I/(\rho N_A)$. Figure 121 shows the final molecular weight distribution as a function of initial molecular weight for various K. To achieve

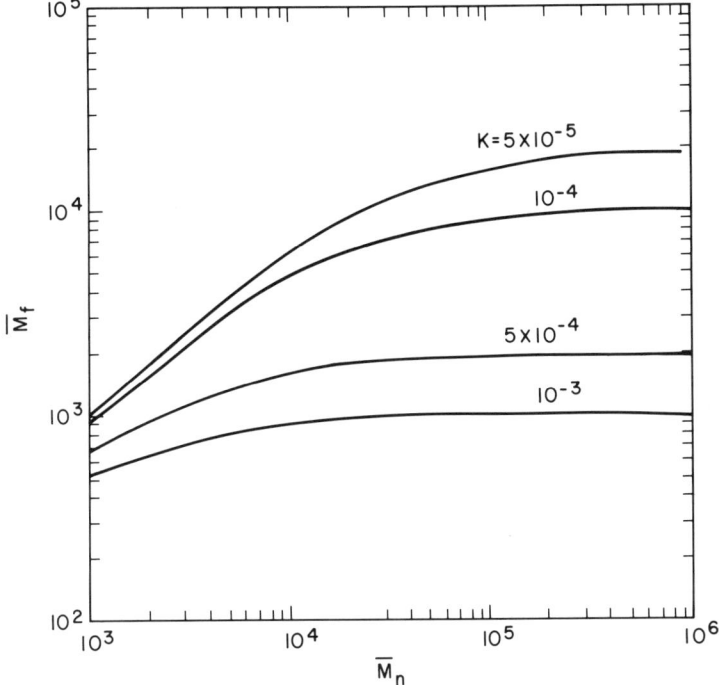

Fig. 121. Plots of $\overline{M}_f$ as a function of $\overline{M}_n$ with $K = gI/\rho N_A$ as a parameter. (From Herzog et al.[364] © 1972 IEEE.)

maximum sensitivity, $\overline{M}_f/\overline{M}_n$ should be small, suggesting a large K factor. For PMMA resist, Greeneich[363] and Neureuther et al.[365] have experimentally determined the removal rate in concentrated developer. Neureuther et al.[365] obtained results for two different PMMA resists, shown in Fig. 122, and they have been fitted by the relation

$$R(I) = R_1[C_m + (I/I_0)]^\alpha, \tag{2.419}$$

where R_1 is the etch rate of unexposed resist, C_m is a constant inversely proportional to $\overline{M}_n$, I_0 is a knee energy defined as the intercept between the low- and the high-dose asymptotes, and α is the asymptotic slope at very high dose. This equation is used in the SAMPLE program.[237]

Kyser and Pyle[348] have used a more complicated relationship between the solubility rate and the absorbed energy, namely,

$$R = (A + BI^n)[1 - \exp(-\alpha z)] + \epsilon(I), \tag{2.420}$$

where z is the depth and A, B, α, and n are fitting constants. The energy dependence of ϵ is given by

$$\epsilon(I) = \epsilon_0 + CI_m, \tag{2.421}$$

where C and m are constants and I is evaluated at $z = 0$. Equation (2.421)

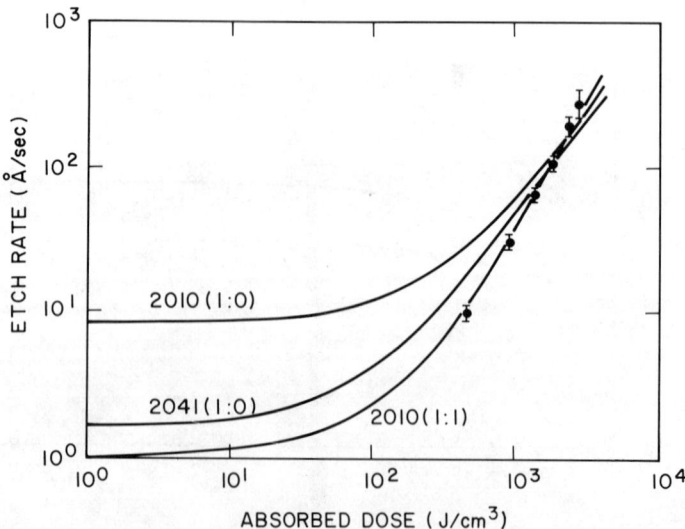

Fig. 122. Resist etch rate versus absorbed dose for PMMA Elvacite 2010 and 2041 for development in a mixture of MIBK and IPA. (From Neureuther et al.[365] This figure was originally presented at the Spring 1978 Meeting of The Electrochemical Society, Inc., held in Seattle, Washington.)

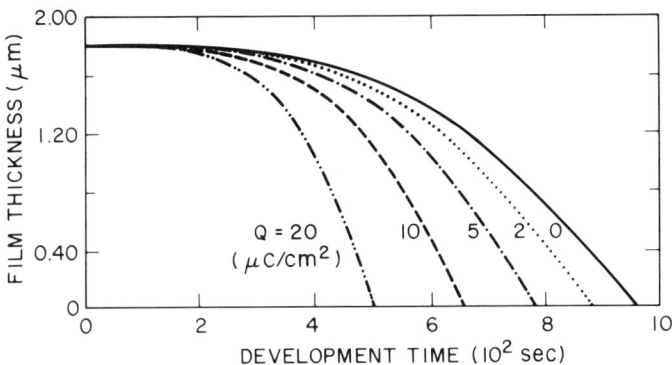

Fig. 123. Thickness of resist remaining versus development time for an etching rate given by Eq. (2.420). (From Kyser and Pyle.[348] Copyright 1980 by International Business Machines Corporation; reprinted with permission.)

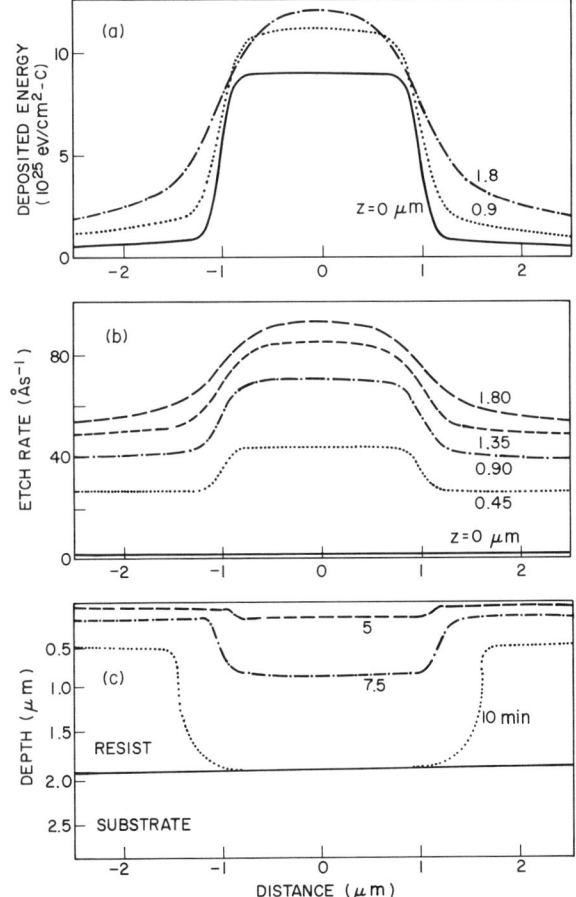

Fig. 124. (a) Lateral distribution of energy deposited within 1.8-μm polumeric resist on Si by 25-keV electrons for a 2-μm rectangular beam. (b) Lateral etch rate distribution for the same case. (c) Development profiles versus time in minutes. The horizontal line indicates the resist–substrate interface. (From Kyser and Pyle.[348] Copyright 1980 by International Business Machines Corporation; reprinted with permission.)

takes into account that even for unexposed resist, a finite dissolution exists $(R(I = 0) = A + \epsilon_0)$.

Figure 123 shows a thickness–time curve for a diazo-type photoresist as fitted by Eq. (2.420) with $A = 50$ Å, $\alpha = 1.4$ μm, $B = 2.5 \times 10^{-18}$, $n = 1.05$, $C = 2 \times 10^{-30}$, $m = 1.5$ and $\epsilon_0 = 0.5$ Å/sec. For large α and $C = 0$, Eq. (2.420) is identical with Eq. (2.411). It offers the advantage of allowing us to model most positive resist materials.

A typical result for the solubility rate is given in Fig. 124. After the point source MC result has been convoluted with a rectangular primary beam profile, one obtains a two-dimensional energy histogram. The curves in Fig. 124a show the results of the convolution for three different depths. The latent image is transformed into a solubility rate distribution where points

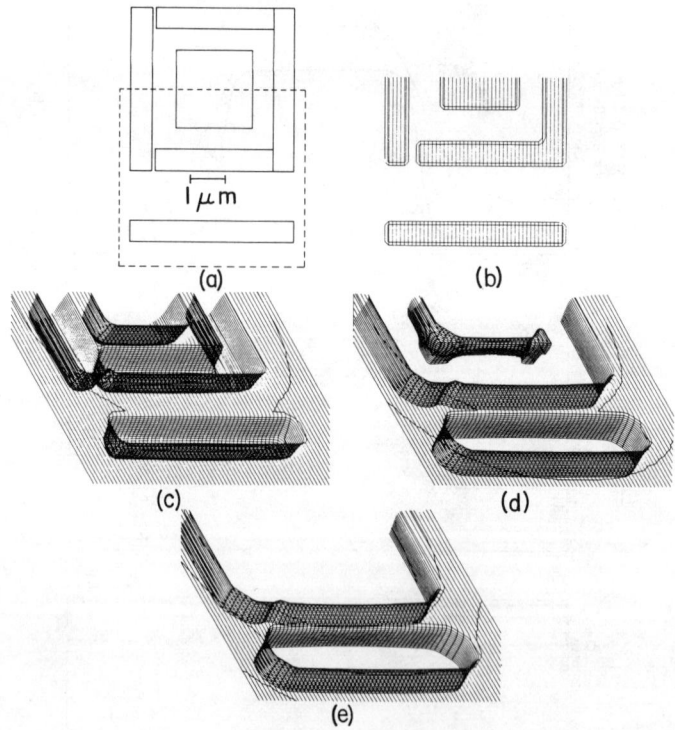

Fig. 125. (a) Pattern layout of connected and separated rectangles. (b) Calculated profiles of exposed patterns after 10 sec of development in a mixture of 1:1 MIBK:IPA. (c) Same after 70 sec. (d) Same after 200 sec. (e) Final developed profiles after 260 sec development. (From Jones and Parasczak.[366] © 1981 IEEE.)

with the same solubility have the same development time (Fig. 124b). With these data the resist can be "developed." Figure 124c shows the time evolution during development for times of 5, 7.5, and 10 min.

Jones and Parasczak[366] have extended the development modeling into the third dimension. Their exposure model is based on the work of Hawryluk et al.[257] A cubic polynomial like the one in Eq. (2.352) was fitted to experimental solubility results. As an example, Fig. 125 shows the development simulation for a 2 × 2-μm square surrounded by rectangles 0.6 μm in width. The region of interest lies inside the dashed line. In Fig. 125b a top view of the developed resist pattern after a 10-sec development time is shown. Vertical lines in the figures are 0.1 μm apart. Figure 125c illustrates the developed pattern after 70 sec. The gap between the 0.6-μm-wide rectangles has more or less vanished into a kink. After 200 sec (Fig. 125d), all directly exposed regions have been fully developed, but a small bridge remains in the central portion of the pattern. Figure 125d gives the final result after 260 sec.

c. *X-Ray Lithography*

X-ray lithography has the potential to produce images of significantly higher resolution than optical or e-beam technology. The application of x-rays to the fabrication of electronic circuits was pioneered by Spears and Smith.[367,368]

Today x-ray lithography has established itself as a potential candidate for submicrometer lithography. The question of which x-ray source is the most appropriate for microelectronic fabrication is beyond the scope of this paper. Spiller and Feder[369] have utilized synchrotron radiation to define small patterns in resist on a wafer. However, no commercial application of synchrotron sources has been reported.

A stationary source in which electrons hit a water-cooled Pd cone has been developed.[370] It has been used to fabricate LSI chips with submicrometer feature sizes.[9] A schematic drawing of this source can be seen in Fig. 126, where all essential elements—the x-ray source, the x-ray mask, and the wafer with the resist—can be identified.

Compared to optical and e-beam lithography, only a few papers have been published describing modeling attempts for x-ray lithography. Heinrich et al.[371] have modeled resist profiles for exposures made either with synchronton radiation or with a Al K_α x-ray tube source. The main differences between the two sources lie in the physical dimensions of the source, the degree of collimation, and the spectral distribution. The monochromatic radiation from the tube has a high degree of convergence, yielding a run-out effect. The finite size of the focus spot causes penumbral blurring. Synchrotron radiation is strongly collimated and covers a certain spectral range.

For parallel monochromatic radiation of wavelength λ and intensity I_0, and an opaque absorber, classical Fresnel diffraction gives for the intensity distribution along the resist

$$I(x) = I_0/2\{[\tfrac{1}{2} + \mathscr{C}(w)]^2 + [\tfrac{1}{2} + \mathscr{S}(w)]^2\}, \qquad (2.422)$$

where $\mathscr{C}$ and $\mathscr{S}$ are the Fresnel integrals and $w = x(2/\lambda s)^{1/2}$ with the proximity gap s.

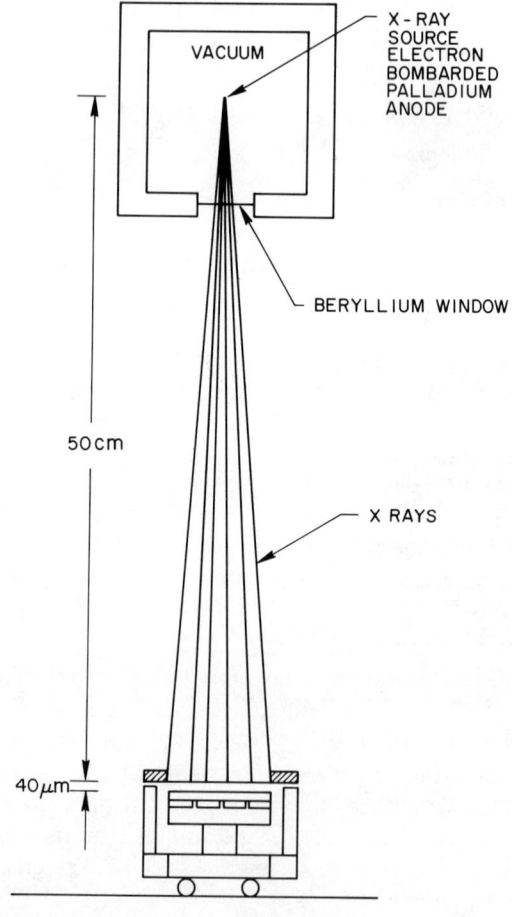

Fig. 126. X-ray system schematic. (From Lepselter et al.[9] © 1983 IEEE.)

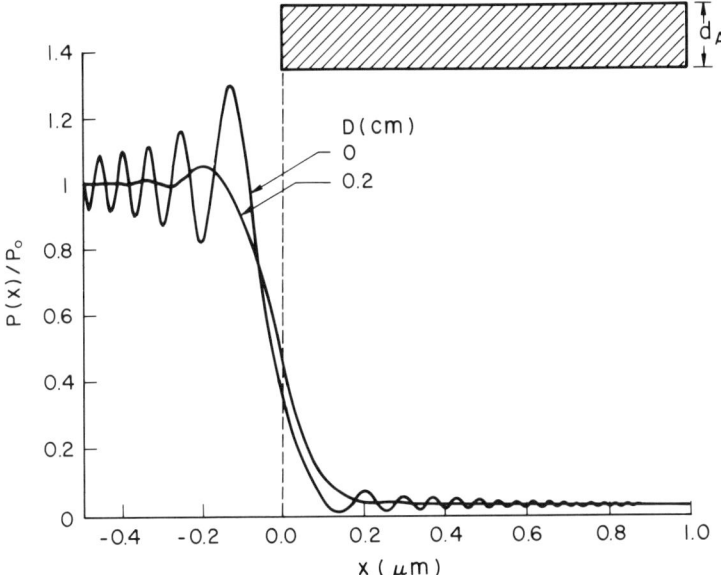

Fig. 127. Influence of the focus parameter on the intensity distribution behind a semitransparent Au absorber due to Fresnel diffraction in case of Al-K_α radiation ($\lambda = 0.834$ μm). The source wafer distance $L = 30$ cm, $s = 30$ μm, $d_A = 0.8$ μm. (From Heinrich et al.[371])

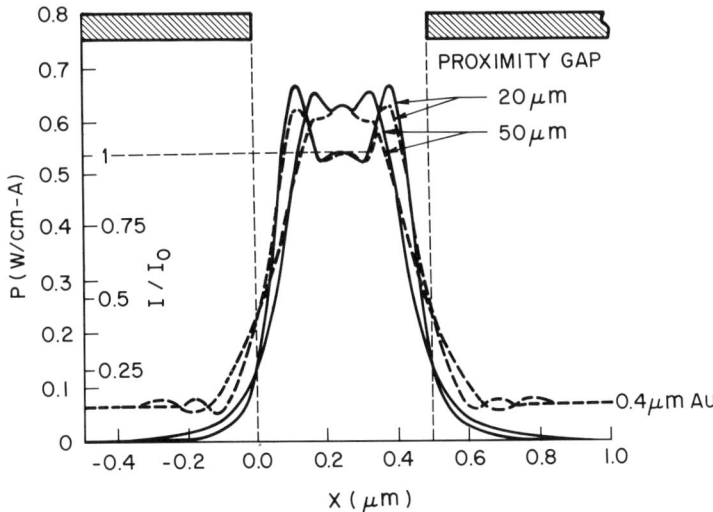

Fig. 128. Fresnel diffraction behind a 0.5-μm slit for synchrotron radiation for a fully opaque (full lines) and a semitransparent absorber (dashed lines) with different proximity gaps. (From Heinrich et al.[371])

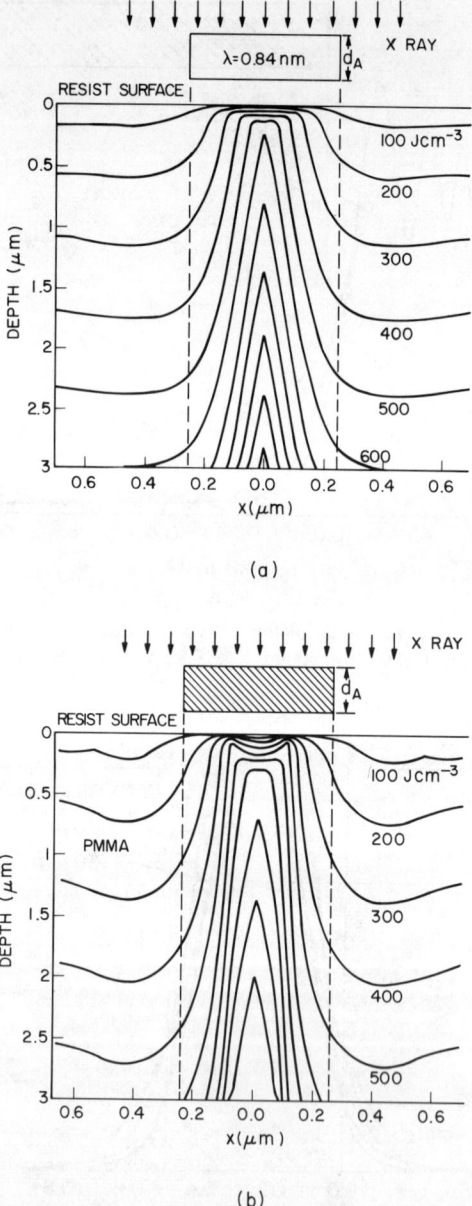

Fig. 129. (a) Resist profiles for x-ray tube exposure at different doses (at the resist surface without absorber). L and s same as in Fig. 127, $d_A = 0.4$ μm, focus diameter $D = 0.2$ cm, development time 4 min. (b) Resist profiles calculated for synchrotron exposure at different doses. $d_A = 0.4$ μm Au on 2 μm Si, $s = 0.4$ μm. (From Heinrich et al.[371])

For a semitransparent absorber with transmitted intensity I_1, complex amplitudes have to be superimposed to yield

$$U(x) = \begin{cases} A_+ - A_-[\mathscr{C}(w) + \mathscr{S}(w)], \\ A_-[\mathscr{C}(w) + \mathscr{S}(w)], \end{cases} \quad (2.423)$$

where $A_+ = I_0^{1/2} + I_1^{1/2}$ and $I(x) = |U(x)|^2$.

Figures 127 and 128 show intensity distributions for a tube source and a synchrotron source exposure. The synchrotron radiation has the maximum power output at 0.8 Å and a divergence of less than 1 mRad. The power absorbed in the resist at depth z is

$$\mathscr{P}_\alpha(x, z) = \int_{-\infty}^{\infty} I(x, \lambda) \alpha_R(\lambda) \exp[-\alpha_R(\lambda) z] \, d\lambda, \quad (2.424)$$

where $I(x, \lambda)$ is the wavelength-dependent Fresnel intensity distribution and α_R is the wavelength-dependent absorption coefficient of the resist.

Equation (2.424) can be approximated by

$$\mathscr{P}_\alpha(x, z) \cong \mathscr{P}(x, 0) \exp[-\beta(x) z], \quad (2.425)$$

where

$$\beta(x) = \frac{\int \alpha_R^2(\lambda) I(x, \lambda) \, d\lambda}{\int \alpha_R(\lambda) I(x, \lambda) \, d\lambda}. \quad (2.426)$$

Figure 129 compares resist profiles calculated for both exposure types. The synchrotron case yields better contrast and better pattern fidelity. It also shows clearly the influence of Fresnel diffraction indicated by the valley at the top of the structure.

d. *Ion-Beam Lithography*

Ion-beam exposure of resists is the least developed of all lithographic techniques. Nevertheless, it offers several advantages because of the reduced scattering of ions if compared to electrons. Furthermore, the sensitivity of classical e-beam resists to ion beams is between one and two magnitudes higher than that for e-beam exposure.[372]

Karapiperis et al.[373] have calculated the ion-beam exposure profiles in PMMA using a technique equivalent to the single-scattering MC approach of Kyser and Murata[317] for e-beam simulations. They used the LSS form for the nuclear scattering cross section, as given in Eqs. (2.44) and (2.45). Figure 130 shows trajectories of 50 H^+ ions implanted at 60 keV in PMMA on three different substrates. These trajectories have been calculated using the MC

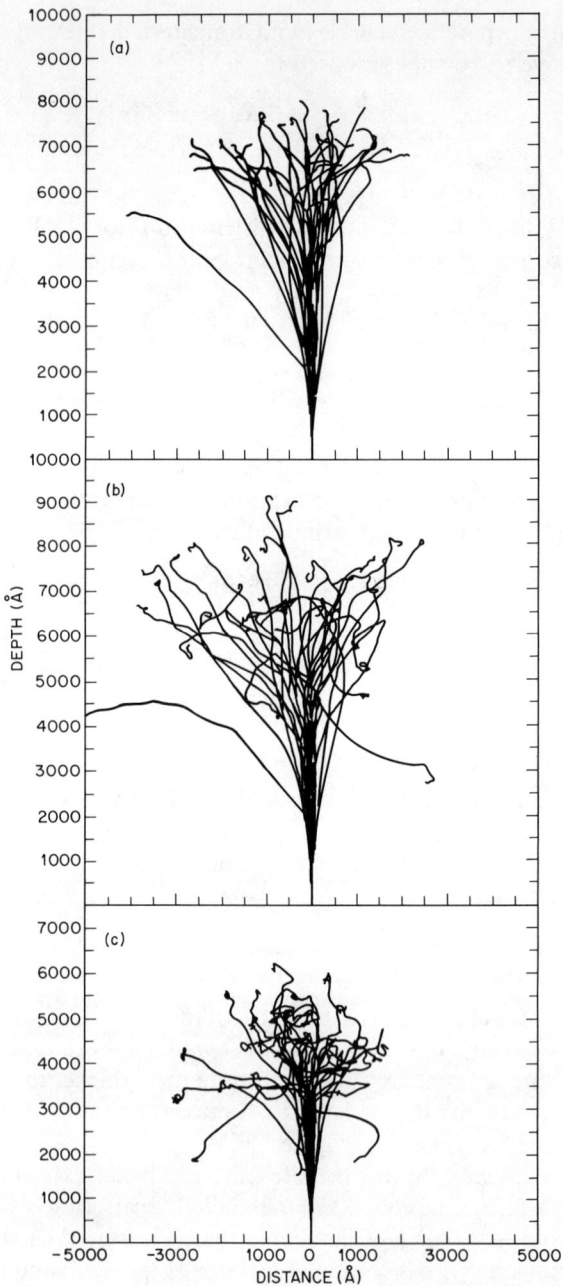

Fig. 130. Trajectories of 50 H⁺ ions implanted into PMMA on various substrates with an energy of 60 keV. (a) PMMA, (b) 0.3 μm PMMA/Si, (c) 0.3 μm PMMA/W.

code described in Petersen et al.[114] These plots and the scales should be compared with e-beam results as in Fig. 107. Backscattering is more or less absent (except for the Au), and the lateral spread of the beam is very small. For all substrates, the ions slow down very fast and stop in the substrate.

The point source response corresponding to these trajectories is shown in Fig. 131 for three different depths in the resist. Even at the resist bottom the absorbed energy drops by one order of magnitude at a lateral distance of 250 Å.

Karapiperis et al.[373] have employed the simple expression in Eq. (2.419) to simulate development profiles. The parameters in this equation have values very close to the ones in e-beam calculations, because the electronic energy loss dominates completely.

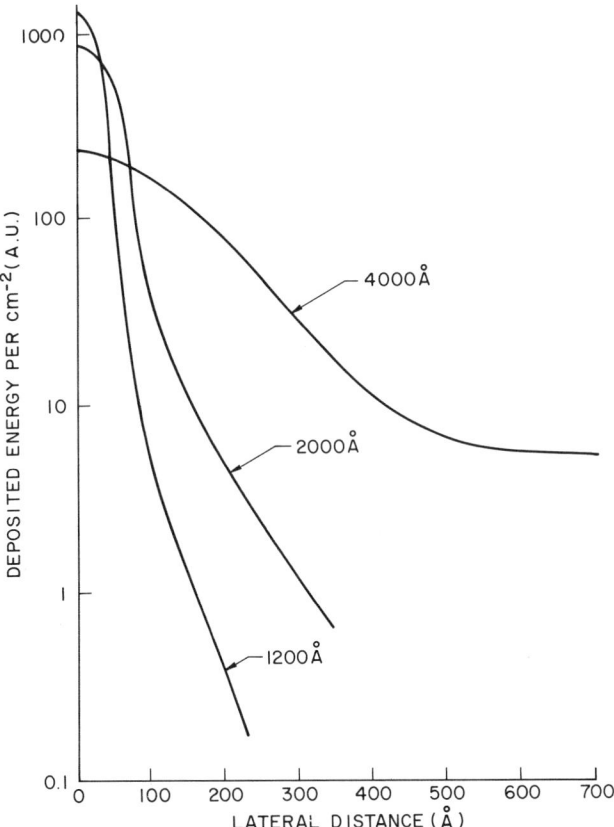

Fig. 131. Lateral variation of absorbed energy density at three characteristic depths for the line response with H^+ ions. (From Karapiperis et al.[373])

For ion-beam exposure calculations, secondary electrons can be ignored. A H^+ ion with 60 keV can maximally transfer about 100 eV, equivalent to a secondary electron range of 100 Å.

5. Etching and Deposition

Etching and deposition are integral process steps in any modern silicon technology. In the schematic wafer process in Table I, many etching and deposition steps occur (e.g., the patterning of resists, and the deposition of polysilicon, insulators, and metal layers).

A complete process simulator has to contain some capabilities for deposition and etching. Contrary to the process steps we have studied up to now, however, no basic physical theories exist that would allow first principles solutions.

In order to circumvent this problem and to obtain some answers through simulation, the modeling of etching and deposition steps can be viewed as a purely geometric problem. The resulting shape of the surface is determined by an initial profile that moves through the medium at a speed that depends on the position and other variables such as etch rates.

In this section we present the basic algorithms used for etching and deposition simulations. Details of physics and chemistry of these process steps are not included, since they are beyond the scope of this review. Interested readers are referred to the relevant chapters in Sze.[374]

a. *Algorithms: Cell, Ray, and String Models*

Three algorithms have been used to simulate the movement of boundaries during etching processes.

The cell removal model was developed by Dill *et al.*[243] to simulate the development of positive resists in photolithography. Jones and Parasczak[366] extended the original concept into three dimensions. In this model the resist is subdivided into cells, as shown in Fig. 132. A matrix characterizes the state of each cell as it develops. The time to remove a cell with only the top surface exposed is given by

$$t = \Delta z/R_{ij}, \qquad (2.427)$$

where R_{ij} is the removal rate associated with cell ij and Δz is the vertical cell dimension. In the case of top and side exposure,

$$t = \frac{\Delta x\, \Delta z}{R_{ij}(\Delta x^2 + \Delta z^2)^{1/2}} \qquad (2.428)$$

with the lateral cell dimension Δx.

PHYSICS OF VLSI PROCESSING AND PROCESS SIMULATION 315

Figure 132 shows detailed results for the etching of an e-beam-exposed resist. For each cell the rate of dissolution is determined by the etch rate.

Rectangular cells can be employed to account for anisotropic etch rate conditions. Although the model is inherently discrete, a continuous surface can be evaluated by interpolation between equally dissolved cells at any point in time. This discreteness of the algorithm can lead to accuracy problems for certain directions where circular profiles tend to converge into octagons.

The ray model[376] works in a manner analogous to Snell's law of optics. A vector perpendicular to the boundary between developed and undeveloped regions (the *etch ray*) is refracted at boundaries between zones of different removal rate (Fig. 133). The effective refractive index of the resist is

$$n_R = R_{max}/R, \tag{2.429}$$

where R_{max} and R are the maximum and the local etch rate, respectively. At $t = 0$, the rays start in a direction perpendicular to the surface of the resist.

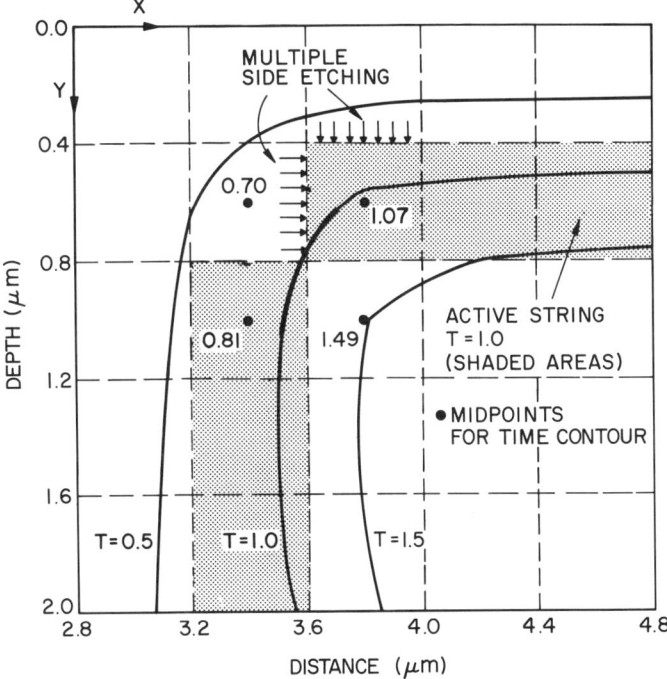

Fig. 132. Cell removal algorithm. Contours shown have been calculated for an e-beam exposure. The numbers in the cells are the timex of dissolution of the cells. The contours are interpolations of these numbers. (From Jewett *et al.*[375])

Subsequent ray trajectories are calculated using the ray equation. The etch front follows the rays that are refracted by the nonuniform etch rate R. The ray algorithm is easily implemented, and it can be extended to three dimensions.

The most promising simulation technique is based on the string model. Jewett[377] has presented a detailed description of this algorithm that forms the basis of etching and deposition calculations in SAMPLE.[237] Figure 134 summarizes the main features of the string model. A polygon consisting of straight line segments forms a string to approximate the etch front. In order to calculate the front position at the next time step, each point along the

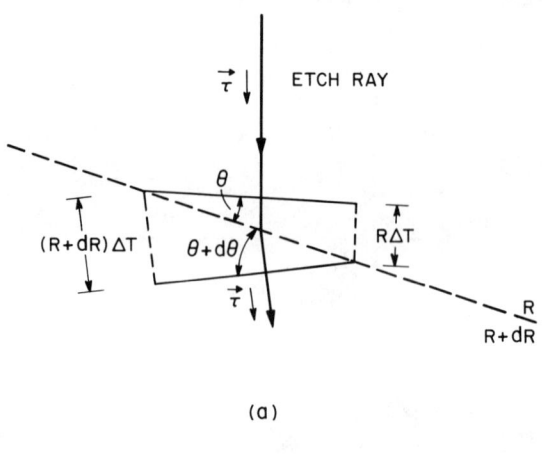

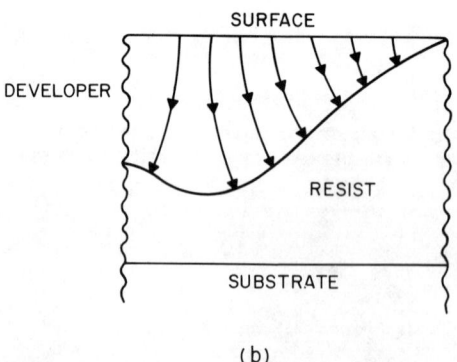

Fig. 133. Ray-tracing algorithm. (a) The etch ray is refracted at a boundary between regions having etch rates R and $R + dR$ according to Snell's law. (b) The rays start at the surface. An etch front is given by a contour joining the end points of the rays. (From Jewett et al.[375])

string advances in a direction perpendicular to the local etch rate (Fig. 134a). This procedure requires occasional reordering of the string point positions to keep the segments at approximately equal length. According to Jewett,[377] one of the major problems in the implementation of the algorithm is the proper choice of the segment length. Another problem is connected with the selection of proper time steps. Since the etch rate normally varies with

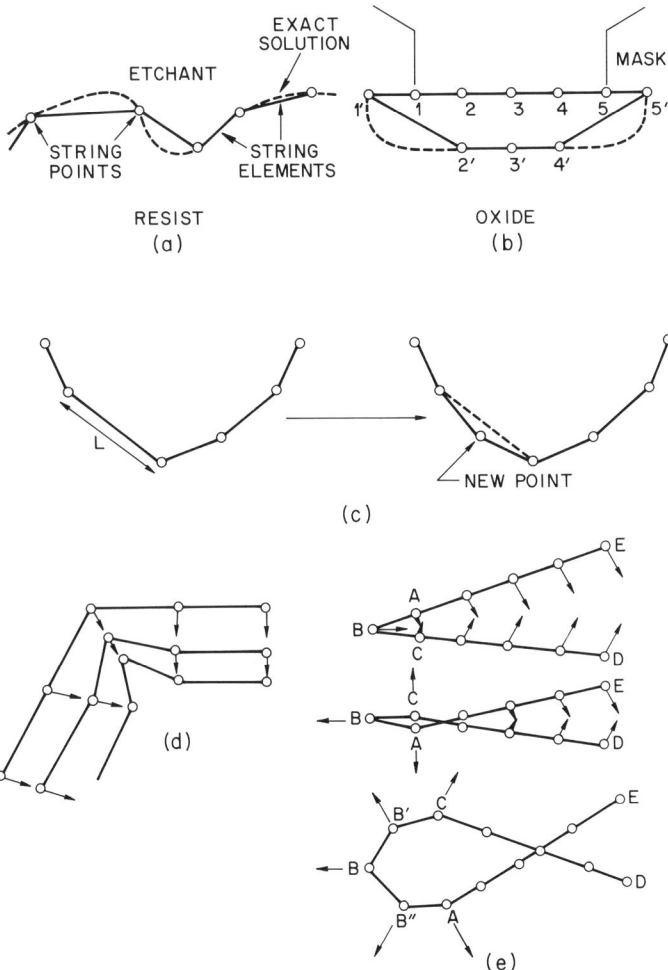

Fig. 134. String algorithm. (a) String model approximation to an etch front. (b) Problems with uniform etching of an oxide. (c) Length reduction by bisection. (d) Strings during contraction. (e) Loop formation, budding, and expansion. (From Jewett.[377])

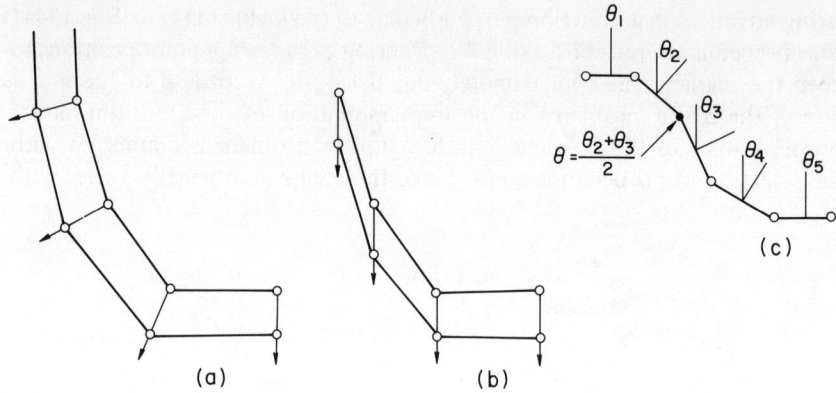

Fig. 135. (a) String sequence for isotropic etching. (b) Same for anisotropic etching. (c) Extraction of local angular orientation from string in ion milling. (From Reynolds.[378])

position, errors can occur. A typical situation is shown in Fig. 134b. During the first time step, points 1 and 5 move under the resist mask due to the etch undercut, and points 2 and 4 move straight down. This has the effect of a large deviation from the accurate result as displayed by the broken line. The solution to this problem as adopted by SAMPLE is to calculate the proper position of points 2 and 4 by advancing them along the angle bisector of the two adjacent segments.

One advantage of the string models is their ability to allow the addition of further points along a string. This is achieved by breaking a large segment into two smaller ones (Fig. 134c). The local curvature is approximated by calculating the angles ϕ and θ formed by the long segment with its two neighbors. The larger these angles are, the more the newly created point should be offset.

For regions where the segments become shorter, contraction can occur. The string model allows the corner point to lag behind (Fig. 134d) until it can be deleted. In same cases, however, a sharp angle can occur, as in Fig. 134e. Segment AB is not allowed to contract before A has not crossed the string between B and D. This process is called loop budding.

The string model is applicable to all important etching variants from isotropic via anisotropic etching to ion milling. Figure 135 gives a schematic sequence of string models for these processes.[378]

b. *Deposition*

Modern technologies with scaled design rules require excellent step coverage of evaporated films. For example, in the metal definition steps of

Table I, electrical continuity has to be achieved for proper circuit operation. Other situations might require complete coverage of conductors with insulating films.

Several different deposition sources are in use today. All simulations of deposition profiles are implemented using the string model algorithm described earlier.[379] The deposition rate is an analytical function and the simulation proceeds through a number of time steps analogous (but in reverse direction) to etching. Blech[380] derived a general model that is applicable to any source geometry and step profile. Sung[379] has extended Blech's model to include surface migration effects at higher temperature.

In this model the following assumptions are made:

(1) The mean free path of atoms is larger than the distance R between source and substrate.
(2) The distance r is large, compared to the step height.
(3) The film grows at a rate proportional to $\cos\theta/r^2$, where θ is the angle between the surface normal and the vapor stream direction. This is called the *cosine growth law*. At the end of this section, new results are presented that challenge this assumption.
(4) The film grows toward the direction of the vapor stream. Roughening effects due to large incidence angles are ignored.
(5) For cold substrates, the sticking coefficient is 1.
(6) For elevated temperatures, a random walk model governs surface migration.[377]

Figure 136 shows the geometries of several important source geometries. For a unidirectional source (a), material arrives at the substrate from one direction only. The growth rate is given as

$$R(x, z) = 0 \qquad (2.430a)$$

if the point at (x, z) is shadowed, and

$$R(x, z) = C(\hat{x}\sin\omega + \hat{z}\cos\omega) \qquad (2.430b)$$

otherwise, where ω is the angle between the z axis and the vapor streams, $\hat{x}$ and $\hat{z}$ are unit vectors in x and z direction, respectively, and C is the growth rate of an unshadowed surface in direction of the vapor stream.

For a dual source evaporator (b), the vapor stream arrives from two different directions under angles ω_1 and ω_2, and the final result is a sum of two terms just like Eq. (2.430).

For a hemispherical source (c), the flux F of the vapor stream is proportional to the area A[380]

$$F = KA, \qquad (2.431)$$

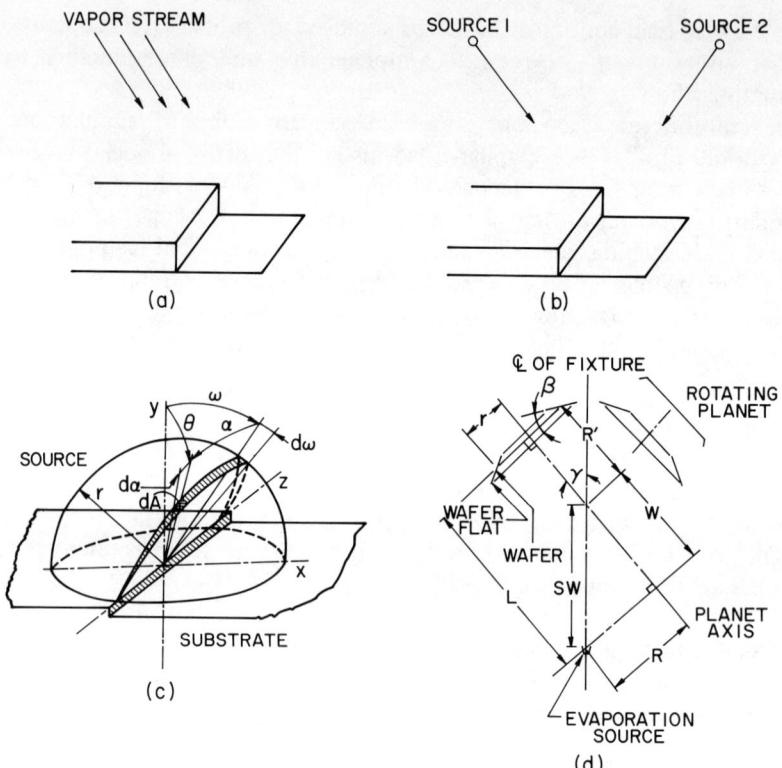

Fig. 136. (a) Geometry for a unidirectional source. (b) Geometry for a dual evaporation source. (From Sung.[379]) (c) Geometrical relations between hemispherical vapor source and substrate. (Reprinted by permission of the publisher from Evaporated film profiles over steps in substrates, by I. A. Blech,[380] *Thin Solid Films*, Vol. 6, p. 113. Copyright 1970 by Elsevier Science Publishing Co., Inc.) (d) Schematic planetary fixture for an e-beam evaporator. (From Blech et al.[381])

where K is a proportionality constant. In Fig. 136c, a flux dF proportional to the area dA will pass through it, namely,

$$dF = Kr^2 \cos\alpha\, d\alpha\, d\omega. \qquad (2.432)$$

Assuming a cosine growth, the change in thickness in direction y due to the entire area enclosed by $d\omega$ is

$$dy = Kr^2 \int_{\alpha=-(\pi/2)}^{\pi/2} \cos\alpha \cos\theta\, d\omega\, d\alpha = 2K \cos\omega\, d\omega \qquad (2.433)$$

and, since assumption 4 holds,

$$dx = dy \tan\omega = 2K \sin\omega\, d\omega \qquad (2.434)$$

PHYSICS OF VLSI PROCESSING AND PROCESS SIMULATION

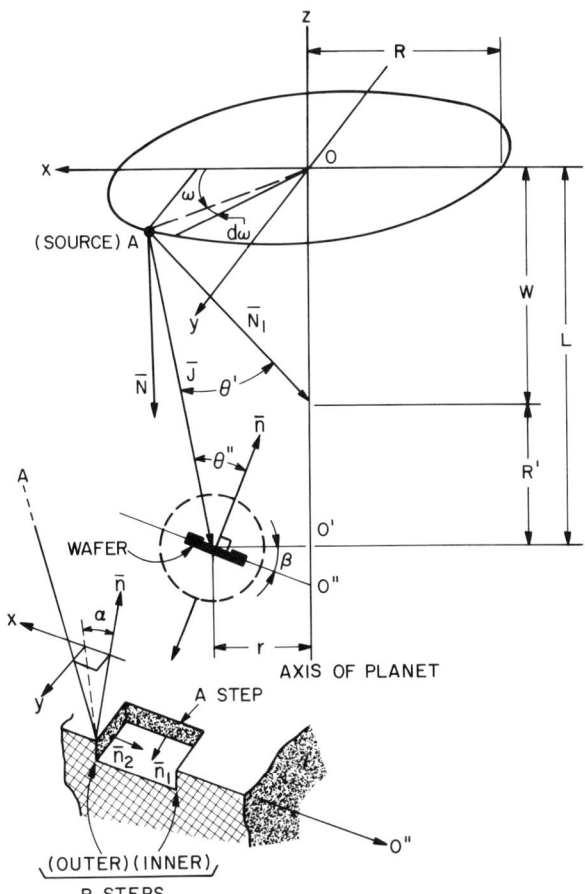

Fig. 137. Geometry and definition of terms. Description of A and B steps. (From Blech et al.[381])

The final deposition rate is calculated by integrating Eqs. (2.433) and (2.434)

$$R(x, y) = 2K[\hat{x}(\cos \omega_1 - \cos \omega_2) + \hat{z}(\sin \omega_2 - \sin \omega_1)] \quad (2.435)$$

Figure 136d shows a schematic of a planetary fixture.[381] The wafers are fixed on the planet, which rotates around its own planetary axis and also around the planetary center line. The rotation around the planet axis does not alter the deposition equations. Blech et al.[381] have solved this problem by assuming the planet to be stationary and the target to rotate. Figure 137 shows a more detailed view of the wafer geometry at position A and the rotating moving source. It also contains a definition of steps A and B.

In case of the outer B step, the amount of material arriving per unit area and rotation $d\omega$ is

$$I(\omega) = \frac{(R/L)[1 + (WL/R^2) - (r/R)^2 - (rL/R^2)\tan(\alpha - \beta)]}{[1 + (W/R)^2]^{1/2}[1 + (R/L)^2 - (r/L)^2 - 2(r^2/L^2) - (2r/L)\tan(\alpha - \beta)]^{3/2}}$$

(2.436)

and the displacements are given by

$$\Delta x(\alpha)\, d\alpha = I(\omega) \cos \theta'' \tan \alpha \, d\omega, \tag{2.437a}$$

$$\Delta z(\alpha)\, d\alpha = I(\omega) \cos \theta'' \, d\omega. \tag{2.437b}$$

The total displacements due to deposition from all angles are obtained by integration of Eq. (2.437) over all $\alpha_{min} \leq \alpha \leq \alpha_{max}$.

Calculated metal deposition profiles for the inner and outer B step are shown in Fig. 138 for angles β from $-5°$ to $30°$. The metal thickness is 1.5 µm and the step height is 1.0 µm. Based on this result, the optimum angle lies around $\beta = 0$. Other results including A step simulations can be found in Blech et al.[381] In Blech and Vander Blas,[382] similar results for a magnetron source have been obtained.

Blech[383] has recently investigated the application of assumption 3, the

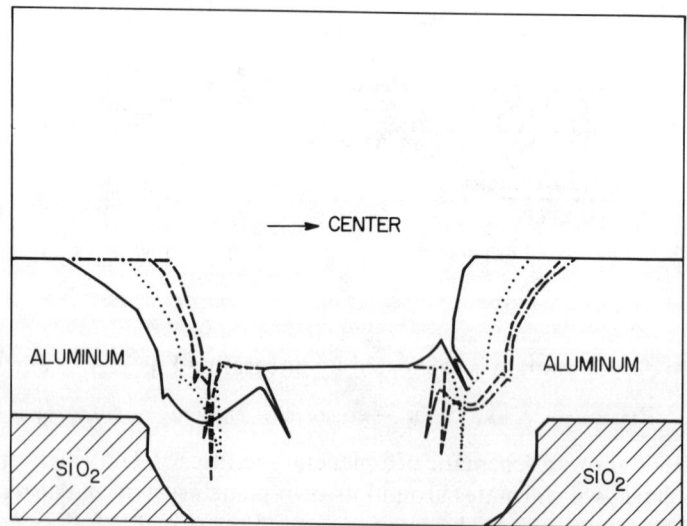

Fig. 138. Dependence of step coverage profiles for step B on the angle β. Beta = 30° ———; beta = 10° ·······; beta = 0° — — — —; beta = $-5°$ —·—·—. (From Blech et al.[381])

PHYSICS OF VLSI PROCESSING AND PROCESS SIMULATION   323

cosine growth law. From his results, it is not at all clear that the cosine law is appropriate for deposition calculations. Several authors[384,385] suggested the use of a tangent law as a better way to model deposition processes. In this case, the film grows at an angle closer to the surface normal and its density is smaller. Figure 139 compares schematically the growth of deposited films according to the (a) cosine and the (b) tangent law. In Fig. 139b, the

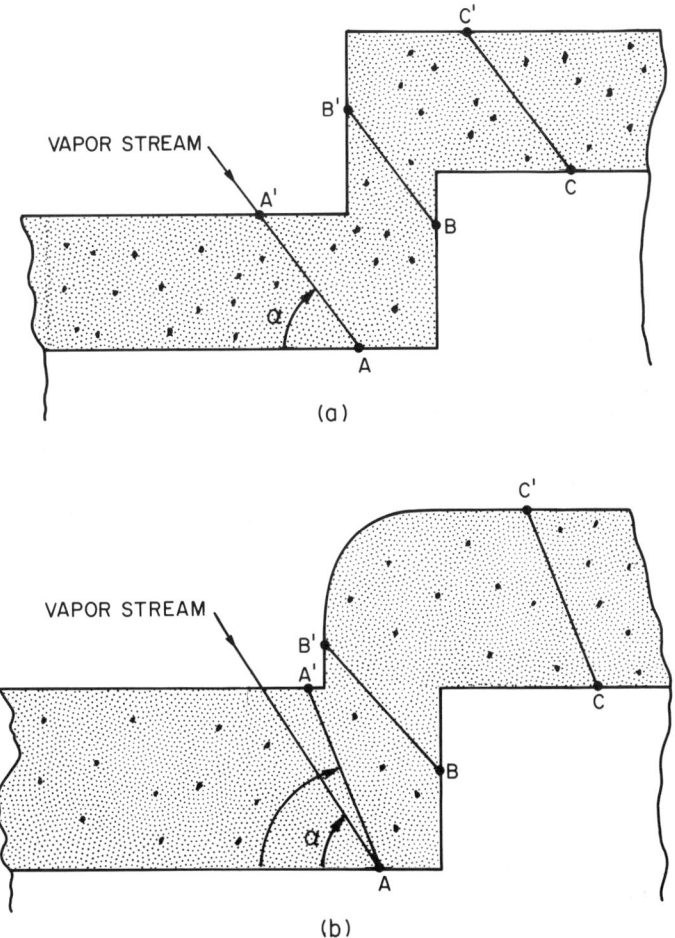

Fig. 139. Growth of a vapor-deposited film according to the (a) cosine and (b) tangent law. (From Blech.[383] Reprinted with permission from *Solid State Technology*, Technical Publishing, a company of Dun & Bradstreet.)

vapor stream arrives under an angle α but the film grows at an angle β to a larger final thickness. Furthermore, the step profile is rounded at the top.

III. Conclusions and the Future

The complexity of modern silicon technology has made it increasingly attractive to use computer simulations for process and technology development and optimization. This chapter was written with two objectives in mind: (1) to review the physical principles underlying the process steps in VLSI technology, and (2) to serve as an introduction and tutorial on process simulation. All major processes have been reviewed emphasizing the theoretical background and the computer implementations.

To keep the text at a reasonable length, several topics have been ignored. The practical aspects of VLSI technology have been left out completely, since an excellent textbook is available[386] that covers this field in detail. Furthermore, computational aspects such as the numerical solution techniques have been ignored since they are beyond the scope of this chapter (deserving at least a book of their own). Interested readers are referred to relevant publications (e.g. Ref.[387]).

The development of future technologies will have to rely more and more on computer simulations to cut development costs. Major efforts will be necessary to make process simulation a fully equal partner to real experiments. While we have witnessed significant progress in our understanding of VLSI physics and simulation, several major problem areas remain that need further attention.

1. Multidimensional Oxidation

Although first steps toward a more complete understanding of the oxidation process have been made (Section II.3.f), no model has been developed yet that would allow a first-principles solution of oxidation phenomena on single crystalline substrates of arbitrary shape. Essentially all work has been restricted to the analysis of bird's-beak geometries. For scaled VLSI devices, however, these local oxidation schemes are no longer appropriate. Buried oxide isolation schemes offer many advantages for different technologies (e.g., trench isolation in CMOS and DRAM storage elements).

Similar comments can be made for the oxidation of polysilicon. At this point, no satisfactory capabilities exist to simulate multidimensional oxidation.

2. Defect Formation and Annealing

The trend toward scaled device structures can only be successful if the annealing steps in a process are kept at the lowest temperature for as short a time as possible. This minimum temperature–time cycle is given by the characteristics of defect formation and annealing for a particular implant situation. Our understanding of the very complicated defect kinetics is rudimentary. This is particularly true for short-time diffusion and annealing effects associated with rapid thermal annealing.

Several other areas exist where future work is necessary to increase our understanding and simulation capabilities. These include etching processes, where first-principles models are completely lacking, and optical lithography, where we need better and faster models to simulate image formation and exposure.

ACKNOWLEDGMENTS

This chapter could not have been written without the help of many colleagues and friends at AT&T Bell Laboratories. It is a particular pleasure to mention E. H. Grosse and W. P. Petersen (MC methods), R. K. Watts (e-beam lithography), B. R. Penumalli (the father of the BICEPS program), S. J. Hillenius, T. E. Seidel, and R. K. Smith for many discussions, G. E. Smith for all his support, and AT&T Bell Laboratories as a whole for providing a unique work environment. D. Kahng has been an extremely helpful and very patient editor. Finally, I do not know how to express my gratitude to my wife Ingrid and my sons Johann and Paul for their love, understanding, and patience.

REFERENCES

1. R. N. Noyce, U.S. Patent 2,981,877 (1961).
2. J. S. Kilby, U.S. Patent 3,138,743 (1964).
3. D. Kahng and M. M. Atalla, *IRE–AIEE Solid-State Device Res. Conf.*, Carnegie Inst. Technol., Pittsburgh, Pa. (1960).
4. D. Kahng, "Silicon–Silicon dioxide Surface Devices," Bell Tel. Lab. Tech. Memo. (1961), available on request.
5. G. E. Moore, *IEEE Spectrum* **16**, 30 (1979), updated.
6. *IEEE Int. Solid State Circuits Conf.*, SESSION VIII: 256K/1Mb DRAMs (1984).
7. W. Fichtner, L. W. Nagel, B. R. Penumalli, W. P. Petersen, and J. L. D'Arcy, *Proc. IEEE* **72**, 96 (1984).
8. R. K. Watts, W. Fichtner, E. N. Fuls, L. R. Thibault, and R. L. Johnston, *IEEE Trans. Electron Devices* **ED-28**, 1338 (1981).
9. M. P. Lepselter, D. S. Alles, H. J. Levinstein, and H. A. Watson, *Proc. IEEE* **71**, 640 (1983).
10. C. O. Thomas, D. Kahng, and R. C. Manz, *J. Electrochem. Soc.* **109**, 1055 (1962).
11. J. J. Grossman, *J. Electrochem. Soc.* **110**, 1065 (1963).
12. W. Rice, *Proc. IEEE* **52**, 284 (1964).
13. A. S. Grove, A. Rhodes, and C. T. Sah, *J. Appl. Phys.* **36**, 802 (1965).
14. T. Abe, Y. Nishi, K. Sato, and N. Oi, *Denki Kagaku oyobi Kogyo Butsuri Kagaku* **35**, 142 (1967).

15. B. A. Joyce, J. C. Weaver, and D. J. Maule, *J. Electrochem. Soc.* **112**, 1100 (1965).
16. W. H. Shepherd, *J. Electrochem Soc.* **115**, 652 (1968).
17. G. Skelly and A. C. Adams, *J. Electrochem. Soc.* **120**, 116 (1973).
18. P. H. Langer and J. I. Goldstein, *J. Electrochem. Soc.* **121**, 563 (1974).
19. D. A. Antoniadis, S. E. Hansen, and R. W. Dutton, "SUPREM II—A Program for IC Process Modeling and Simulation," Stanford Electron. Lab. Tech. Rep. No. 5019-2 (1978).
20. D. A. Antoniadis and R. W. Dutton, *in* "Process and Device Modelling for Integrated Circuit Design" (F. van de Wiele, W. L. Engl, and P. G. Jespers, eds.), pp. 837–868. Noordhoff, Leiden, 1977.
21. J. D. Meindl, R. W. Dutton, K. C. Saraswat, J. D. Plummer, T. I. Kamins, and B. E. Deal, *in* "Process and Device Modelling for Integrated Circuit Design" (F. van de Wiele, W. L. Engl, and P. G. Jespers, eds.), pp. 57–114. Noordhoff, Leiden, 1977.
22. R. Reif, T. I. Kamins, and K. C. Saraswat, *J. Electrochem. Soc.* **126**, 644 (1979).
23. R. Reif, T. I. Kamins, and K. C. Saraswat, *J. Electrochem. Soc.* **126**, 653 (1979).
24. R. Reif and R. W. Dutton, *J. Electrochem. Soc.* **128**, 909 (1981).
25. R. Reif, *J. Electrochem. Soc.* **129**, 1122 (1982).
26. R. Reif and M. Vanzi, *J. Electrochem. Soc.* **128**, 2187 (1981).
27. R. Reif, T. I. Kamins, and K. C. Saraswat, *J. Electrochem. Soc.* **125**, 1860 (1978).
28. M. M. Faktor and I. Garrett, "Growth of Crystals from the Vapor." Chapman & Hall, London, 1974.
29. D. W. Shaw, *in* "Crystal Growth—Theory and Techniques" (C. H. L. Goodman, ed.), Vol. 1, pp. 1–48. Plenum, New York, 1975.
30. W. Rice, *Proc. IEEE* **52**, 284 (1964).
31. J. Crank, "The Mathematics of Diffusion." Oxford Univ. Press (Clarendon), London and New York, 1965 and 1975.
32. B. Tuck, "Introduction to Diffusion in Semiconductors," IEE Monograph Series, No. 16. Peregrinus, London, 1974.
33. H. S. Carslaw and J. C. Jaeger, "Conduction of Heat in Solids." Oxford Univ. Press (Clarendon), London and New York, 1959.
34. M. Abramowitz and I. A. Stegun, "Handbook of Mathematical Functions." Dover, New York, 1972.
35. N. Bohr, *Philos. Mag.* **25**, 10 (1913).
36. N. Bohr, *Mat.-Fys. Medd.—K. Dan. Vidensk. Selsk* **18**, 8 (1948).
37. N. Bohr and J. Lindhard, *Mat.-Fys. Medd.—K. Dan. Vidensk. Selsk.* **28**, 7 (1954).
38. J. Lindhard, *Mat.-Fys. Medd.—K. Dan. Vidensk. Selsk.* **28**, 8 (1954).
39. J. Lindhard, V. Nielssen, and M. Scharff, *Mat.-Fys. Medd.—K. Dan. Vidensk. Selsk.* **36**, 10 (1968).
40. J. Lindhard, M. Scharff, and H. E. Schiott, *Mat.-Fys. Medd.—K. Dan. Vidensk. Selsk.* **33**, 14 (1963).
41. J. Lindhard, V. Nielssen, M. Scharff, and P. V. Thomsen, *Mat.-Fys. Medd.—K. Dan. Vidensk. Selsk.* **33**, 18 (1963).
42. O. B. Firsov, *Sov. Phys.—JETP (Engl. Transl.)* **5**, 1192 (1957).
43. O. B. Firsov, *Sov. Phys.—JETP (Engl. Transl.)* **6**, 534 (1953).
44. O. B. Firsov, *Sov. Phys.—JETP (Engl. Transl.)* **34**, 308 (1958).
45. O. B. Firsov, *Sov. Phys.—JETP (Engl. Transl.)* **36**, 1076 (1959).
46. C. Carter and J. C. Colligan, "Ion Bombardment of Solids." Heinemann, London, 1968.
47. J. F. Gibbons, *Proc. IEEE* **56**, 295 (1968); **60**, 1062 (1972); *also in* "Handbook on Semiconductors" (S. P. Keller, ed.), Vol. 3, pp. 599–640. North-Holland Publ., Amsterdam, 1980.

48. H. Ryssel and I. Ruge, "Ionen Implantation." Teubner, Stuttgart, 1978.
49. H. Maes, W. Vandervorst, and R. van Overstraeten, in "Impurity Doping Processes," (F. F. Y. Wang, ed.), pp. 443–638. North-Holland Publ., Amsterdam, 1981.
50. L. D. Landau and E. M. Lifshitz, "Mechanics." Pergamon, New York, 1976.
51. H. Goldstein, "Classical Mechanics." Addison-Wesley, Reading, Massachusetts, 1950.
52. P. Sigmund, *Rev. Roum. Phys.* **17,** 823 (1972).
53. C. Lehmann, "Interaction of Radiation with Solids and Elementary Defect Production." North-Holland Publ., Amsterdam, 1977.
54. I. A. Torrens, "Interatomic Potentials." Academic Press, New York, 1972.
55. K. Winterbon, P. Sigmund, and J. B. Sanders, *Mat.-Fys. Medd.—K. Dan. Vidensk. Selsk.* **37,** 14 (1970).
56. J. Gibbons, W. S. Johnson, and S. Mylroie, "Projected Range Statistics," 2nd Ed. Wiley, New York, 1975.
57. B. Smith, "Ion Implantation Range Data for Silicon and Germanium Device Technologies." Research Studies Press, Beaverton, Oregon, 1977.
58. D. K. Brice, "Ion Implantation Range and Energy Deposition Distributions." IFI/Plenum, New York, 1975.
59. S. Kalbitzer and H. Oetzmann, *Radiat. Eff.* **47,** 57 (1980).
60. W. D. Wilson, L. G. Haggmark, and J. P. Biersack, *Phys. Rev. B* **15,** 2458 (1977).
61. G. Moliere, *Z. Naturforsch., A* **2A,** 133 (1947).
62. U. Littmark and J. F. Ziegler, *Phys. Rev. A* **23,** 64 (1981).
63. U. Littmark and J. F. Ziegler, "Handbook of Range Distribution for Energetic Ions in All Elements." Pergamon, New York, 1980.
64. H. Bethe, *Ann. Phys. (Leipzig)* **5,** 325 (1930).
65. K. O. Nielsen, in "Electromagnetically Enriched Isotopes and Mass Spectroscopy" (M. L. Smith, ed.), p. 68. Academic Press, New York, 1956.
66. J. W. Mayer, L. Eriksson, and J. A. Davies, "Ion Implantation in Symiconductors; Silicon and Germanium." Academic Press, New York, 1970.
67. I. M. Cheshire, G. Dearnally, and J. M. Poate, *Proc. R. Soc. London, Ser. A* **311,** 47 (1969).
68. M. Cheshire, G. Dearnally, and J. M. Poate, *Phys. Lett. A* **27A,** 304 (1968).
69. A. P. Pathak, *J. Phys. F* **4,** 1883 (1974).
70. A. P. Pathak, *J. Phys. C* **7,** 3239 (1974).
71. J. S. Briggs and A. P. Pathak, *J. Phys. C* **7,** 1929 (1974).
72. F. H. Eisen, *Can. J. Phys.* **46,** 561 (1968).
73. S. Furukawa and H. Ishiwara, in "Ion Implantation in Semiconductors" (S. Namba, ed.), pp. 143–154. Plenum, New York, 1975.
74. W. S. Johnson and J. F. Gibbons, "Projected Range Statistics in Semiconductors." Stanford Univ. Bookstore, 1969. Stanford, California.
75. D. K. Brice, *Proc. Int. Conf., 1st, Thousand Oaks, Calif., 1970* p. 101 (1971).
76. S. Furukawa, H. Matsumura, and H. Ishiwara, *Jpn. J. Appl. Phys.* **11,** 134 (1972).
77. S. Mylroie and J. F. Gibbons, *Ion Implantation Semicond. Other Mater., Proc. Int. Conf., 3rd, Yorktown Heights, New York, 1972* p. 243 (1973).
78. P. Sigmund and J. B. Sanders, *Proc. Int. Conf. Appl. Ion Beams Semicond. Technol., Grenoble, Fr.* p. 215 (1978).
79. K. B. Winterbon, *Radiat. Eff.* **13,** 215 (1972); **30,** 199 (1976).
80. D. H. Smith and J. F. Gibbons, in "Ion Implantation in Semiconductors 1976" (F. Chernow, J. A. Borders, and D. K. Brice, eds.), pp. 333–346. Plenum, New York, 1977.
81. L. A. Christel, J. F. Gibbons, and S. Mylroie, *J. Appl. Phys.* **51,** 6176 (1980).
82. L. A. Christel, J. F. Gibbons, and S. Mylroie, *Nucl. Instrum. Methods* **182,** 187 (1981).

83. M. D. Giles and J. F. Gibbons, *Nucl. Instrum. Methods* **209/210**, 33 (1983).
84. D. K. Brice, *Appl. Phys. Lett.* **16**, 103 (1970).
85. T. Tsurushima and H. Tanoue, *J. Phys. Soc. Jpn.* **31**, 1695 (1971).
86. S. Furukawa and H. Ishiwara, *J. Appl. Phys.* **43**, 1268 (1972).
87. J. P. Biersack and L. G. Haggmark, *Nucl. Instrum. Methods* **174**, 257 (1980).
88. A. T. Macrander, Cornell Univ. Materials Science Center Rep. No. 4234 (1979).
89. H. E. Schiott, *Mat.-Fys. Medd.—K. Dan. Vidensk. Selsk.* **35**, 9 (1966).
90. K. B. Winterbon, "Ion Implantation Range and Energy Deposition Distributions. Vol. 2: Low Incident Ion Energies." IFI/Plenum, New York, 1975.
91. D. K. Brice, *Radiat. Eff.* **6**, 77 (1970).
92. O. S. Oen and M. T. Robinson, *J. Appl. Phys.* **35**, 2515 (1964).
93. O. S. Oen, D. K. Holmes, and M. T. Robinson, *J. Appl. Phys.* **34**, 302 (1963).
94. O. S. Oen and M. T. Robinson, *Nucl. Instrum. Methods* **132**, 647 (1976).
95. M. T. Robinson and O. S. Oen, *Appl. Phys. Lett.* **2**, 30 (1962).
96. M. T. Robinson and O. S. Oen, *Phys. Rev.* **132**, 64 (1976).
97. M. T. Robinson and I. M. Torrens, *Phys. Rev. B* **9**, 5008 (1973).
98. M. T. Robinson, "Marlowe-Binary Collision Cascade Simulation Program Version 11," Oak Ridge Natl. Lab. PSR-137 (1978).
99. T. Ishitani, R. Shimizu, and K. Murata, *Jpn. J. Appl. Phys.* **11**, 125 (1972); *Phys. Status Solidi B* **50**, 681 (1972).
100. J. E. Robinson, *Radiat. Eff.* **23**, 29 (1974).
101. K. Günther, *Z. Naturforsch. A* **26A**, 1290 (1971).
102. K. Günther, H. Ewald, and H. Schmidt, *Radiat. Eff.* **13**, 111 (1972).
103. D. K. Hutchence and S. Hontzeas, *Nucl. Instrum. Methods* **116**, 217 (1974).
104. D. A. Eastham, *Nucl. Instrum. Methods* **125**, 277 (1975).
105. Y. Miyagawa and S. Miyagawa, *J. Appl. Phys.* **54**, 7124 (1983).
106. E. M. Lifshitz and L. P. Pitaevskii, "Physical Kinetics." Pergamon, New York, 1981.
107. T. Hirao, K. Inoue, S. Takayanagi, and Y. Yaegashi, *J. Appl. Phys.* **50**, 193 (1979).
108. W. K. Hofker, D. P. Oesthoek, N. J. Koeman, and H. A. M. de Grefte, *Radiat. Eff.* **24**, 223 (1975).
109. D. K. Brice, *J. Appl. Phys.* **46**, 3385 (1975).
110. G. H. Kinchin and R. S. Pease, *Rep. Prog. Phys.* **18**, 2 (1955).
111. D. Sigmund, *Appl. Phys. Lett.* **14**, 114 (1969).
112. J. C. North and W. M. Gibson, in "Ion Implantation" (F. H. Eisen and L. T. Chadderton, eds.), pp. 143–144. Gordon & Breach, New York, 1971.
113. M. Kendall and A. Stuart, "The Advanced Theory of Statistics. Vol. 1: Distribution Theory." Macmillan, New York, 1977.
114. W. P. Petersen, W. Fichtner, and E. H. Grosse, *IEEE Trans. Electron Devices* **ED-30**, 1011 (1983).
115. W. P. Petersen, W. Fichtner, and E. H. Grosse, *IEEE Trans. Comput. Aided Design*, to be published.
116. K. B. Winterbon, *Appl. Phys. Lett.* **42**, 205 (1983).
117. H. Runge, *Phys. Status Solidi* **39**, 595 (1977).
118. R. B. Fair, in "Silicon Integrated Circuits," Part B (D. Kahng, ed.), pp. 1–108. Academic Press, New York, 1981.
119. R. B. Fair, in "Impurity Doping Processes in Silicon" (F. F. Y. Wang, ed.), pp. 315–441. North-Holland Publ., Amsterdam, 1981.
120. A. F. W. Willoughby, in "Impurity Doping Processes in Silicon" (F. F. Y. Wang, ed.), pp. 1–52. North-Holland Publ., Amsterdam, 1981.
121. A. F. W. Willoughby, *J. Phys. D* **10**, 455 (1977).

122. S. M. Hu, *in* "Atomic Diffusion in Semiconductors" (D. Shaw, ed.), pp. 217–350. Plenum, New York, 1973.
123. S. M. Hu, *J. Appl. Phys.* **43**, 2015 (1972).
124. R. K. Jain and R. J. van Overstraeten, *J. Electrochem. Soc.* **122**, 553 (1975).
125. R. K. Jain and R. J. van Overstraeten, *IEEE Trans. Electron Devices* **ED-21**, 155 (1974).
126. R. van Overstraeten, H. de Man, and R. Mertens, *IEEE Trans. Electron Devices* **ED-20**, 290 (1973).
127. R. P. Mertens, R. van Overstraeten, and H. de Man, *Adv. Electron. Electron Phys.* **55**, 77 (1981).
128. E. O. Kane, *Phys. Rev.* **131**, 79 (1963).
129. T. N. Morgan, *Phys. Rev.* **139**, A343 (1965).
130. R. K. Jain, *Indian J. Pure Appl. Phys.* **19**, 662 (1981).
131. S. M. Hu and S. Schmidt, *J. Appl. Phys.* **39**, 4272 (1968).
132. S. M. Hu, *Phys. Rev.* **180**, 773 (1969).
133. J. R. Lowney, *Ext. Abstr. Electrochem. Soc., Spring Meet., St. Louis, Mo.* Abstr. 139 (1980).
134. F. F. Morehead, *Ext. Abstr. Electrochem. Soc., Spring Meet., St. Louis, Mo.* Abstr. 138 (1980).
135. M. Y. Tsai, F. G. Morehead, and J. E. E. Baglin, *J. Appl. Phys.* **51**, 3230 (1980).
136. B. R. Penumalli, *IEEE Trans. Electron Devices* **ED-30**, 986 (1983).
137. S. Zaromb, *IBM J. Res. Dev.* **1**, 57 (1957).
138. J. F. Ziegler, G. W. Cole, and J. E. E. Baglin, *Appl. Phys. Lett.* **21**, 177 (1972).
139. J. R. Lowney and R. D. Larrabee, *IEEE Trans. Electron Devices* **ED-27**, 1795 (1980).
140. U. Gösele and T. Y. Tan, *in* "Defects in Semiconductors" (E. Sirtl and T. Goorissen, eds.), pp. 17–36. Electrochem. Soc., Princeton, New Jersey, 1982.
141. N. A. Stolwijk, B. Schuster, J. Holzl, H. Mahrer, and W. Frank, *Proc. Int. Conf. Defects Semicond., 12th, Amsterdam, 1983*
142. F. Morehead, W. Stolwijk, W. Meyberg, and U. Gösele, to be published.
143. W. P. Wilcox, T. J. La Chapelle, and D. H. Forbes, *J. Electrochem. Soc.* **111**, 1377 (1964).
144. H. Kitagawa, K. Hashimoto, and M. Yoshida, *Jpn. J. Appl. Phys.* **21**, 276 (1982).
145. D. A. Antoniadis, *J. Electrochem. Soc.* **129**, 1093 (1982).
146. T. Y. Tan, U. Gösele, and F. F. Morehead, *Appl. Phys., A* **31**, 97 (1983).
147. D. A. Antoniadis and I. Moskovitz, *in* "Aggregation Phenomena of Point Defects in Silicon" (E. Sirtl and J. Goorissen, eds.), p. 1. Electrochem. Soc., Princeton, New Jersey, 1983.
148. D. A. Antoniadis and I. Moskovitz, *J. Appl. Phys.* **53**, 6788 (1982).
149. U. Gösele and W. Frank, *in* "Defects in Semiconductors" (J. Narayan and T. Y. Tan, eds.), p. 55. North-Holland Publ., Amsterdam, 1981.
150. W. Frank, A. Seeger, and U. Gösele, *in* "Defects in Semiconductors" (J. Narayan and T. Y. Tan, eds.), p. 31. North-Holland Publ., Amsterdam, 1981.
151. U. Gösele and T. Y. Tan, *in* "Defects in Semiconductors II" (S. Mahajan and J. W. Corbett, eds.), pp. 45–60. North-Holland Publ., Amsterdam, 1983.
152. R. B. Fair, *in* "Defects in Semiconductors II" (S. Mahajan and J. W. Corbett, eds.), pp. 61–74. North-Holland Publ., Amsterdam, 1983.
153. K. Taniguchi and D. A. Antoniadis, *in* "Defects in Semiconductors" (W. M. Bullis and L. C. Kimerling, eds.), pp. 315–324. Electrochem. Soc., Princeton, New Jersey, 1983.
154. T. Y. Tan, F. Morehead, and U. Gösele, *in* "Defects in Semiconductors" (W. M. Bullis and L. C. Kimerling, eds.), pp. 325–336. Electrochem. Soc., Princeton, New Jersey, 1983.
155. R. Tielert, *IEEE Trans. Electron Devices* **ED-27**, 1479 (1980).
156. H. Ryssel, K. Haberger, K. Hoffmann, G. Prinke, R. Dümcke, and A. Sachs, *IEEE Trans. Electron Devices* **ED-27**, 1484 (1980).

157. T. Y. Tan and U. Gösele, *Appl. Phys. Lett.* **39**, 86 (1981).
158. P. S. Dobson, *Philos. Mag.* **24**, 567 (1971).
159. S. M. Hu, *J. Appl. Phys.* **45**, 1567 (1974).
160. R. Francis and P. S. Dobson, *J. Appl. Phys.* **50**, 280 (1979).
161. T. Y. Tan and U. Gösele, *Appl. Phys. Lett.* **40**, 616 (1982).
162. S. Prussin, *J. Appl. Phys.* **43**, 2850 (1972).
163. E. Sirtl, *in* "Semiconductor Silicon 1977" (H. R. Huff and E. Sirtl, eds.), pp. 4–17. Electrochem. Soc., Princeton, New Jersey, 1977.
164. T. Y. Tan and U. Gösele, to be published.
165. S. Mizuo and H. Higuchi, *Jpn. J. Appl. Phys.* **20**, 739 (1981).
166. D. A. Antoniadis, A. M. Lin, and R. W. Dutton, *Appl. Phys. Lett.* **33**, 1030 (1978).
167. T. Y. Tan and B. J. Ginsberg, *Appl. Phys. Lett.* **42**, 448 (1983).
168. A. M. Lin, D. A. Antoniadis, and R. W. Dutton, *J. Electrochem. Soc.* **128**, 1131 (1981).
169. D. P. Kennedy and P. C. Murley, *Proc. IEEE* **59**, 335 (1971).
170. R. O. Schwenker, E. S. Pan, and R. F. Lever, *J. Appl. Phys.* **42**, 3195 (1971).
171. J. S. Sandhu and J. L. Reuter, *IBM J. Res. Dev.* **15**, 464 (1971).
172. E. Guerrero, H. Pötzl, R. Tielert, M. Grasserbauer, and G. Stingeder, *J. Electrochem. Soc.* **129**, 1826 (1982).
173. S. Solmi, G. Celloti, D. Nobili, and P. Negrini, *J. Electrochem. Soc.* **123**, 654 (1976).
174. F. N. Schwettmann and D. L. Kendall, *Appl. Phys. Lett.* **19**, 218 (1971).
175. F. N. Schwettmann and D. L. Kendall, *Appl. Phys. Lett.* **20**, 2 (1972).
176. M. Yoshida, E. Arai, H. Nakamura, and Y. Terunuma, *J. Appl. Phys.* **45**, 1498 (1974).
177. M. Yoshida, *J. Appl. Phys.* **48**, 2169 (1977).
178. M. Yoshida, *Jpn. J. Appl. Phys.* **18**, 479 (1979).
179. R. B. Fair and J. C. C. Tsai, *J. Electrochem. Soc.* **124**, 1107 (1977).
180. S. M. Hu, P. Fahey, and R. W. Dutton, *J. Electrochem. Soc.* **54**, 6912 (1983).
181. T. Y. Tan and U. Gösele, *Proc.—Electrochem. Soc.* **84**, 151–175 (1984).
182. B. E. Deal and A. S. Grove, *J. Appl. Phys.* **36**, 3770 (1965).
183. C. P. Ho and J. D. Plummer, *J. Electrochem. Soc.* **126**, 1516, 1523 (1979).
184. R. W. Dutton and D. A. Antoniadis, *in* "Moving Boundary Problems" (D. G. Wilson, A. D. Solomon, and P. T. Boggs, eds.), pp. 233–247. Academic Press, New York, 1978.
185. A. S. Grove, O. Leistiko, and C. T. Sah, *J. Appl. Phys.* **35**, 2695 (1964).
186. M. Av-Ron, M. Shatzkes, P. J. Burkhardt, and I. Cadoff, *J. Appl. Phys.* **47**, 3159 (1976).
187. P. R. Wilson, *Solid-State Electron.* **15**, 961 (1972).
188. S. Wagner, *J. Electrochem. Soc.* **119**, 1570 (1972).
189. G. Masetti, P. Negri, S. Solmi, and G. Sondhi, *Alta Freq.* **42**, 346 (1973).
190. J. L. Prince and F. W. Schwettmann, *J. Electrochem. Soc.* **121**, 705 (1974).
191. R. H. Krambeck, *J. Electrochem. Soc.* **121**, 588 (1974).
192. H. Guckel and L. A. Hall, *Solid-State Electron.* **18**, 99 (1975).
193. H. Guckel and L. A. Hall, *Solid-State Electron.* **18**, 875 (1975).
194. R. Kraft, *in* "Advances in Computer Methods for Partial Differential Equations" (R. Vichnevetsky, ed.), pp. 328–333. AICA, Dep. Comput. Sci., Rutgers Univ., New Brunswick, New Jersey, 1975.
195. D. A. Antoniadis, S. H. Hansen, R. W. Dutton, and A. G. Gonzalez, "SUPREM I—A Program for IC Process Modeling and Simulation," Stanford Electron. Lab. Tech. Rep. SEL-77-006 (1977).
196. R. W. Dutton, H. G. Lee, and S. Y. Oh, *in* "Numerical Analysis of Semiconductor Devices and Integrated Circuits" (B. T. Browne and J. J. H. Miller, eds.), pp. 3–33. Boole Press, Dublin, 1981.

197. D. Chin, M. Kump, and R. W. Dutton, "SUPRA—Stanford University Process Analysis Program," Stanford Electron. Lab. Tech. Rep. (1981).
198. J. Huang and L. Welliver, *J. Electrochem. Soc.* **117**, 1577 (1970).
199. H. G. Lee, R. W. Dutton, and D. A. Antoniadis, *J. Electrochem. Soc.* **126**, 2001 (1979).
200. B. R. Penumalli, in "Numerical Analysis of Semiconductor Devices and Integrated Circuits" (B. T. Browne and J. J. H. Miller, eds.), pp. 264–269. Boole Press, Dublin, 1981.
201. L. O. Wilson, *J. Electrochem. Soc.* **129**, 831 (1982).
202. L. I. Rubinstein, "The Stefan Problem." Am. Math. Soc., Providence, Rhode Island, 1971.
203. G. Charitat and A. Martinez, *J. Appl. Phys.* **55**, 909 (1984).
204. E. P. Eernisse, *Appl. Phys. Lett.* **30**, 290 (1977).
205. R. H. Doremus, "Glass Science." Wiley, New York, 1973.
206. D. Chin, R. W. Dutton, S. Y. Oh, J. L. Moll, and S. M. Hu, *Tech. Dig.—Int. Electron Devices Meet.* p. 228 (1982).
207. D. Chin, S. Y. Oh, S. M. Hu, R. W. Dutton, and J. L. Moll, *IEEE Trans. Electron Devices* **ED-30**, 744 (1983).
208. D. Chin, S. Y. Oh, and R. W. Dutton, *IEEE Trans. Electron Devices* **ED-30**, 993 (1983).
209. H. Matsumoto and M. Fukuma, *Tech. Dig.—Int. Electron Devices Meet.* p. 39 (1983).
210. E. Nicollian and J. R. Brews, "MOS (Metal Oxide Semiconductor) Physics and Technology." Wiley, New York, 1982.
211. D. P. Kennedy and R. R. O'Brien, *IBM J. Res. Dev.* **9**, 179 (1965).
212. D. I. Cherednichenko, H. Gruenberg, and T. K. Sarkes, *Solid-State Electron.* **17**, 315 (1974).
213. L. C. Parillo, R. S. Payne, R. E. Davis, G. W. Reutlinger, and R. L. Field, *Tech. Dig.—Int. Electron Devices Meet.* p. 752 (1980).
214. B. J. Lin, in "Fine-Line Lithography" (R. Newman, ed.), pp. 105–232. North-Holland Publ., Amsterdam, 1980.
215. W. G. Heitman and P. M. van der Berg, *Can. J. Phys.* **53**, 1305 (1975).
216. F. L. Neerhof and G. Mur, *Appl. Sci. Res.* **28**, 73 (1973).
217. B. J. Lin, *J. Opt. Soc. Am.* **62**, 976 (1972).
218. A. Sommerfeld, *Math. Ann.* **47**, 317 (1896). A nice derivation of the Sommerfeld result can be found in reference 219, Chap. XI, Sect. 11.5.
219. M. Born and E. Wolf, "Principles of Optics-Electromagnetic Theory of Propagation, Interference and Diffraction of Light." Pergamon, New York, 1959.
220. L. Levi, "Applied Optics—A Guide to Optical System Design," Vol. 1. Wiley, New York, 1968.
221. Reference 219, Chap. VIII, Sect. 8.3.3.
222. Reference 219, Chap. VIII, Sect. 8.3.2.
223. B. J. Lin, *Proc. Int. Conf. Microlithogr.—Microcircuit Eng. 81 Lausanne, Switz.* p. 47, (1981).
224. J. Bruning, *Electrochem. Soc. Spring Meet., Montreal* 47–75 (1982).
225. H. H. Hopkins, *Proc. R. Soc. London, Ser. A* **208**, 263 (1951); **217**, 408 (1953); *Proc. R. Soc. London, Ser. B* **69**, 562 (1956); *J. Opt. Soc. Am.* **47**, 508 (1957).
226. E. Wolf, *Proc. R. Soc. London, Ser. A* **225**, 96 (1954); **230**, 246 (1955).
227. R. E. Kinzly, *J. Opt. Soc. Am.* **55**, 1002 (1965); **46**, 9 (1956).
228. B. M. Watrasiewicz, *Opt. Acta* **12**, 391 (1965).
229. E. C. Kintner, *Appl. Opt.* **17**, 2747 (1978).
230. A. Offner and J. Meiron, *Appl. Opt.* **8**, 183 (1969).
231. R. Hershel, *Kodak Microelectron. Semin. Proc. Interface* **178**, 62 (1978); *Proc. Soc. Photo-Opt. Instrum. Eng.* **135**, 24 (1978).

232. D. Nyssonen, *Appl. Opt.* **16**, 2223 (1977).
233. P. S. Considine, *J. Opt. Soc. Am.* **56**, 1001 (1966).
234. M. Lacombat and G. M. Dubroeucq, *Proc. Soc. Photo-Opt. Instrum. Eng.* **174**, 28 (1979).
235. P. Tigreat, *Proc. Soc. Photo-Opt. Instrum. Eng.* **174**, 37 (1979).
236. M. M. O'Toole and A. R. Neureuther, *Proc. Soc. Photo-Opt. Instrum. Eng.* **174**, 22 (1979).
237. "SAMPLE User's Guide," Electron. Res. Lab., Dep. Electr. Eng. Comput. Sci., Univ. of California, Berkeley, 1982.
238. J. W. Goodman, "Introduction to Fourier Optics," McGraw-Hill, New York, 1968.
239. M. J. Beran and G. B. Parent, "Theory of Partial Coherence." Prentice-Hall, Englewood Cliffs, New Jersey, 1964.
240. M. C. King, *in* "VLSI Electronics: Microstructure Science" (N. G. Einspruch, ed.), pp. 42–81. Academic Press, New York, 1981.
241. B. J. Lin, *IEEE Trans. Electron Devices* **ED-27**, 931 (1980).
242. F. H. Dill, W. P. Hornberger, P. S. Hange, and J. M. Shaw, *IEEE Trans. Electron Devices* **ED-22**, 445 (1975).
243. F. H. Dill, A. R. Neureuther, J. A. Tuttle, and E. J. Walker, IBM Res. Rep. RC 5162 (1975).
244. W. S. deForest, "Photoresist Materials and Processes," McGraw-Hill, New York, 1975.
245. P. H. Berning, *Phys. Thin Films* **1**, 69 (1963).
246. J. D. Cuthbert, *Solid State Technol.* **20**, Aug., 59 (1977).
247. A. R. Neureuther and F. H. Dill, *Symp. Ser.—Polytechnic Inst. N.Y., Microwave Res. Inst.* **23**, 233–249 (1975).
248. M. A. Narasimhan and J. H. Carter, *Proc. Soc. Photo-Opt. Instrum. Eng.* **135**, 2 (1978).
249. J. M. Shaw and M. Hatzakis, *IEEE Trans. Electron Devices* **ED-25**, 425 (1978).
250. A. R. Neureuther and W. Wong, "Fundamentals of High-Resolution Lithography," Univ. of California, Berkeley (1982).
251. M. A. Narashimhan and J. B. Lownsbury, *Proc. Soc. Photo-Opt. Instrum. Eng.* **135**, 57 (1978).
252. R. K. Watts, *in* "Very Large Scale Integration (VLSI)—Fundamentals and Applications" (D. F. Barbe, ed.), pp. 42–88. Springer-Verlag, Berlin and New York, 1980.
253. N. D. Wittels, *in* "Fine-Line Lithography" (R. Newman, ed.), pp. 1–104. North-Holland Publ., Amsterdam, 1980.
254. T. L. Brewer, *in* "Electron-Beam Technology in Microelectronic Fabrication" (G. R. Brewer, ed.), pp. 2–258. Academic Press, New York, 1980.
255. D. L. Herriot and G. R. Brewer, *in* "Electron-Beam Technology in Microelectronic Fabrication" (G. R. Brewer, ed.), pp. 142–216. Academic Press, New York, 1980.
256. P. R. Thornton, *Adv. Electron. Electron Phys.* **48**, 272 (1979).
257. R. J. Hawryluk, A. M. Hawryluk, and H. I. Smith, *J. Appl. Phys.* **45**, 2551 (1974); R. J. Hawryluk, Ph.D. Thesis, MIT, Cambridge, Massachusetts, 1974.
258. W. Bothe, *Handb. Phys.* **22**, 1–74 (1933).
259. E. J. Williams, *Proc. R. Soc. London, Ser. A* **169**, 531 (1939); *Phys. Rev.* **58**, 292 (1940); *Rev. Mod. Phys.* **17**, 217 (1945).
260. H. A. Bethe, M. E. Rose, and L. P. Smith, *Proc. Am. Phys. Soc.* **78**, 573 (1938).
261. H. W. Lewis, *Phys. Rev.* **78**, 526 (1950).
262. H. Meister, *Z. Naturforsch., A* **13A**, 809 (1958).
263. J. W. Weymouth, *Phys. Rev.* **84**, 766 (1951).
264. G. Moliere, *Z. Naturforsch., A* **3A**, 78 (1948).
265. F. Lenz, *Z. Naturforsch., A* **9A**, 185 (1954).
266. G. Wentzel, *Z. Phys.* **40**, 590 (1927).

267. B. P. Nigarn, M. K. Sundaresan, and T. Y. Wu, *Phys. Rev.* **115**, 491 (1959).
268. N. F. Mott, *Proc. R. Soc. London, Ser. A* **125**, 259 (1930).
269. E. Krefting and L. Reimer, *in* "Quantitative Analysis With Electron Microprobe and Secondary Ion Mass Spectroscopy" (E. Preuss, ed.), pp. 114–121. Zentralbibliothek Kernforschungsanlage, Jülich, FRG, 1974.
270. S. Ichimura, M. Aratama, and R. Shimizu, *J. Appl. Phys.* **5**, 2853 (1980).
271. W. Bühring, *Z. Phys.* **212**, 61 (1968).
272. L. Reimer and E. Krefting, *in* "Use of Monte Carlo Calculations in Electron Probe Microanalysis and Scanning Electron Microscopy" (K. F. J. Heinrich, D. E. Newbury, and H. Yakowitz, eds.), NBS Spec. Publ. 460, pp. 45–60. U.S. Gov. Print. Off., Washington, D.C., 1976.
273. W. J. Byatt, *Phys. Rev.* **104**, 1298 (1958).
274. T. Tietz, *Nuovo Cimento* **87**, 1365 (1965).
275. M. Kotera, K. Murata, and K. Nagami, *J. Appl. Phys.* **52**, 997 (1981); **52**, 7403 (1981).
276. I. Adesida, R. Shimizu, and T. E. Everhart, *Appl. Phys. Lett.* **33**, 849 (1978).
277. J. S. Greeneich, *in* "Electron-Beam Technology in Microelectronic Fabrication" (G. R. Brewer, ed.), pp. 60–141. Academic Press, New York, 1980.
278. V. E. Cosslett and R. N. Thomas, *Br. J. Appl. Phys.* **15**, 883 (1964).
279. V. E. Cosslett and R. N. Thomas, *Br. J. Appl. Phys.* **15**, 1283 (1964).
280. V. E. Cosslett and R. N. Thomas, *Br. J. Appl. Phys.* **15**, 235 (1964).
281. S. Goudsmit and J. L. Saunderson, *Phys. Rev.* **57**, 24 (1940).
282. L. V. Spencer, *Phys. Rev.* **98**, 1507 (1955).
281. M. J. Berger, *Methods Comput. Phys.* **1**, 135 (1963).
284. J. Henoc and F. Maurice, *in* "Use of Monte Carlo Calculations in Electron Probe Microanalysis and Scanning Electron Microscopy," (K. F. J. Heinrich, D. E. Newbury, and H. Yakowitz, eds.), NBS Spec. Publ. 460, pp. 61–95. U.S. Gov. Print. Off., Washington, D.C., 1976.
285. H. A. Bethe, *Phys. Rev.* **89**, 1256 (1953).
286. B. P. Nigam and V. S. Mathur, *Phys. Rev.* **121**, 1577 (1962).
287. N. F. Mott and H. S. W. Massey, "The Theory of Atomic Collisions." Oxford Univ. Press, London and New York, 1965.
288. B. Rossi and K. Greisen, *Rev. Mod. Phys.* **13**, 240 (1941).
289. W. T. Scott, *Phys. Rev.* **76**, 212 (1949).
290. R. W. Nosker, *J. Appl. Phys.* **40**, 1872 (1969).
291. H. A. Bethe and J. Ashkin, *Exp. Nucl. Phys.* **1**, 166–357 (1953).
292. R. D. Birkhoff, *Handb. Phys.* **34**, 53 (1958).
293. M. J. Berger and S. M. Seltzer, "Tables of Energy Losses and Ranges of Electrons and Positrons," pp. 1–12. NTIS N65-12506. U.S. Dep. Commer., Washington, D.C.
294. D. B. Brown and R. E. Ogilvie, *J. Appl. Phys.* **37**, 4429 (1966).
295. L. D. Landau, *J. Phys. (USSR)* **8**, 201 (1944).
296. O. Blunck and S. Leisegang, *Z. Phys.* **128**, 500 (1950).
297. R. Shimizu, T. Ikuta, T. E. Everhart, and W. J. deVore, *J. Appl. Phys.* **46**, 1581 (1975).
298. R. Shimizu, Y. Kataoka, T. Ikuta, T. Koshikawa, and H. Hasimoto, *J. Phys. D* **9**, 101 (1976).
299. M. Gryzinski, *Phys. Rev.* **138**, A336 (1965).
300. L. Vriens, *in* "Case Studies in Atomic Collision Physics I" (E. W. McDaniel and M. R. C. McDowell, eds.), pp. 337–398. North-Holland Publ., Amsterdam, 1976.
301. A. E. Kingston, *Proc. Phys. Soc., London* **87**, 193 (1966).
302. R. Shimizu and T. E. Everhart, *Appl. Phys. Lett.* **33**, 784 (1978).

303. E. Eggartner, *J. Chem. Phys.* **62**, 833 (1975).
304. R. D. Evans, "The Atomic Nucleus." McGraw-Hill, New York, 1955.
305. C. Møller, *Z. Phys.* **70**, 786 (1931).
306. K. Murata, D. F. Kyser, and C. H. Ting, *J. Appl. Phys.* **52**, 4396 (1981).
307. R. Castaing, P. Deschamps, and J. Philibert, "X-Ray Optics and Microanalysis." Hermann, Paris. 1966.
308. H. E. Bishop, *in* "Use of Monte Carlo Calculations in Electron Probe Microanalysis and Scanning Electron Microscopy" (K. F. J. Heinrich, D. E. Newbury, and H. Yakowitz, eds.), NBS Spec. Publ. 460, pp. 5–14. U.S. Gov. Print. Off., Washington, D.C., 1976.
309. M. Green, *Proc. R. Soc. London, Ser. A* **82**, 204 (1963).
310. H. E. Bishop, *Proc. Phys. Soc., London* **85**, 855 (1965).
311. R. Shimizu, T. Ikuta, and K. Murata, *J. Appl. Phys.* **43**, 4233 (1972).
312. B. Pascal, *in* "X-Ray Optics and Microanalysis" (G. Mollenstedt and K. H. Gaukler, eds.), pp. 137–152. Springer-Verlag, Berlin and New York, 1969.
313. J. Henoc and F. Maurice, *Proc. Int. Conf. X-Ray Opt. Microanal., 6th, Osaka, 1971* pp. 113–121 (1972).
314. K. Murata, T. Matsukawa, and R. Shimizu, *Jpn. J. Appl. Phys.* **10**, 678 (1971).
315. L. Reimer, *Optik* **27**, 86 (1968).
316. N. Saitou, *Jpn. J Appl. Phys.* **12**, 941 (1973).
317. D. F. Kyser and K. Murata, *Electron Ion Beam Sci. Technol.* **6**, 205–225 (1974).
318. S. Horiguchi, M. Suzuki, T. Kobayashi, H. Voshino, and Y. Sakakibara, *Appl. Phys. Lett.* **39**, 512 (1981).
319. R. J. Hawryluk, A. M. Hawryluk, and H. I. Smith, *J. Appl. Phys.* **53**, 5985 (1982).
320. M. S. C. Chung and K. L. Tai, *Electron Ion Beam Sci. Technol.* **8**, 242–255 (1978).
321. D. O. Schneider and D. V. McCormack, *Radiat. Res.* **11**, 418 (1959).
322. J. S. Greeneich and T. van Duzer, *IEEE Trans. Electron Devices* **ED-21**, 286 (1974).
323. J. S. Greeneich and T. van Duzer, *J. Vac. Sci. Technol.* **10**, 1056 (1973).
324. T. E. Everhart, *J. Appl. Phys.* **31**, 1483 (1960).
325. J. S. Greeneich and T. van Duzer, *IEEE Trans. Electron Devices* **ED-20**, 598 (1973).
326. M. Parikh, *Electron Ion Beam Sci. Technol.* **8**, 382–392 (1978).
327. M. Parikh, *J. Vac. Sci. Technol.* **15**, 931 (1978).
328. M. Parikh and D. E. Schreiber, *Electron Ion Beam Sci. Technol.* **9**, 304–313 (1980).
329. M. Parikh, *J. Appl. Phys.* **50**, 4371 (1979).
330. W. Grobman, A. J. Speth, and T. H. P. Chang, *Electron Ion Beam Sci. Technol.* **9**, 314–325 (1980).
331. D. P. Kern, *Electron Ion Beam Sci. Technol.* **9**, 326–339 (1980).
332. J. C. H. Phang and H. Ahmed, *J. Vac. Sci. Technol.* **16**, 1754 (1979).
333. P. Nehmiz, H. Bohlen, J. Greschner, E. Bretscher, and P. Vettiger, *J. Vac. Sci. Technol.* **19**, 1291 (1981).
334. A. M. Carroll, *J. Vac. Sci. Technol.* **19**, 1296 (1981).
335. H. Sewell, *J. Vac. Sci. Technol.* **15**, 927 (1978).
336. N. D. Wittels and C. I. Youngman, *Electron Ion Beam Sci. Technol.* **8**, 361–370 (1978).
337. N. Sugiyama and K. Saitoh, *Electron Ion Beam Sci. Technol.* **9**, 272–281 (1980).
338. N. Sugiyama, *Trans. IECE Jpn.* **E63**, 310 (1980).
339. M. Parikh and D. E. Schreiber, *IBM J. Res. Dev.* **24**, 530 (1980).
340. A. Kikuchi, A. Kanamura, N. Okazaki, Y. Nakone, and K. Tsubai, *J. Vac. Sci. Technol.* **16**, 1764 (1979).
341. E. Kratschmer, *J. Vac. Sci. Technol.* **19**, 1261 (1981).
342. W. W. Molzen and W. D. Grobman, *J. Vac. Sci. Technol.* **19**, 1300 (1981).

343. K. Murata, E. Nomura, K. Nagami, T. Kato, and H. Nakata, *Jpn. J. Appl. Phys.* **17**, 1851 (1978).
344. E. Nomura, K. Murata, and K. Nagami, *Jpn. J. Appl. Phys.* **18**, 1353 (1979).
345. T. H. P. Chang, *J. Vac. Sci. Technol.* **12**, 1271 (1975).
346. M. Parikh and D. F. Kyser, *J. Appl. Phys.* **50**, 1104 (1979).
347. D. F. Kyser, D. E. Schreiber, and C. H. Ting, *Electron Ion Beam Sci. Technol.* **9**, 255–271 (1982).
348. D. F. Kyser and R. Pyle, *IBM J. Res. Dev.* **24**, 426 (1980).
349. N. Aizaki, *J. Vac. Sci. Technol.* **16**, 1726 (1979).
350. D. F. Kyser and C. H. Ting, *J. Vac. Sci. Technol.* **16**, 1759 (1979).
351. C. H. Shaw, *J. Vac. Sci. Technol.* **19**, 1286 (1981).
352. J. S. Greeneich, *J. Vac. Sci. Technol.* **16**, 1749 (1979).
353. J. S. Greeneich, *J. Vac. Sci. Technol.* **19**, 1269 (1981).
354. J. S. Greeneich, *Electron Ion Beam Sci. Technol.* **9**, 282–303 (1980).
355. T. Kato, Y. Watakabe, and N. Nakata, *J. Vac. Sci. Technol.* **19**, 1279 (1981).
356. S. K. S. Ma, M. Parikh, and W. Ward, *J. Vac. Sci. Technol.* **19**, 1275 (1981).
357. R. D. Heidenreich, J. P. Ballantyne, and L. F. Thompson, *J. Vac. Sci. Technol.* **12**, 1284 (1975).
358. R. D. Heidenreich, L. F. Thompson, E. D. Feit, and C. M. Melliar-Smith, *J. Appl. Phys.* **44**, 4039 (1973).
359. L. H. Lin, *J. Vac. Sci. Technol.* **12**, 1289 (1975).
360. H. Nakata, T. Kata, K. Murata, Y. Hirai, and K. Nagami, *J. Vac. Sci. Technol.* **19**, 1248 (1981).
361. N. Nakayama, Y. Machida, S. Furuya, S. Yamamoto, and T. Hisatsugu, *Electron Ion Beam Sci. Technol.* **10**, 252–259 (1983).
362. J. S. Greeneich, *J. Electrochem. Soc.* **121**, 1669 (1974); **122**, 970 (1975).
363. J. S. Greeneich, *J. Appl. Phys.* **45**, 5264 (1974).
364. R. F. Herzog, J. S. Greeneich, T. E. Everhart, and T. van Duzer, *IEEE Trans. Electron Devices* **ED-19**, 635 (1972).
365. A. Neureuther, D. F. Kyser, K. Murata, and C. H. Ting, *Electron Ion Beam Sci. Technol.* **8**, 265–275 (1978).
366. F. Jones and J. Parasczak, *IEEE Trans. Electron Devices* **ED-28**, 1544 (1981).
367. D. L. Spears and H. I. Smith, *Electron. Lett.* **8**, 102 (1972).
368. D. L. Spears and H. I. Smith, *Solid State Technol.* **15**, 21 (1972).
369. E. Spiller and R. Feder, *in* "X-Ray Optics" (H.-J. Queisser, ed.), pp. 35–92. Springer-Verlag, Berlin and New York.
370. J. R. Maldonado, M. E. Poulsen, T. E. Saunders, F. Vratny, and A. Zacharias, *J. Vac. Sci. Technol.* **16**, 1942 (1979).
371. K. Heinrich, H. Betz, A. Heuberger, and S. Pongratz, *J. Vac. Sci. Technol.* **19**, 1254 (1981).
372. H. Ryssel, H. Kranz, K. Haberger, and J. Bosch, *Proc. Microcircuit 1980, Delft, Neth.*, 293–300 (1980).
373. L. Karapiperis, I. Adesida, C. A. Wolf, and E. D. Wolf, *J. Vac. Sci. Technol.* **19**, 1259 (1981).
374. C. J. Mogab, *in* "VLSI Technology" (S. M. Sze, ed.), pp. 303–346. McGraw-Hill, New York, 1983.
375. R. E. Jewett, P. I. Hagouel, A. R. Neureuther, and T. van Duzer, *Polym. Eng. Sci.* **16**, 381 (1977).
376. P. I. Hagouel, Ph.D. Thesis, Univ. of California, Berkeley, 1976.
377. R. E. Jewett, Memo. UCB/ERL M79/68, Univ. of California, Berkeley (1979).

378. J. L. Reynolds, Memo. UCB/ERL M81/2, Univ. of California, Berkeley (1981).
379. C. Sung, Memo. UCB/ERL M81/1, Univ. of California, Berkeley (1981).
380. I. A. Blech, *Thin Solid Films* **6,** 113 (1970).
381. I. A. Blech, D. B. Fraser, and S. E. Hasko, *J. Vac. Sci. Technol.* **15,** 13 (1978).
382. I. A. Blech and H. A. Vander Blas, *J. Appl. Phys.* **54,** 3489 (1983).
383. I. A. Blech, *Solid State Technol.* **26,** Sept., 123 (1983).
384. J. M. Niewenhuizen and H. B. Haanstra, *Philips Tech. Rev.* **27,** 87 (1966).
385. A. G. Dirks and H. J. Leamy, *Thin Solid Films* **47,** 219 (1977).
386. S. M. Sze, "VLSI Technology." *McGraw-Hill, New York,* **1983**.
387. *IEEE Trans. Electron Devices* **ED-30,** Sept. (1983).

Author Index

Numbers in parentheses are reference numbers and indicate that an author's work is referred to although the name is not cited in the text. Numbers in italics show the page on which the complete reference is listed.

A

Abe, H., 76(5), *116*
Abe, T., 126, *325*
Abramowitz, M., 136(34), *326*
Adachi, Y., 77(18), 78, 81(21), 82(21), 84(21), 86, 87, 88, 93, 94, *116*
Adams, A. C., 95(29), *117*, 126, *326*
Adda, L., 36(86), *71*
Adesida, I., 270, 276, 311(373), 313(373), *333*, *335*
Ahmed, H., 294(332), *334*
Aizaki, N., 301, *335*
Alexander, F. B., 95(29), *117*
Alles, D. S., 122(9), 307(9), 308(9), *325*
Anderson, C. H., Jr., 55(116), *72*
Anthony, T., 38(91), *71*
Antoniadis, D. A., 126(19, 20), 183(145), 184 (147, 148, 153), 185(19), 186, 187, 188, 189, 190, 197(19), 200(184, 195), 201, 202, 203, 204, 205(199), 206(199), *326*, *329*, *330*, *331*
Appleton, B. R., 17(43), 18(43), 26(65), *69*, *70*
Arai, E., 197(176), *330*
Arai, T., 28(73), *70*
Aratama, M., 269(270), 270(270), 276(270), *333*
Aritome, H., 77(13), *116*
Asamaki, T., 77(8), *116*
Ashkin, J., 275, *333*
Aspnes, D. E., 12(33), 13, *68*, *69*
Atalla, M. M., 119, *325*
Atwater, H. A., 56(120), *72*
Auston, D. H., 6(18), 11, *68*

Auvert, G., 20(53), *69*
Av-Ron, M., 200(186), 203(186), 220, *330*

B

Baeri, P., 6, 9, 18, 19, 20(49), *68*, *69*
Bagley, B. G., 29, *70*
Baglin, J. E., 19(46), *69*
Baglin, J. E. E., 176(135), 182(138), 194(135), 195(135), 196(135), *329*
Bailey, P., 11(25), *68*
Ballantyne, J. P., 302(357), *335*
Bardet, R., 76(2), *116*
Baumgart, H., 26(67), 27, 44(97), 56(119), 61 (137), 62(137), *70*, *71*, *72*, *73*
Bayazitov, R. M., 4(2), *67*
Bayruns, R. J., 57(123), *72*
Bean, J. C., 6(13), *68*
Bean, K. E., 62(142), *73*
Bensahel, D., 20(53), *69*
Bentini, G. G., 37(87), *71*
Benton, J. L., 14, 48, *69*, *72*
Beran, M. J., 245(239), *332*
Berger, M. J., 273(283), 275(293), 278(283), *333*
Berkowitz, H. L., 28(70), *70*
Berning, P. H., 260, *332*
Bersin, R. L., 76(7), *116*
Bethe, H. A., 268, 273, 275, *327*, *332*, *333*
Betz, H., 307(371), 309(371), 310(371), *335*
Bhattacharyya, A., 6(19), *68*
Biegelsen, D. K., 39(93), 61(138), *71*, *73*

AUTHOR INDEX

Biersack, J. P., 138, 141(60), 142(60), 146(87), 151, *327, 328*
Birkhoff, R. D., 275, *333*
Bishop, H. E., 278, *334*
Blech, I. A., 59(128), *72*, 319, 320, 321, 322, 323, *336*
Bloembergen, N., 12(29, 30), 19(45, 47), *68, 69*
Blunck, O., 276, *333*
Bohlen, H., 295(333), *334*
Bohr, N., 137, 275, *326*
Bokor, J., 12(31), *68*
Bomke, H. A., 28(70), *70*
Born, M., 234(219), 245, 253(219), *331*
Bosch, J., 311(372), *335*
Bothe, W., 268, 272, 273(258), *332*
Bottoms, W. R., 77(14), *116*
Bresnock, F. J., 80(20), 81(20), *116*
Bretscher, E., 295(333), *334*
Brewer, G. R., 267, *332*
Brewer, T. L., 267(254), *332*
Brews, J. R., 219, *331*
Brice, D. K., 140(58), 146(75, 84), 147, 148, 149, 157, *327, 328*
Briggs, J. S., 144(71), *327*
Brown, D. B., 276(294), *333*
Brown, W. L., 20(55), 21, 22(59), 25(63), 26(63), *70*
Bruning, J., 243(224), 244, 245, *331*
Bucksbaum, P. H., 12(31), *68*
Bühring, W., 270, *333*
Burggraaf, P. S., 60(130), *72*
Burkhardt, P. J., 200(186), 203(186), 220(186), *330*
Butler, A. L., 15(41), 16(41), 17(41), *69*
Byatt, W. J., 270, *333*

C

Cadoff, I., 200(186), 203(186), 220(186), *330*
Campisano, S. U., 6(15), 9(15), 18(44), *68, 69*
Capio, C. D., 95(29), *117*
Carey, P., 55(115), *72*
Carroll, A. M., 295(334), *334*
Carslaw, H. S., 135(33), 216(33), *326*
Carter, C., 47(106), *71*, 137(46), *326*
Carter, J. H., 261(248), 262(248), *332*
Castaing, R., 278(307), *334*
Celler, G. K., 6(13), 11, 12, 13(33), 14(36, 40), 20(55), 22(59), 23(61), 28(61), 33(84), 34, 38(92), 39, 44(97), 48(109), 56(118, 119, 121, 122), 57, 60(133, 136), 62, 63, *68, 69, 70, 71, 72, 73*
Celloti, G., 196(173), 197(173), *330*
Chang, C. C., 61(137), 62, *72*
Chang, T. H. P., 294(330), 295(330), 297, 301(330), *334, 335*
Chapman, R. L., 20(51, 54), *69, 70*
Charitat, G., 213, *331*
Chem, K., 55(115), *72*
Chen, H. S., 29, *70*
Cheng, J., 61(137), 62(137), *72*
Cherednichenko, D. I., 221, *331*
Cheshire, I. M., 144(67, 68), 145, *327*
Chew, N. G., 19(48, 49), 20(49), 21(56), *69, 70*
Chew, R., 59(128), *72*
Chiang, S. Y., 55(115), *72*
Child, C. D., 79, *116*
Chin, D., 204, 206(197), 207, 213, 214, 215, *331*
Chiou, H. D., 37(89), *71*
Christie, W. H., 10(23), *68*
Christel, L. A., 146(81, 82), 150(81, 82), 154(81), 155, 156, 157, *327*
Chu, W. K., 47(107), *71*
Chuang, T. J., 87(25), *116*
Chung, M. S. C., 284, *334*
Cleland, J. W., 10(24), *68*
Cline, H. E., 37(88), 38(91), *71*
Coburn, J. W., 87(24, 25), *116*
Cohen, R. L., 28, *70*
Cohen, S. A., 51(111), 52(113), *72*
Cole, G. W., 182(138), *329*
Colligan, J. C., 137(46), *326*
Collins, R. J., 6(16), *68*
Considine, P. S., 245, *332*
Consoli, T., 76(1, 2), *116*
Correra, L., 28(72), 37(87), *70, 71*
Cosslett, V. E., 272, 273, *333*
Crank, J., 135(31), 216(31), *326*
Csepregi, L., 23, *70*
Cullis, A. G., 11(25), 18(44), 19, 20, 21(56), 66(147), *68, 69, 70, 73*
Cuomo, J. J., 80(20), 81(20), *116*
Cuthbert, J. D., 260(246), 264(246), *332*

D

Daly, J. A., 15(41), 16(41), 17(41), *69*
D'Arcy, J. L., 121(7), 122(7), 123(7), 285(7), *325*

AUTHOR INDEX

Das, S. C., 10(24), *68*
Davies, J. A., 144(66), 145(66), *327*
Davis, R. E., 223(213), *331*
Deal, B. E., 126(21), 199, *326*, *330*
Dearnally, G., 144(67, 68), 145(68), *327*
de Forest, W. S., 258(244), *332*
de Grefte, H. A. M., 156(108), 157(108), 162(108), *328*
Delfino, M., 58(124), *72*
Deline, V. R., 20(52), 24(62), 25(62), 44(100), 45(100), 51(100), 52(100, 113), 54(100), *69, 70, 71, 72*
Delne, V. R., 51(111), *72*
de Man, H., 170(126, 127), *329*
Deschamps, P., 278(307), *334*
Diehl, H. T., 44(99), 46(99), *71*
Dill, F. H., 258(242, 243), 259, 260(243), 261(243, 247), 262, 263(242, 243), 264, 265, 314, *332*
Dirks, A. G., 323(385), *336*
Dobson, P. S., 186(158, 160), *330*
Doerschel, C. J., 37(90), *71*
Doherty, C. J., 14(40), 56(119), *69, 72*
Donin, V. I., 4(3), *67*
Donnelly, V. M., 87(23), *116*
Doremus, R. H., 213(205), *331*
Downey, D. F., 33(82), *71*, 77(14), *116*
Durbroeucq, G. M., 245, *332*
Dutton, R. W., 126(19, 20, 21, 24), 128, 132, 133, 134, 135, 185(19), 187(166), 188, 197(19, 180), 198(180), 199(180), 200(184, 195, 196), 201, 202, 203, 204, 205(199), 206, (197, 199), 207, *326, 330, 331*

E

Eastham, D. A., 151(104), *328*
Eby, R. E., 10(24), *68*
Eernisse, E. P., 213, *331*
Eggartner, E., 278(303), *304*
Ehara, K., 113(35), 114, 115, *117*
Eisen, F. H., 145, *327*
Enomoto, T., 76(5), *116*
Eriksson, L., 144(66), 145(66), *327*
Evans, C. A., Jr., 20(52), 24(62), 25(62), *69, 70*
Evans, R. D., 278(304), *334*
Everhart, T. E., 270(276), 276, 292, 303, *333, 334, 335*
Ewald, H., 151(102), *328*

F

Fahey, P., 197(180), 198(180), 199(180), *330*
Fair, R. B., 26(64), 51, 70, 72, 168(118, 119), 174(118), 184, 185(118), 197, *328, 329, 330*
Faktor, M. M., 129, *326*
Fan, J. C. C., 20(51, 54, 55), 22(59), 60(135), *69, 70, 72*
Feder, R., 307, *335*
Feit, E. D., 302(358), *335*
Feldman, L. C., 28(69), *70*
Fennell, L. E., 61(138), *73*
Ferris, S. D., 14(40), *69*
Fichtner, W., 121, 122(7, 8), 123, 164(114, 115), 165(115), 166(115), 285, 313, *325, 328*
Field, R. L., 223(213), *331*
Finstad, T., 47(107), *71*
Firsov, O. B., 137, 139(43), 145, *326*
Flamm, D. L., 87(23), *116*
Forbes, D. H., 183(143), 184(143), *329*
Foti, G., 6(15), 9(15), 18(44), 19(50), 20(55), 22(59), *68, 69, 70*
Francis, R., 186(160), *330*
Frank, W., 183(141), 184(141, 149, 150), *329*
Fraser, D. B., 320(381), 321(381), 322(381), *336*
Frye, R. C., 56(118), *72*
Fukuma, M., 214, *331*
Fulks, R. T., 33(82), 58, 59, *71, 72*
Fuls, E. N., 122(8), 123(8), *325*
Furukawa, S., 146(73, 76, 86), 165(76), *327, 328*
Furuya, S., 302(361), *335*

G

Gale, R. P., 20(51, 54), *69, 70*
Galvin, G. J., 21(56), *70*
Galyatudinov, M. F., 4(2), *67*
Garrett, I., 129, *326*
Gat, A., 24, 25, 53(114), 54(114), 58(126), 59(126), *70, 72*
Geis, M. W., 60 (135), *72*
Geller, R., 76(2), 77(15), 99(32), *116, 117*
Gibbons, J. F., 6(17), 14(38), 20(52), 24(62), 25(62), 26(66), 140(56), 146(56, 74, 77, 80, 81, 82, 83), 150, 154, 155(81), 156(81),

157(81), 158, 159, 160, 162, *68*, *69*, *70*, *327*, *328*
Gibson, W. M., 157, 158, *328*
Giles, G. E., 6(20), *68*
Giles, M. D., 146(83), 150, 154, 158, 159, *328*
Gilmer, G. H., 20(55), 22(58, 59), *70*
Ginsberg, B. J., 187, 188, 189, *330*
Gold, R. B., 14(38), 20(52), *69*
Goldsmith, A., 6(8), *68*
Goldstein, H., 137(51), *327*
Goldstein, J. I., 125, 126, *326*
Golovchenko, J. A., 6(18), 11(28), *68*
Gonzalez, A. G., 200(195), 201, *330*
Goodman, J. W., 245(238), 247(238), *332*
Gösele, U., 183(140, 142), 184, 185, 186(161, 164), 187(146), 188(146), 189(146), 199, *329*, *330*
Goudsmit, S., 278, *333*
Grasserbauer, M., 191(172), 192(172), 194 (172), *330*
Green, M., 278, *334*
Greeneich, J. S., 272, 288, 291, 292, 293, 301, 302, 303, 304, *333*, *334*, *335*
Greenwald, A. C., 10(22), *68*
Gregory, R. B., 44(99), 46(99), *71*
Greisen, K., 273, *333*
Greschner, J., 295(333), *334*
Grobman, W., 294(330), 295, 301, *334*
Grobman, W. D., 295, 302(342), *334*
Grosse, E. H., 164(114, 115), 165(115), 166 (115), 313, *328*
Grossman, J. J., 125, *325*
Grove, A. S., 125, 199, 200(185), *325*, *330*
Gruenberg, H., 221(212), *331*
Gryzinski, M., 276(299), *333*
Guckel, H., 200(192, 193), *330*
Guerrero, E., 191, 192, 194, *330*
Günther, K., 151(101, 102), *328*

H

Haanstra, H. B., 323(384), *336*
Haberger, K., 185(156), 311(372), *329*, *335*
Haggmark, L. G., 138, 141(60), 142(60), 146 (87), 151, *327*, *328*
Hagoule, P. I., 315(375, 376), 316(375), *335*
Hall, L. A., 200(192, 193), *330*
Hall, R. B., 76(1), *116*
Hange, P. S., 258(242), 263(242), *332*

Hanley, P. R., 77(14), *116*
Hansen, S. E., 126(19), 185(19), 197(19), 200 (195), 201, 203(19), *326*, *330*
Harmatz, M., 28(70), *70*
Harper, J. M. E., 80(20), 81(20), *116*
Hashimoto, K., 183(144), 184(144), *329*
Hasimoto, H., 276(298), *333*
Hasko, S. E., 320(381), 321(381), 322(381), *336*
Hatzakis, M., 261(249), 262(249), 264(249), *332*
Haszko, S. E., 95(27), *116*
Hawkins, W. G., 39(93), *71*
Hawryluk, A. M., 267(257), 279(257), 281 (257), 282(257, 319), 283(257), 287(257), 288(257), 292(257), 293(257), 307(257), *332*, *334*
Hawryluk, R. J., 267(257), 279, 281, 282, 283, 287, 288, 292, 293, 307, *332*, *334*
Heidenreich, R. D., 302, *335*
Heinrich, K., 307, 309, 310, *335*
Heitman, W. G., 227, 228, 230, 232, *331*
Henoc, J., 273(284), 279(313), *333*, *334*
Herriot, D. L., 267, *332*
Hershel, R., 244, 261(231), 264(231), *331*
Herzog, R. F., 303, *335*
Hess, K., 6(19), *68*
Hess, L. D., 28(74), 29(74), 30(74), 31(77), *70*
Heuberger, A., 307(371), 309(371), 310, *335*
Higuchi, H., 187, 188, 189, 190, *330*
Hill, C., 3, 15, 16, 17, 32(1), *67*, *69*
Hirai, Y., 302(360), *335*
Hiramoto, T., 28(73), *70*
Hiraro, T., 155, *328*
Hirlimann, C., 12(32), *68*
Hisatsugu, T., 302(361), *335*
Ho, C. P., 199, *330*
Hodgson, R. T., 19(45, 46), 44, 45, 51, 52, 54, *69*, *71*
Hoffmann, K., 185(156), *329*
Hofker, W. K., 44, 51(101), *71*, 156, 157, 162, *328*
Holden, S. C., 33(82), *71*
Holland, O. W., 26(65), *70*
Holmes, D. K., 151(93), *328*
Holzl, J., 183(141), 184(141), *329*
Hong, J. D., 26(66), *70*
Hontzeas, S., 151(103), *328*
Hopfgarten, N., 99(32), *117*
Hopkins, H. H., 244, *331*
Horiguchi, S., 282, 284, *334*

AUTHOR INDEX

Horiike, Y., 76(6), 77(12), *116*
Hornberger, W. P., 258(242), 263(242), *332*
Hosokawa, M., 76(3), *116*
Hosokawa, N., 77(8), *116*
Hu, S. M., 169(122, 123), 173, 174, 175, 176, 179, 181, 186(159), 191(122), 195(122), 196(122), 197, 198, 199, 213(206, 207), 214(206, 207), 215(206), *329, 330, 331*
Huang, J., 204, *331*
Hutchence, D. K., 151(103), *328*

I

Ichimura, S., 269, 270, 276, *333*
Igarashi, T., 28(73), *70*
Ikegami, H., 76(3), *116*
Ikeji, H., 76(3), *116*
Ikuta, T., 276(297, 298), 279(312), *333, 334*
Inoue, K., 155(107), *328*
Ishitani, T., 151(99), *328*
Ishiwara, H., 146(73, 76, 86), 165(76), *327, 328*

J

Jackson, K. A., 5, 6(7), 39(94), *68, 71*
Jacquot, B., 99(32), *117*
Jacquot, C., 99(32), *117*
Jacobson, D. C., 21(56), 48(109), *70, 72*
Jaeger, J. C., 135(33), 216(33), *326*
Jain, R. K., 170(124, 125), 171, 172, *329*
Jellison, G. E., 6, 7, 8, 10(24), *68*
Jewett, R. E., 315, 316, 317, 319(377), *335, 336*
Johnson, N. M., 14(38, 39), 56(120), 61(138), *69, 72, 73*
Johnson, W. S., 140(56), 146(56, 74), 159(56), 160(56), 162(56), *327*
Johnston, R. L., 122(8), 123(8), *325*
Jones, F., 306, 307, 314, *335*
Joyce, B. A., 126, *326*

K

Kachurin, G. A., 4(4), *67*
Kadono, K., 77(12), *116*
Kahng, D., 119, 125(10), *325*
Kalbitzer, S., 141, *327*
Kalish, R., 51(111), 52(113), *72*
Kamins, T. I., 126(21, 22, 23), 127(27), 128(22), *326*
Kanamura, A., 295(340), 302(340), *334*
Kane, E. O., 170, *329*
Kanomata, I., 77(17), *116*
Karapiperis, L., 311, 313, *335*
Kata, T., 302(360), *335*
Kataoka, Y., 276(298), *333*
Kato, T., 295(343), 296(343), 302(355), *335*
Kaufman, H. R., 80(20), 81(20), *116*
Keigler, A. L., 56(120), *72*
Kelley, M. J., 36(86), *71*
Kemp, R. F., 100(33), *117*
Kendall, D. L., 196, *330*
Kendall, M., 162(113), *328*
Kennedy, D. P., 191(169), 220, 222, *330, 331*
Kennedy, E. F., 23(60), *70*
Kern, D. P., 294(331), *334*
Khaibullin, I. B., 4(2), *67*
Kikuchi, A., 295, 302(340), *334*
Kilby, J. S., 119, *325*
Kimerling, L. C., 11(26), 12(26), 14(36, 37, 40), 48(109), *68, 69, 72*
Kinchin, G. H., 157, *328*
King, M. C., 249(240), 250(240), *332*
Kingston, A. E., 276(301), *333*
Kintner, E. C., 244, 245, 247, 249, *331*
Kinzly, R. E., 243, 244, *331*
Kirkpatrick, A. R., 10(22), *68*
Kirscht, F. G., 37(90), *71*
Kitagawa, H., 183(144), 184, *329*
Kiuchi, M., 95(31), 96, 97, 100, 103, 104, 105, 106, 107, 108, *117*
Klimenko, A. G., 4(3), *67*
Klimenko, E. A., 4(3), *67*
Knapp, J. A., 60(134), *72*
Knoell, R. W., 45(103), 46(103), 47(103), 48(103), 49(103), *71*
Kobayashi, T., 282(318), 284(318), *334*
Koeman, N. J., 156(108), 157(108), 162(108), *328*
Koike, H., 77(17), *116*
Kokorowski, S. A., 28(74), 29(74), 30(74), *70*
Koshikawa, T., 276(298), *333*
Kotera, M., 270, 271, *333*
Kraft, R., 200, *330*
Krambeck, R. H., 200(191), *330*
Kranz, H., 311(372), *335*
Kratschmer, E., 295, 302(341), *334*
Krefting, E., 269, 270, 276, *333*

Kronenberg, S., 28(70), *70*
Kump, M., 204(197), 206(197), 207(197), *331*
Kurz, H., 12(29, 30), 19(47), *68*, *69*
Kwong, D. W., 44(98), 51(98), *71*
Kwor, R., 44(98), 51, *71*
Kyser, D. F., 278(306) 279, 280, 281, 282, 283, 288, 294, 298, 299, 300, 301, 304, 305, 311, *334*, *335*

L

La Chapelle, T. J., 181(143), 183(143), 184 (143), *329*
Lacombat, M., 245, *332*
Lam, H. W., 61(139), 62(139), *73*
Lancaster, L., 36(86), *71*
Landau, L. D., 137(50), 276, 284(295), *327*, *333*
Landi, E., 55(117), *72*
Lanford, W. A., 95(28), *116*
Langer, P. H., 125, 126, *326*
Larrabee, R. D., 182, *329*
Lasky, J. B., 44, 45(103), 46, 47, 48, 49, 53, *71*
Lau, S. S., 33(83), 36(83), 37(83), 38(83), *71*
Leamy, H. J., 6(13), 9(21), 10, 14(40), 20(55), 22(58, 59), 25(63), 26(63, 67), 27(67), 56 (118, 119), 60, 61(137), 62, *68*, *69*, *70*, *72*, *73*, 323(385), *336*
Leary, P. A., 80(20), 81(20), *116*
Lee, H. G., 200(196), 204, 205, 206, *330*, *331*
Lehmann, C., 137(53), *327*
Leisegang, S., 276, *333*
Leistiko, O., 200(185), *330*
Lemons, R. A., 61(137), 62(137), *72*
Lenz, F., 268, 272, *332*
Lepselter, M. P., 122(9), 307(9), 308, *325*
Levatter, J. I., 10(23), *68*
Lever, R. F., 191(170), *330*
Levi, L., 234(220), 245(220), *331*
Levinstein, H. J., 95(27), *116*, 122(9), 307(9), 308(9), *325*
Lewis, H. W., 268, 273, *332*
Lietoila, A., 6(17), 14(38), 26(66), *68*, *69*, *70*
Lifshitz, E. M., 137(50), 154(106), *327*, *328*
Lin, A. M., 187(166), 188, *330*
Lin, B. J., 226(214), 231, 233, 234(214), 235, 236, 238, 242, 252, 253, 255, *331*, *332*
Lin, L. H., 302, *335*
Lindhard, J., 137, 139(39, 40, 41), 144(39), *326*

Lischner, D. J., 33(83, 84), 34, 36(83, 86), 37 (83), 38(83, 92), 39(92, 94), 44(97), 48 (109), 51, 55(112), 60(136), 62(143, 144, 145), 63(145), *71*, *72*, *73*
Littmark, U., 142, *327*
Liu, J., 47(106, 107), 51(110), *71*, *72*
Liu, J. M., 12(29, 30), 19(47), *68*, *69*
Liu, P. L., 19(45), *69*
Liu, T. M., 47(105), *71*
Logan, J. S., 87(22), *116*
Lowndes, D. H., 6, 8, 10(24), *68*
Lowney, J. R., 174, 175, 177, 182, *329*
Lownsbury, J. B., 264(251), *332*
Lue, J. T., 28(71), *70*
Luthy, W., 6(14), *68*
Lux, R., 28(70), *70*

M

Ma, S. K. S., 302(356), *335*
Maby, E. W., 56(120), *72*
McCormack, D. V., 286, *334*
Macfarlane, G. G., 6(11), 8(11), *68*
Machida, Y., 302(361), *335*
McLean, T. P., 6(11), 8(11), *68*
Macrander, A. T., 146(88), *328*
Mader, S. M., 44(100), 45(100), 51(100), 52 (100), 54(100), *71*
Maes, H., 137(49), *327*
Magee, T. J., 20(52), 24(62), 25(62), 26(66), *69*, *70*
Mahrer, H., 181(141), 183(141), 184(141), *329*
Maldonado, J. R., 307(370), *335*
Manz, R. C., 125(10), *325*
Martinez, A., 213, *331*
Masetti, G., 200(189), *330*
Massey, H. S. W., 273, *333*
Maszara, W., 47(106), *71*
Mathur, V. S., 273, *333*
Matsui, S., 77(13), *116*
Matsukawa, T., 279(314), *334*
Matsumoto, H., 214, *331*
Matsumura, H., 146(76), 165(76), *327*
Matsuo, S., 77(9, 10, 11, 18), 78, 80(21), 81 (21), 82(21), 84(21), 86, 87, 88, 90(26), 91, 92, 93, 94, 95(31), 96, 97, 100, 103, 104, 105), 106, 107, 108, 109(34), 110(34), 111(34), 112(34), 113(35), 114(35), 115 (35, 36), *116*, *117*

Matsuzaki, R., 77(8), *116*
Mauer, J. L., 87(22), *116*
Maule, D. J., 126(15), *326*
Maurice, F., 273(284), 279(313), *333, 334*
Mayer, J. W., 21(56), 23(60), 59(129), *70, 72,* 144, 145(66), *327*
Meek, R. L., 66(147), *73*
Meindl, J. D., *326*
Meiron, J., 244, *331*
Meister, H., 268, *332*
Melliar-Smith, C. M., 302(358), *335*
Mertens, R. P., 170(126, 127), *329*
Merz, J. L., 26(68), *70*
Meyberg, W., 181(142), 183(142), 184(142), *329*
Miller, G. L., 14(36), 48(109), *69, 72*
Minakata, M., 114(36), *117*
Minnucci, J. A., 10(22), *68*
Miyagawa, S., 151(105), *328*
Miyagawa, Y., 151(105), *328*
Mizuo, S., 187, 188, 189, 190, *330*
Mizuta, M., 26(68), *70*
Modine, F. A., 6, 7, *68*
Mogab, C. J., 314(374), *335*
Moliere, G., 141, 268, 272, 273, *327, 332*
Moll, J. L., 213(206, 207), 214(206, 207), 215(206), *331*
Møller, C., 277, *334*
Molzen, W. W., 295, 302(342), *334*
Moore, G. E., 120, *325*
Morehead, F. F., 44(100), 45(100), 51(100), 52(100), 54(100), *71*, 176(134), 181, 182, 183(142), 184, 185(146), 187(146), 188(146), 189(146), *329*
Morehead, F. G., 176(135), 194(135), 195(135), 196(135), *329*
Morgan, T. N., 171(129), *329*
Morimoto, T., 113(35), *117*
Moskovitz, I., 184(147, 148), 186, 189, 190, *329*
Mott, N. F., 269, 273, *333*
Moyer, M. D., 61(138), *73*
Mur, G., 231, *331*
Muramoto, S., 113(35), *117*
Murarka, S., 59, *72*
Murata, K., 151(99), 207(275), 271(275), 279, 280, 281, 282, 283, 288, 294, 295, 296, 297(344), 302(360), 304(365), 311, *328, 333, 334, 335*
Murley, P. C., 191(169), *330*

Myers, E., 47(107), *71*
Mylroie, S., 140(56), 146(56, 77, 81, 82), 150(81, 82), 155(81), 156(81), 157(81), 159(56), 160(56), 162(56), *327*

N

Nagami, K., 270(275), 271(275), 295(343, 344), 296(343, 344), 297(344), 302(360), *333, 335*
Nagel, L. W., 121(7), 122(7), 123(7), 285(7), *325*
Nakamura, H., 197(176), *330*
Nakata, H., 295(343), 296(343), 302, *335*
Nakayama, N., 302, *335*
Nakone, Y., 295(340), 302(340), *334*
Namba, S., 77(13), *116*
Narashimhan, M. A., 264(251), *332*
Narasimhan, M. A., 261(248), 262(248), *332*
Narayan, J., 10(23), 10(24), 12(34), 13(35), 26(65), 31(77), 47(108), *68, 69, 70, 72*
Neerhof, F. L., 231, *331*
Negri, P., 200(189), *330*
Negrini, P., 55(117), *72*, 196(173), 197(173), *330*
Nehmiz, P., 295(333), *334*
Neureuther, A. R., 245, 250, 251, 252, 258(243), 259(243), 260(243), 261(243, 247), 262(243, 250), 263, 264(243), 265(243), 304, 314(243), 315(375), 316(375), *332, 335*
Ng, K. K., 56, 57(121, 122, 123), 62(140, 141), *72, 73*
Nicollian, E., 219, *331*
Nielsen, K. O., 143, *327*
Nielssen, V., 137(39), 139(39, 41), 144(39), *326*
Niewenhuizen, J. M., 323(384), *336*
Nigam, B. P., 273, *333*
Nigarn, B. P., 269, *333*
Nilson, J. A., 10(24), *68*
Nishi, Y., 126(14), *325*
Nobili, D., 196(173), 197(173), *330*
Nomura, E., 295, 296, 297, *335*
North, J. C., 157, 158, *328*
Nosker, R. W., 274, 275, 292, *333*
Noyce, R. N., 119, 325
Nyssonen, D., 245, 249(232), *332*

O

O'Brien, R. R., 220, 222, *331*
Oehrlein, G. S., 51(111), 52(113), *72*
Oen, O. S., 151, *328*
Oesthoek, D. P., 156(108), 157(108), 162(108), *328*
Oetzmann, H., 141, *327*
Offner, A., 244, *331*
Ogilvie, R. E., 276(294), *333*
Oh, S. Y., 200(196), 204, 213(206, 207, 208), 214(206, 207, 208), 215(206), *330*, *331*
Oi, N., 126(14), *325*
Okamoto, Y., 77(16), *116*
Okazaki, N. 295(340), 302(340), *334*
Oldham, W. G., 47(105), *71*
Olson, G. L., 28, 29, 30, 31, *70*
Ono, T., 81(21), 82, 84, 92, 94, 109(34), 110, 111, 112, *116*, *117*
Ormond, R., 20(52), *69*
O'Toole, M. M., 245, 250, 251, 252, *332*
Ozguz, V., 47(106), *71*

P

Pai, C. S., 33(83), 36(83), 37(83), 38(83), *71*
Palm, B. J., 20(54), *69*, *70*
Pan, E. S., 191(170), *330*
Pan, J.-D. T., 59(128), *72*
Parasczak, J., 306, 307, 314, *335*
Parent, G. B., 245(239), *332*
Parikh, M., 294, 295, 298, 299, 300, 301, 302 (356), *334*, *335*
Parillo, L. C., 223, *331*
Pascal, B., 279(312), *334*
Pathak, A. P., 144(69, 70, 71), *327*
Paulson, W. M., 44(99), 46(99), *71*
Payne, R. S., 223(213), *331*
Pearce, C. W., 66, *73*
Pease, R. S., 157, *328*
Pedula, L., 28(72), *70*
Peercy, P. S., 21(56), *70*
Peng, J., 20(52), 24(62), 25(62), 26(66), *69*, *70*
Penumalli, B. R., 121(7), 122(7), 123(7), 178 (136), 185(136), 206, 207, 209, 285(7), *325*, *329*, *331*
Peters, D., 55(115), *72*
Petersen, W. P., 121(7), 122(7), 123(7), 164, 165, 166, 285(7), 313, *325*, *328*
Phang, J. C. H., 294(332), *334*

Philibert, J., 278(307), *334*
Phillipp, F., 26(67), 27(67), *70*
Picraux, S. T., 60(134), *72*
Pinizzotto, R. F., 61(139), 62(139), *73*
Pitaevskii, L. P., 154(106), *328*
Plummer, J. D., 126(21), 199, *326*, *330*
Poate, J. M., 6(13), 11(26), 12(26, 33), 13(33), 18(44), 19(49, 50), 20(49), 21(56), 23(61), 25(63), 26(63), 28(61), 59(129), *68*, *69*, *70*, *72*, 144(67, 68), 145(68), *327*
Poli, G., 45(103), 46(103), 47(103), 48(103), 49(103), *71*
Pongratz, S., 307(371), 309(371), 310(371), *335*
Pötzl, H., 191(172), 192(172), 194(172), *330*
Poulsen, M. E., 307(370), *335*
Povilonis, E. I., 56(118, 121, 122), 57(121, 122, 123), *72*
Powell, R. A., 58(125), 59(128), *72*
Pridachin, N. B., 4(4), *67*
Prince, J. L., 200(190), *330*
Prussin, S., 186, *330*
Pyle, R., 301(348), 304, 305, *335*

Q

Quarrington, J. E., 6(11), 8(11), *68*
Quintana, G., 95(27), *116*

R

Ramani, R., 58(126), 59(126), *72*
Rand, M. J., 95(28), *116*
Redfield, D., 6(12), 8(12), *68*
Regolini, J. L., 26(66), *70*
Reif, R., 126, 127, 128, 132, 133, 134, 135, *326*
Reifsteck, T. A., 58(124), *72*
Reimer, L., 269, 270, 276, 279, *333*, *334*
Reuter, J. L., 191, *330*
Reutlinger G. W., 223(213), *331*
Reynolds, J L., 318, (378), *336*
Rhodes, A., 125(13), *325*
Rice, W., 125, 135, *325*, *326*
Rimini, E., 6(15), 9(15), *68*
Roberts, V., 6(11), 8(11), *68*
Robinson, D. A. H., 14(36), *69*
Robinson, J. E., 151(100), *328*
Robinson, McD., 38(92), 39(92, 94), 44(97), 48(109), 60(136), 62(140, 143, 144, 145), 63(145), *71*, *72*, *73*

Robinson, M. T., 151, *328*
Rodgers, J. W., 25(63), 26(63), *70*
Rose, M. E., 268(260), 273(260), *332*
Rossi, B., 273, *333*
Roth, J. A., 28(74), 29(74), 30(74), 31(77), *70*
Rothe, D. E., 10(23), *68*
Rousseau, D., 25(63), 26(63), *70*
Rozgonyi, G. A., 6(13), 12(33), 13(33), 23(61), 25(63), 26(63), 28(61), 47(106, 107), *68, 69, 70, 71*
Rubinstein, L. I., 208(202), *331*
Ruge, I., 137(48), *327*
Runge, H., 165, 167, 168, *328*
Runyan, W. R., 62(142), *73*
Russo, C. J., 33(82), *71*
Ryssel, H., 137(48), 185(156), 311(372), *327, 329, 335*

S

Sadana, D. K., 47(106, 107), 53, 54, *71, 72*
Sah, C. T., 125(13), 200(185), *325, 330*
Saitoh, K., 295(337), *334*
Saitou, N., 279, *334*
Sakakibara, Y., 282(318), 284(318), *334*
Sakamoto, Y., 76(4), *116*
Sakuto, N., 77(17), *116*
Salmi, S., 55(117), *72*
Sanders, J. B., 140(55), 146(78), 147, *327*
Sandku, J. S., 191, *330*
Saraswat, K. C., 126(21, 22, 23), 127(27), 128(22), *326*
Sarkes, T. K., 221(212), *331*
Sato, K., 126(14), *325*
Saunders, T. E., 307(370), *335*
Saunderson, J. L., 278, *333*
Scharff, M., 137(39), 139(39, 40, 41), 144(39), *326*
Schiott, H. E., 137(40), 139(40), 146, *326, 328*
Schmidt, H., 151(102), *328*
Schmidt, S., 173(131), 179, 181, *329*
Schneider, D. O., 286, *334*
Schreiber, D. E., 294(328), 295, 300(347), *334, 335*
Schultz, J. C., 6(16), *68*
Schuster, B., 183(141), 184(141), *329*
Schwartz, B., 45(103), 46(103), 47(103), 48(103), 49(103), *71*
Schwartz, G. S., 87(22), *116*

Schwenker, R. O., 191(170), *329*
Schwettmann, F. N., 196, 200(190), *330*
Scott, W. T., 274(289), *333*
Sedgwick, T. O., 32(78), 51(111), 52, *71, 72*
Seeger, A., 184(150), *329*
Seidel, T. E., 32(79), 33(83), 36(83), 37, 38, 42(95), 43, 44(79), 45(103), 46, 47(103), 48, 49, 51, 55(112), 65(146), 66(147), *71, 72, 73*
Sellen, J. M., 100(33), *117*
Seltzer, S. M., 275(293), *333*
Sewell, H., 295(335), *334*
Shank, C. V., 12(32), *68*
Shatas, A. G., 34(85), *71*
Shatas, S., 51(111), 52(113), 53(114), 54(114), 58, 59, *72*
Shatzkes, M., 200(186), 203(186), 220(186), *330*
Shaw, C. H., 301, *335*
Shaw, D. W., 129, *326*
Shaw, J. M., 258(242), 261(249), 262(249), 263(242), 264(249), *332*
Shellnutt, J. A., 25(63), 26(63), *70*
Sheng, N. H., 26(68), *70*
Sheng, T. T., 12(33), 13(33), 23(61), 25(63), 26(63), 28(61), 44(97), 62(144), *68, 69, 70, 71, 73*
Shepherd, W. H., 126, *326*
Shibagaki, M., 76(6), 77(12), *116*
Shimizu, R., 151(99), 269(270), 270(270, 276), 276, 278, 279, 280, *328, 333, 334*
Shtyrkov, E. I., 4(2), *67*
Sigmon, T. W., 23(60), 26(66), *70*
Sigmund, D., 158, *328*
Sigmund, P., 137(52), 140(55), 146(78), 147, *327*
Simons, A. L., 6(18), 11(28), *68*
Sinha, A. K., 95(27), *116*
Siregar, M. R. T., 6(14), *68*
Sirtl, E., 186, *330*
Skelly, G., 126, *326*
Slusher, R. E., 6(18), 11(28), *68*
Smirnov, L. S., 4(4), *67*
Smith, B., 140(57), 146(57), *327*
Smith, D. H., 146(80), 150, *327*
Smith, H. I., 267(257), 279(257), 281(257), 282(257, 319), 283(257), 287(257), 288(257), 292(257), 293(257), 307(257, 367, 368), *332, 334, 335*
Smith, L. P., 268(260), 273(260), *332*

Smith, P. R., 11(28), *68*
Smith, R. E., 95(29), *117*
Smith, T. E., 95(27), *116*
Solmi, S., 196(173), 197, 200(189), *330*
Sommerfeld, A., 231, *331*
Sondhi, G., 200(189), *330*
Sonobe, Y., 76(5), *116*
Spaepen, F., 29(76), *70*
Spears, D. L., 307, *335*
Spencer, L. V., 273, *333*
Speth, A. J., 294(330), 295(330), 301(330), *334*
Spiller, E., 307, *335*
Stacy, W. T., 58(125), *72*
Stegun, I. A., 136(34), *326*
Stevie, F. A., 45(103), 46(103), 47(103), 48(103), 49(103), *71*
Stingeder, G., 191(172), 192(172), 194(172), *330*
Stolwijk, N. A., 183(141), 184, *329*
Stolwijk, W., 181(142), 183(142), 184(142), *329*
Streetman, B. G., 6(19), 47(104), *68, 71*
Stuart, A., 162(113), *328*
Sugiyama, N., 295(337, 338), *334*
Summa, G. M., 80(20), 81(20), *116*
Sung, C., 319, 320, *336*
Surko, C. M., 6(18), 11(28), *68*
Suzuki, M., 282(318), 284(318), *334*
Sze, S. M., 56(118, 122), 57(122) *72*, 324(386), *336*

T

Tai, K. L., 284, *334*
Takahashi, C., 109(34), 110(34), 111(34), 112(34), *117*
Takayama, K., 76(3), *116*
Takayanagi, S., 155(107), *328*
Takehara, Y., 77(9), *116*
Tamagawa, H., 77(16), *116*
Tan, T. Y., 19(46), *69*, 183(140), 184, 185, 186 (161, 164), 187, 188, 189, 199, 211(157), *329, 330*
Tanaka, S., 76(3), *116*
Taniguchi, K., 184(153), *329*
Tanoue, H., 146(85), 219(85), *328*
Tasch, A. F., 61(139), 62(139), *73*
Tauan, H. C., 61(138), *73*
Taylor, G. W., 57(123), *72*

Terunuma, Y., 197(176), *330*
Thibault, L. R., 122(8), 123(8), *325*
Thomas, C. O., 125, *325*
Thomas, R. N., 272, 273, *333*
Thompson, L. F., 302(357, 358), *335*
Thompson, M. O., 21(56), *70*
Thomsen, P. V., 137(41), 139(41), *326*
Thornton, P. R., 267, *332*
Thridandam, C., 59(128), *72*
Tielert, R., 185(155), 191(172), 192(172), 194 (172), *329, 330*
Tietz, T., 270, *333*
Tigreat, P., 245, 249(235), *332*
Ting, C. H., 300(347), 301, 304(365), 278(306) *334, 335*
Tokiguchi, K., 77(17), *116*
Torrens, I. A., 137(54), 139(54), *327*
Torrens, I. M., 151(97), *328*
Towner, J. J., 60(131), *72*
Trimble, L. E., 38(92), 39(92, 94), 56(119, 121, 122), 57(121, 122, 123), 62(140, 141, 145), 63(145), *71, 72, 73*
Tsai, J. C. C., 26(64), *70*, 197, *330*
Tsai, M. Y., 47(104), *71*, 176(135), 194, 195, 196, *329*
Tsaur, B.-Y., 44(98), 51(98), 55(116), 60(135), *71, 72*
Tsu, R., 19(46), *69*
Tsubai, K., 295(340), 302(340), *334*
Tsuchimoto, T., 95(30), *117*
Tsurushima, T., 146(85), 219(85), *328*
Tu, K. N., 59(129), *72*
Tuck, B., 135(32), *326*
Turnbull, D., 29(76), *70*
Tuttle, J. A., 258(243), 259(243), 260(243), 261 (243), 262(243), 263(243), 264(243), 265 (243), 314(243), *332*

V

van der Berg, P. M., 227, 228, 230, 232, *331*
Vander Blas, H. A., 322, *336*
van der Leeden, G. A., 10(23), *68*
Vandervorst, W., 137(49), *327*
van der Ziel, J. P., 6(13), *68*
van de Ven, E. P., 60(131), *72*
van Duzer, T., 288, 291, 292, 293, 303, 315 (375), 316(375), *334, 335*

van Overstraeten, R. J., 137(49), 170(124, 125, 126, 127), 171, *327*, *329*
Van Vechten, J. A., 11(27), *68*
Vanzi, M., 127(26), *326*
Venkatesan, T. N. C., 6(18), 11(28), *68*
Vettiger, P., 295(333), *334*
von Allmen, M., 6(14), *68*
Voshino, H., 282(318), 284(318), *334*
Vratny, F., 307(370), *335*
Vriens, L., 276(300), *333*

W

Wagner, S., 200(188), *330*
Walker, E. J., 258(243), 259(243), 260(243), 261(243), 262(243), 263(243), 264(243), 265(243), 314(243), *332*
Ward, W., 302(356), *335*
Watakabe, Y., 302(355), *335*
Watrasiervicz, B. M., 244, *331*
Watson, H. A., 122(9), 307(9), 308(9), *325*
Watts, R. K., 122(8), 123, 267(252), *325*, *332*
Weakliem, H. A., 6(12), 8(12), *68*
Weaver, J. C., 126(15), *326*
Webber, H. C., 11(25), 19(48, 49), 20(49), *68*, *69*
Welliver, L., 204, *331*
Wentzel, G., 268, 272, *332*
West, K. W., 28(69), *70*
Westbrook, R. D., 10(24), *68*
Weymouth, J. W., 268, *332*
White, C. W., 12(13), 17, 18(42, 43), *69*
Wilcox, W. P., 183(143), 184, *329*
Williams, E. J., 268, *332*
Williams, J. S., 6(13), 25, 26(63), 28(69), *68*, *70*
Williams, P., 24(62), 25(62), *70*
Willoughby, A. F. W., 165(121), 166(121), 168(120, 121), 178, *328*
Wilson, L. O., 208, 209, 212, 213, *331*

Wilson, P. R., 200(187), *330*
Wilson, S. R., 17(43), 18(43), 44, 46, *69*, *71*
Wilson, W. D., 141, 142, *327*
Winterbon, K. B., 140, 146(79), 147, 148, 162, 164, *327*, *328*
Winters, H. F., 87(24, 25), *116*
Witt, A. F., 5, 6(7), *68*
Wittels, N. D., 267(253), 295, 301, *332*, *334*
Wolf, C. A., 311(373), 313(373), *335*
Wolf, E. D., 234(219), 244, 245, 253(219), 311(373), 313(373), *331*, *335*
Wong, W., 262(250), 263, *332*
Wood, R. F., 6(20), 10(23, 24), *68*
Wortman, J. J., 47(106, 107), 51(110), *71*, *72*

Y

Yaegashi, Y., 155(107), *328*
Yamaguchi, E., 115(36), *117*
Yamamoto, S., 302(361), *335*
Yamato, T., 77(13), *116*
Yen, R., 12(29, 32), 19, *68*, *69*
Yoshida, M., 183(144), 184(144), 197, *329*, *330*
Young, F. W., Jr., 17(43), 18(43), *69*
Young, R. T., 10(23), 12(34), *68*, *69*
Youngman, C. I., 295(336), *334*

Z

Zacharias, A., 307(370), *335*
Zaleckas, V. J., 66, *73*
Zaripov, M. M., 4(2), *67*
Zaromb, S., 178(137), *329*
Zeiger, H. J., 20(51, 54), *69*, *70*
Ziegler, J. F., 142, 182(138), *327*, *329*
Zielinski, L. B., 87(22), *116*

Subject Index

A

Absorbtance, sublayer, 261
Absorption coefficient
  energy density threshold as function of, 6–7, 9
  as function of temperature, 6, 8
  versus photon energy, 6–7
Absorption constant, resist, 259
Activity coefficients, 173–174
Adiabatic annealing, 2–23
  advantages, 22
  amorphization of Si and other rapid regrowth phenomena, 18–22
    events during explosive crystallization, 21
    explosive crystallization, 20
    inner and outer radii of amorphous ring, 19
    irradiated Si layers cross sections, 19–20
  average adiabatic temperature rise, 6
  depth profile of deposited energy, 8–9
  disadvantages, 22–23
  electron–hole pairs generated by one electron, 9–10
  energy density threshold for surface melting as function of absorption coefficient, 6–7, 9
  equipment, 10–11
  microstructure and dopant incorporation, 11–18
    As depth profiles, 11–12
    Bi segregation coefficients as function of liquid–solid interface velocity, 18
    defect-free material, laser melting, 12–13
    depth profiles, 17–18
    detection of molten layer, 11–12
    DLTS spectrum and capacitance versus temperature, 14
    effect of laser energy on redistribution of As implant, 15
    emitter base region fabrication, 16–17
    overlapped annealing, 12–13
  optical absorption coefficient
    as function of temperature, 6, 8
    versus photon energy, 6–7
  output parameters of commercial high-power lasers, 5
  photon and electron absorption, 4, 6–10
  physical properties of silicon, 5
Al etching, 85–86
  ion extraction voltage and etch rates, 86–87
Al_2O_3 deposition, characteristics, 112–113
Aluminum contact sintering, 59–60
Amorphization, silicon, 18–22
  inner and outer radii of amorphous ring following laser pulse, 19
Annealing, 325, *see also* Adiabatic annealing; Isothermal rapid annealing; Thermal flux annealing
  furnace laser, arsenic concentration profiles, 24–25
  overlapped, 13
  solid-phase laser, arsenic concentration profiles, 24–25
Arrhenius reaction rate formula, 88–89
Arsenic diffusion, 48–53, 176, 191–196
  α as function of electrically active concentration, 192–193
  carrier concentration
    at annealing temperature, 195
    as function of time, 196
  clustering coefficient, 194

Arsenic diffusion (*continued*)
  cluster model comparison, 194
  electrical neutrality, 192
  electrical solubility as function of temperature, 192
  enhanced diffusion, 51
  fit of oxidation-enhanced diffusion data, 187
  flux, 177
  following RTA using one-sided tungsten–halogen furnace, 49–50
  kinetic equation of clustering, 195
  law of mass action, 191
  rapid thermal annealing using arc lamp, 51–52
  reaction for clustering/declustering, 191
  sheet resistance, 49–50
  total arsenic concentration, 191, 195

B

Backscattering contribution, 292
Bessel function, first kind of order one, 253
Bipolar transistor, emitter–base region fabrication, 16–17
Blech's model, assumptions, 319
Boron
  fit of oxidation-enhanced diffusion data, 188
  impurity profile under zero surface concentration boundary condition, 180–181
Boron diffusion, 44–51
  dislocations, 47–48
  electrically active hole-concentration versus depth, 44, 47
  enhanced diffusion, 51
  profile
    arc lamp pulses annealing, 44–45
    As-implanted, 208–209
    calculated numerically and analytically, 205
    graphite rapid thermal annealing, 44, 46
    implantation through thin SiO_2 layer, 46, 48–49
  n^+p-junction region, 223–224
  oxidized, 208–209
  removal of precipitation effects, 53–54
  under field oxide, 208–209

shallow junction formation, 45–46
sheet resistance, 49–50
Bulk gas phase, 126–127

C

CASPER, *see* Computer-aided semiconductor processing and epitaxial redistribution
Cell removal algorithm, 314–315
CF_4 plasma, 76
CMOS, *see* Complementary metal–oxide–semiconductor technology
Complementary metal–oxide–semiconductor technology
  inverter, 60
  Twin-Tub inverter, 223–225
Computer-aided semiconductor processing and epitaxial redistribution, 126
Contact and proximity printing, 226–241
  configuration, N-1 layers and semi-infinite screen, 227–228
  diffraction pattern comparison, 235–237
  evolution of diffracted wave from aperture plane, 231, 233
  exact Kirchoff integral for scalar wave function, 234
  Fresnel approximation, 235
  Helmholtz equation, 234
  intensity distribution of proximity image of contact hole, 237–241
  near-field diffraction of infinite slit, 231
  normalized power flow density, 230–232
  orthogonal tangential magnetic field, 226–228
  power flow density, 229–230
  Poynting vector, 231
  spectral parameters, 237
  transverse electric polarization, 226–228
Continuity equations, charged particles, 169
Creeping-flow equation, 213
Cross-coefficient function, 249
Crystallization, explosive, 20
  sequence of events, 21

D

Deep-level transient spectroscopy spectrum, versus temperature, 14

SUBJECT INDEX 351

Defect, formation and annealing, 325, *see also* Annealing
Delta function
  geometry for computing scattering response, 291
  point source, energy density profiles, 285, 288
Density of states functions, versus electron energy, 170
Deposition, 318–324
  amount of material arriving per unit area, 322
  Blech's model, assumptions, 319
  dependence of step coverage profiles on angle, 322
  geometry and definition of terms, 319–321
  growth rate, 319
  vapor-deposited film growth, 323
  vapor stream flux, 319
Diffusion, 168–225
  analytical solutions, 216–223
    constant surface concentration, 216–217
    diffusion from thick film, 218
    impurity redistribution during oxidation, 219–220
    infinite composite medium, 218–219
    instantaneous source, 217, 222–223
    surface evaporation condition, 217–218
    two-dimensional diffusion of impurities near mask edge, 220–221
  anomalies, 191–199
    arsenic, *see* Arsenic diffusion
    phosphorus, 196–199
  anomalous diffusion phenomena, 185–191
    concentration of self-interstitials, 190
    fit of As oxidation-enhanced diffusion data, 187
    fit of B oxidation-enhanced diffusion data, 188
    fit of P oxidation-enhanced diffusion data, 188
    fit of Sb oxidation-reduced diffusion data, 186, 189
    local point defect equilibrium, 186
    normalized diffusion as function of self-interstitial supersaturation ratio, 186–187
    normalized Sb diffusivity during oxidation, 189–191

  oxidation and segregation, 199–216
    boron profile calculated numerically and analytically, 205
    boron profile under field oxide, 208–209
    creeping-flow equation, 213
    discretization method of $Si–SiO_2$ region, 203
    equation of impurity conservation, 201
    equilibrium segregation coefficient, 201
    final shape for local oxide, 216
    flow rate induced by stress, 212–213
    flux across interface, 201–202
    gas–oxide interface, 210
    growth rate of SiO_2 on top of Si, 199
    hydrodynamic equation, 213
    implicit approximations of integrand, 202
    interface boundary condition, 205
    isolation oxide growth, simulated results, 214
    linear rate constant, 199–200
    model configuration and definition of variables, 209–210
    oxide layer with initial wall angle of 60°, 211–212
    percentage error versus diffusivity ratio, 203–204
    $Si–SiO_2$ interface, 210
    simulation regions and coordinate transformation, 207
    stress distribution along silicon interface, 214–215
    Stanford University Process Analysis program result for metal–oxide–semiconductor process simulation, 206–207
    Tan–Gösele model, 211–213
    two-dimensional Stefan problem, 210–211
    vacancy concentrations, 200
    velocity field distribution, 214–215
    Twin-Tub complementary metal–oxide–semiconductor inverter, 223–225
    two-dimensional numerical example, 223–225
Diffusivity, 183
Divergent magnetic field plasma extraction, 97–102
  electric field, 98
  gas pressure and negative potential, 99, 101

Divergent magnetic field plasma extraction (*continued*)
  intensity distribution, 97
  magnetic moment, 98
  negative potential generation as function of distance, 99–100
  plasma potential variation along plasma stream, 100–101
  plasma stress, 99
  potential, 98
Dopant, *see* Adiabatic annealing, microstructure and dopant incorporation

E

Elastic scattering, probability, 271–272, 288
Electric field, concentration as function of depth, 172–173
Electron-beam lithography, 266–307
  deposited energy distribution, 267
  development simulations, 302–307
    degraded average number molecular weight, 303
    lateral distribution of energy deposited, 305–307
    number of scission events, 303
    pattern layout of connected and separated rectangles, 306–307
    resist etch rate versus absorbed dose, 304
    thickness of resist remaining versus development time, 303–304
  electron–electron scattering and energy loss, 275–278
  energy deposition function calculation, *see* Energy, deposition functions
  nuclear scattering, *see* Nuclear scattering
  process schematic, 266–267
  proximity functions and effect correction techniques, 294–302
    backward-scattering component, exposure intensity distribution function, 299–300
    deposited energy density variation, 296–297
    dose compensation curve, two Gaussian model, 301–302
    energy deposition due to backward-scattered electrons, 299–300
    exposure intensity distribution function, 297, 299

  forward-scattering, two Gaussian approximation, 301
  intrashape and intershape proximity effects, 294
  isoenergy contours, 296
  pattern shape and configuration layout, 295–296
  radial distribution of energy distribution, 298–299
  two-Gaussian expression, 299–300
Electron cyclotron resonance, 76; *see also* Plasma deposition; Reactive ion-base etching
Electron–electron scattering, 275–278
  comparison of differential cross sections for inelastic scattering, 277–278
  energy straggling, 276
  Gryzinski's core-electron cross section, 276
  secondary electron production, 278
Energy
  deposition functions, 278–294
    analytical methods, 291–294
    Monte Carlo method, *see* Monte Carlo method
  distribution function, 291
  straggling, 276
Etching, 314–320
  cell removal algorithm, 314–315
  effective refractive index of resist, 315
  rate data, 264, 266
    versus absorbed dose, 304
  ray-tracing algorithm, 315–316
  string algorithm, 316–317
  string sequence for isotropic and anisotropic etching, 320
Exposure intensity distribution function, 297, 299
  backward-scattered component, 299–300
  dose compensation curve, 301–302
  forward-scattering term, 301
  parameters, 299
  radial distribution, 298–299
  two Gaussian expression, 299–300
Extinction coefficient, 259

F

Fick's second law, solution, 131, 133
Focus error, 250–251

SUBJECT INDEX

Focus parameter, 309, 311
Forward-scattering contribution, 292
Fresnel approximation, 235
Fresnel coefficient, 260
Fresnel diffraction, 309, 311
   intensity distribution, 308

G

Gas-phase boundary layer, 127–128
Grain-boundary diffusion, 56–57
   output characteristics comparison, 57
Gryzinski's core-electron cross section, 276

H

Helmholtz equation, 234
Hydrodynamic equation, 213

I

ICs, see Integrated circuits
Illumination, partially coherent, 243–244
Image
   amplitude, 250
   calculations, see Optical lithography
   in and out of focus, 254–257
   intensity for incoherent illumination, 249
Implantation, fast evaluation, 55
Impurity
   concentration
      continuous Gaussian band, 171
      contours of constant concentration around mask edge, 222
         as function of depth, 172–173
         within oxide, 220
         within silicon, 220
      coordinate system used to describe redistribution, 219
   diffusion, 183
   flux, 173
   redistribution, 130
      during oxidation, 219–220
      two-dimensional diffusion near mask edge, 220–221
Index of refraction
   complex, photoresist, 259–260
   effective, resist, 315
Inelastic scattering, 279
   comparison of differential cross sections, 277–278
   probability, 288
Integrated circuits
   history, 119–120
   increase in chip component complexity, 120
   process steps, 123–124
   shallow junctions, 55–56
Intensity
   coherent image, 250
   destruction of inhibitor, 258
   distribution
      Fresnel diffraction, 308
      proximity image, 237–241
   image plane, 246–247
   mutual, 253
   optical, 261
   profile, 261–262
Ion extraction
   characteristics
      as function of extraction voltage, 82–83
      as function of microwave power, 79
   shielded single-grid, 80–84
Ion implantation, 137–168
   as-deposited energy deposition profiles, 156–157
   Boltzmann transport equation
      calculations, 152, 154–159
      distribution function, 154
   characteristic energies of elements in Si, 143–144
   classical scattering theory, 137–139
   classification of theories, 146–150
   comparison of LSS and transport equation calculations, 155–157
   comparison of RBS and TE results, 157–158
   continuous slowing down approximation, 143
   contours of constant damage density, 148–149
   damage energy
      deposition distributions as function of depth, 148
      distribution, 147
   electronic stopping, 142–145
      measured versus theoretical power, 145
   energy and length parameters, 139
   implanted profile, 158–159

Ion implantation (*continued*)
  ion concentration as function of depth, 152–153
  Lindhard–Scharff relation, 143
  Monte Carlo model
    advantages, 150
    calculations, 150–152
    constants values based on Moliere potential, 152
    radius of curvature, 151
    scattering triangle, 151
  nuclear stopping
    power, 143–144
    and scattering cross section, 139–142
  probability density function, 146
  range distributions and profile construction, 159–168
    Chebycheff–Hermite polynomials, 161
    coordinate system and schematic geometry, 167
    Edgeworth distribution, 161
    equidensities, 167–168
    Gaussian probability function, 166
    Gram–Charlie distribution, 161–162
    joined half-Gaussian distribution, 159–161
    ordinates for Gaussian distribution, 159
    parameter K as function of energy, 164–165
    peak impurity density, 159
    Pearson distribution, 162–165
    simulated Monte Carlo profile, 165–166
    standard deviation for joined half-Gaussian distribution, 161
    third moment, 160
    two-dimensional profile construction, 165–168
  recurrence relations, 147
  scattering function
    as function of reduced energy, 141–142
    Thomas–Fermi potential, 140–141
  screened Coulombic Thomas–Fermi potential, 139
  total electronic cross section, 143
  total reduced nuclear cross section, 140
  two-particle scattering, 138
  Z_1 oscillations, 145
Ion-beam lithography, 311–314
  ion trajectories, 311–312
  lateral variation of absorbed energy density, 313

Isothermal rapid annealing, 32–55
  arsenic diffusion, 48–53
    enhanced diffusion, 51
    following rapid thermal annealing using one-sided tungsten-halogen furnace, 49–50
    rapid thermal annealing using arc lamp, 51–52
    sheet resistance, 49–50
  boron diffusion, *see* Boron diffusion
  dislocation modification and removal, 53–55
  dopant activation, 40–44
    classification according to dopant incorporation, 41–42
    sheet resistance versus temperature, 40–43
  equipment and general uses, 32–35
    electrical activation of implanted boron, 32
    equipment summary, 35
    heating sources, 32–33
    heating system, 33–34
  temperature determination and stress effects, 35–40
    calibration of equipment, 36
    coexistence of solid and liquid areas on surface, 39
    doping effect on heating rates, 36–37
    normalized blackbody-like spectrum, 37–38
    polished, single-crystalline Si surface after melting and recrystallization, 38–39
    radial temperature gradients, 37–38
    temperature assignment error, 36
    thermocouple voltage as function of time, 36–37

K

Kaufman-type ion source, introduction of reactive gases, 77

L

Lambert–Beer law, optical absorption, 258
Langmuir–Child's law, 79

SUBJECT INDEX

Lasers
  adiabatic annealing, 10–11
  gettering of impurities, 64–66
    denuded zone, 64, 66
    front side view of wafer with laser-damaged segment on back, 65–66
  output parameters, 5
  thermal flux annealing, 23–24
Lithography
  electron-beam, see Electron-beam lithography
  ion-beam, 311–314
  optical, see Optical lithography
  x-ray, 307–311

M

Modulation transfer function, 242–243
Molecular weight, degraded average number, 303
Monte Carlo method, 278–291
  advantages, 150
  angle of scattering, 281
  angular distributions, 279
  calculations, 150–152
  constants values based on Moliere potential, 152
  donut-shaped geometry, 285, 287
  elastic and inelastic scattering probabilities, 288
  electron energy versus path length, 286, 289
  electron path, 282–283
  electron trajectories, 279–280
  energy density profiles, delta-function point source, 285, 288
  free pathlength, 282
  geometry and step sequence, 281–283
  inelastic scattering, 279
  lateral distribution of absorbed energy density, 290
  mean free path, 279–280
    approximation, 280–281
    from different models, 284
  point source, 284–285
  primary and secondary electron trajectories, 288–289
  radius of curvature, 151
  random azimuthal angle, 282
  scattering triangle, 151
  simulated particle trajectories, 284–286

Metal–oxide–semiconductor, process simulation, 206–207

N

NMOS process, see n-type-channel metals–oxide–semiconductor process
n-type-channel metal–oxide–semiconductor process, 122–123
Nuclear cross section, total reduced, 140
Nuclear scattering, 268–275
  angular distribution, 272
  comparison of Mott and Rutherford cross sections, 270–271
  delta function response for forward-scattered electrons, 274–275
  electron diffusion, 268
  mean free path, 274
  multiple scattering, 268
    angular distribution, 273
  plural scattering, 268, 272
  probability for elastic scattering, 271–272
  ratio of Mott to Rutherford cross section as function of scattering angle, 270
  Rutherford expression for scattering of charged particle, 269
  single scattering, 268
  spatial probability distribution function, 274
  total scattering cross section, 269
Numerical aperture, 242

O

Optical absorption
  coefficient
    as function of temperature, 6, 8
    versus photon energy, 6–7
  Lambert–Beer law, 258
Optical lithography, 226–266
  contact and proximity printing, see Contact and proximity printing
  one-dimensional image calculations, 247–252
    cross-coefficient function, 249
    edge slope versus focus error, 251
    effect of focus error, 250
    image amplitude, 250
    image intensity for incoherent illumination, 249

SUBJECT INDEX

Optical lithography (*continued*)
  intensity of coherent image, 250
    toe intensity versus focus error, 252
  projection printing, *see* Projection printing
  resist exposure, 258–266
    ABC data summary, 261–262
    development rate, AZ1350J photoresist, 264, 266
    edge profiles, 265–266
    etch rate data summary, 264, 266
    exposure parameter dependence on prebake conditions, 261, 263
    Poynting vector magnitude, 261
    simulated intensity and inhibitor profile, 261–262
  two-dimensional image calculations, 252–258
    Bessel function of first kind of order 1, 253
    coordinate system, 253
    image in and out of focus, 254–257
    mutual intensity of light incident on object, 253
    transfer function, 253
Oxidation, *see also* Diffusion, oxidation and segregation
  multidimensional, 324

P

Phosphorus
  diffusion anomalies, 196–199
    consumption of two vacancies, 198
    generation of Si self-interstitial, 198
    profiles, 196–197
  fit of oxidation-enhanced diffusion data, 188
  profile
    n^+p-junction region, 223–224
    surface and contour plots, 224–225
Phosphorus glass flow, 58
Plasma deposition, 95–116
  apparatus, 95–97
    with material supply by sputtering, 110
  applications, 113–116
    Si_3N_4-InP metal-insulator-semiconductor diode, 115–116
    surface planarization for large-scale integration fabrications, 113–114
  deposition characteristics, 102–109
    ion incidence effects, 108–109

    Si_3N_4 deposition, 102–105
    SiO_2 deposition, 105–108
    wafer temperature rise during deposition, 102
  divergent magnetic field plasma extraction, *see* Divergent magnetic field plasma extraction
  material supply by sputtering, 109–113
    Al_2O_3 deposition characteristics, 112–113
    deposition mechanism, 110–111
    Ta_2O_3 deposition characteristics, 113
    target current characteristics as function of target voltage, 110–111
Point defect
  generation mechanism, 185–186
  local, equilibrium, 186
  neutral, concentration, 183
Poisson's equation, 169
Potential, electrostatic, 173
Poynting vector, 231
  magnitude, 261
Probability density function, 146
Processing and process simulation
  advanced *n*-type-channel metal–oxide–semiconductor process, 122–123
  coupling between simulation fields, 121–122
  epitaxy, 124–137
    adsorbed layer, 128–135
    analytical expression, 135–137
    boundary layer, enlarged view, 127–128
    bulk gas phase, 126–127
    complementary error function, 136
    doping profiles from decreasing step change in arsine flow, 135
    doping profiles used to determine K, 134
    flux on silicon surface, 130–131
    gas-phase boundary layer, 127–128
    horizontal reactor tube schematic, 126
    impurity distribution, 124–125
    impurity redistribution, 130
    input partial pressure versus silicon growth rate, 132
    numerical technique used to solve Fick's second law, 132–133
    out-diffusion of dopant, 125–126
    reactor schematic, 125–126
    sequence of steps occurring on surface, 128–129
    solution of Fick's second law, 131, 133

SUBJECT INDEX 357

integrated circuit process steps, 123–124
ion implantation, see Ion implantation
simulation and experiment, 121
wafer process, 122–124
Projection printing, 241–247
  Cartesian coordinates, 245
  effective f/number, 242
  entrance and exit pupils of imaging system, 242–243
  geometrical coordinates. 245
  image intensity profile, 243, 245
  intensity in image plane, 246–247
  modulation transfer function, 242–243
  numerical aperture, 242
  partial coherence, 243
  partially coherent illumination, interpretation, 243–245
  refractive-type projection mask aligner elements, 241–242
  transmission cross coefficient, 247
  transmission function, 245–246
Proximity effects, see Electron-beam lithography, proximity functions and effect correct techniques

Q

Quasi-neutrality condition, 173–174

R

Rapid isothermal annealing, 3
Rapid thermal annealing, classification, 2–3
Rapid thermal processes, 1–67
  adiabatic annealing, see Adiabatic annealing
  aluminum contact sintering, 59–60
  fast evaluation of implantation, 55
  grain-boundary diffusion, 56–57
  isothermal rapid annealing, see Isothermal rapid annealing
  laser gettering of impurities, 64–66
  phosphorus glass flow, 58
  recrystallization of Si on insulator, 60–64
    basic approaches, 60–61
    crystallized over nonplanar oxide, 63–64
    lamp-recrystallized Si, 62–63
    pattern of Si before irradiation, 61
    recrystallized with scanned strip heater, 61–62
    seeded crystallization, 61–62
  shallow junctions for integrated circuits, 55–56
  silicide formation, 58–59
  thermal flux annealing, see Thermal flux annealing
Ray-tracing algorithm, 315–316
Reactive ion-beam etching, 76–95
  broad-beam electron cyclotron resonance ion source, 77–80
    ion extraction characteristics as function of microwave power, 79
    schematic, 78
    shielded single-grid ion extraction system, 80–81
  etching characteristics
    Al, 85–86
    Arrhenius reaction rate formula, 88–89
    dependence of etch rates on angle of ion incidence, 89–90
    ion current density and etch rates, 85–86
    ion extraction voltage and etch rates, 86–87
    reaction mechanism, 86–90
    Si, 87
    SiO_2, 85–88
  pattern formation process
  Al patterns, 94–95
    analytical treatment, 90–93
    characteristic curve, 90–91
    etch rate dependence on angle of ion incidence, 92–93
    fine-pattern fabrications, 93–95
    pattern engraved into fused quartz, 94–95
    profile, 92–93
    profile formation with mask, 90–91
    SiO_2 pattern, 93–95
  RTA, see Rapid thermal annealing
  shielded single-grid ion extraction, 80–84
    effect of shield electrode, 82–83
    influence of grid potential on plasma, 81, 83
    ion current density distribution, 83
    ion extraction characteristics as function of extraction voltage, 82–83
    system, 84–85
Reactive sputter etching, 76–77

Recrystallization, Si on insulator, 60–64
  base approaches, 60–61
  complementary metal–oxide–semiconductor inverter, 60
  lamp-recrystallization over buried square island of SiO_2, 62–63
  over nonplanar oxide, 63–64
  patterning of Si before irradiation, 61
  scanned strip heater, 61–62
  seeded crystallization, 61–62
Reflection coefficient, 260
Resist
  complex index of refraction, 259–260
  development rate, 264, 266
  distribution of deposited energy, 305–307
  edge profiles, 265–266
  effective refractive index, 315
  etch rate versus absorbed dose, 304
  exposure, 258–266
    ABC data summary, 261–262
    absorbtance of sublayer, 261
    absorption constant of resist, 259
    complex index of refraction of photoresist, 259–260
    destruction of inhibitor by light intensity, 258
    development rate, AZ1350J photoresist, 264, 266
    edge profiles, 265–266
    etch rate data summary, 264, 266
    exposure parameter dependence on prebake conditions, 261, 263
    extinction coefficient, 259
    Fresnel coefficient, 260
    Lambert–Beer law, optical absorption, 258
    optical phase thickness of layer, 260
    Poynting vector magnitude, 261
    reflectance and transmittance of substrate, 260
    reflection and transmission coefficient, 260
    simulated intensity and inhibitor profile, 261–262
    substrate and resist cross section, 259–260
  isoenergy contours, 296
  profiles, x-ray tube and synchrotron exposure, 310–311
  thickness versus development time, 303–304
Rf discharge plasma, 75

S

Sb, fit of oxidation-reduced diffusion data, 189
Scattering, see also Nuclear scattering
  classical theory, 137–139
  electron–electron, 275–278
Scission events, 303
Self-diffusion coefficient, 183
Self-interstitial
  concentration, 190
  ratio, 185–187
Si etching
  etch rate dependence on angle of ion incidence, 89–90
  ion extraction voltage and etch rates, 86–87
  reaction mechanism, 87
Silicide
  formation, 58–59
  resistivities, 59
Silicon
  beam-annealed, defects, 26–28
  physical properties, 5
Si_3N_4 deposition, 102–105
  deposition rate and refractive indices, 102–103
  film etch rates, 105
  films deposited partially masked, 108–109
  gas pressure, 103
  infrared absorption spectrum, 104–105
  internal stress, 104
Si_3N_4–InP metal–insulator–semiconduction diode, 115–116
  C–V characteristics, 116
SiO_2
  deposition, 105–108
    deposition rates and refractive indices, 106
    film etch rates, 107–108
    gas pressure, 106–107
    infrared absorption spectrum, 107
  etching, 85
    etch rate dependence on angle of ion incidence, 89–90

ion current density and etch rates, 85–86
ion extraction voltage and etch rates, 86–87
reaction mechanism, 86–87
growth of new layer, 211–212
Si–SiO$_2$ interface, normal component of flux, 210
Spatial probability distribution function, 274
Stanford University Process Analysis program, MOS process simulation, 206–207
Stanford University Process Engineering Models program, 126
  discretization method, 203
Stefan problem, two-dimensional, 210–211
Stress
  distribution along silicon interface, 214–215
  flow rate induced by, 212–213
String algorithm, 316–317
  sequence for isotropic and anisotropic etching, 318
SUPRA, see Stanford University Process Analysis Program
SUPREM, see Stanford University Process Engineering Models program
Surface planarization, 113–114
  buried Al line, 114–115

T

Ta$_2$O$_3$ deposition, characteristics, 112
Tan-Gösele model, 211–213
Thermal flux annealing, 3, 23–32
  arsenic concentration profiles, 24–25
  defects in beam-annealed Si, 26–28
  diffusion profiles, 25–26
  equipment, 23–24
  glancing-angle RBS random and aligned spectra of As-implanted Si, 25–26
  Si surface annealed with cw argon laser, 26–27
  solid-phase epitaxy rate measurements, 28–32

rate dependence on temperature, 29–31
schematic view, 28–29
velocity of interface, 23
Thomas–Fermi potential, screened Coulombic, 139
Thomas–Whiddington law, energy-range relationship, 292
Toe intensity, 252
Transmission coefficient, 260
Transmission cross coefficient, 247
Transmission function, 245–246
Twin-Tub complementary metal–oxide–semiconductor inverter, 223–225
Two Gaussian model, 299–301

V

Vacancy supersaturation ratio, 185

W

Wafer
  heating and cooling, 2
  process, 122–124
Wave function, scalar, exact Kirchhoff integral, 234

X

X-ray lithography, 307–311
  Fresnel diffraction
    behind slit, 309, 311
    intensity distribution, 308
  influence of focus parameter on intensity distribution, 309, 311
  resist profiles for x-ray tube and synchrotron exposure, 310–311
  system schematic, 307–308